Advances in Plant Auxin Biology

Advances in Plant Auxin Biology

Guest Editors

Dongwei Di
Verena Kriechbaumer

MDPI

Basel • Beijing • Wuhan • Barcelona • Belgrade • Novi Sad • Cluj • Manchester

Guest Editors
Dongwei Di
Institute of Soil Science
Chinese Academy of Sciences
Nanjing
China

Verena Kriechbaumer
Department of Biological and Medical Sciences
Oxford Brookes University
Oxford
UK

Editorial Office
MDPI AG
Grosspeteranlage 5
4052 Basel, Switzerland

This is a reprint of the Special Issue, published open access by the journal *Plants* (ISSN 2223-7747), freely accessible at: https://www.mdpi.com/journal/plants/special_issues/P6226I7780.

For citation purposes, cite each article independently as indicated on the article page online and as indicated below:

Lastname, A.A.; Lastname, B.B. Article Title. *Journal Name* **Year**, *Volume Number*, Page Range.

ISBN 978-3-7258-7741-6 (Hbk)
ISBN 978-3-7258-7742-3 (PDF)
https://doi.org/10.3390/books978-3-7258-7742-3

Contents

About the Editors

Dongwei Di

Dongwei Di is an Associate Professor at the Institute of Soil Science, Chinese Academy of Sciences, and a member of the State Key Laboratory of Soil & Sustainable Agriculture. His research focuses on the molecular mechanisms underlying plant responses to ammonium (NH_4^+) nutrition, with a particular interest in the regulatory networks governing futile NH_4^+ efflux in plants. In recent years, he has extended his research scope to explore the regulatory functions of auxin metabolism and signaling pathways in plant responses to NH_4^+ stress and other abiotic stresses. To date, he has served as principal investigator for more than ten research projects, including those funded by the National Natural Science Foundation of China (NSFC), and has published over 40 SCI-indexed papers in related fields.

Verena Kriechbaumer

Verena Kriechbaumer is a Professor of Plant Sciences and Biotechnology at Oxford Brookes University, where she leads the Endomembrane Structure and Function research group. Prof. Kriechbaumer's research lies at the intersection of cell biology, molecular biology, and advanced imaging, with a focus on the organization and function of the plant endoplasmic reticulum, membrane proteins and auxin biosynthesis. Her interdisciplinary approach bridges fundamental research with practical applications, such as engineering plants for methane detoxification and production of human therapeutics in plant systems.

Preface

We are pleased to present this Reprint, "Advances in Plant Auxin Biology", developed from the Special Issue published in *Plants*. The articles collected here reflect the remarkable progress made in understanding auxin, one of the most influential signaling molecules in plant biology. Long recognized as a key regulator of growth and development, auxin is now understood to function through a far more sophisticated and interconnected regulatory system than previously assumed.

Recent years have witnessed a significant shift in the field. The traditional view centered primarily on nuclear transcriptional signaling has expanded to include rapid cellular responses, complex feedback regulation, metabolic control, spatial transport dynamics, and extensive crosstalk with other hormonal and environmental signaling pathways. Together, these discoveries have transformed auxin biology into one of the most vibrant and integrative areas of plant science.

The contributions in this Reprint provide a multi-scale perspective on these developments. They explore auxin biosynthesis, conjugation, degradation, transport, and signal transduction, while also addressing how these processes shape plant architecture, developmental transitions, and stress adaptation. Several authoritative reviews synthesize current knowledge on IAA biosynthetic pathways, GH3-mediated regulation, and lineage-specific auxin metabolism in Brassicaceae. Original research articles further demonstrate how auxin influences diverse biological phenomena, including bamboo shoot development, metabolic reprogramming, and signaling variation in polyploid crops.

This collection also highlights methodological innovation. Techniques such as Non-invasive Micro-test Technology are enabling real-time analyses of auxin transport and associated signaling events, helping bridge the gap between molecular processes and whole-plant physiology. These approaches are opening new opportunities to understand how auxin responses are coordinated in time and space.

Taken together, the studies assembled in this Reprint provide a coherent and contemporary overview of auxin as a central organizer of plant growth, adaptation, and evolution. We hope this volume will serve as a valuable resource for researchers, teachers, and advanced students, and that it will stimulate further discoveries in this rapidly advancing field.

Dongwei Di and Verena Kriechbaumer
Guest Editors

plants

Editorial

Recent Advances and Future Perspectives in Plant Auxin Biology

Dong-Wei Di [1,2,*] and Verena Kriechbaumer [3,*]

1 State Key Laboratory of Soil and Sustainable Agriculture, Institute of Soil Science, Chinese Academy of Sciences, Nanjing 210008, China
2 University of Chinese Academy of Sciences, Beijing 100049, China
3 School of Biological and Medical Sciences, Faculty of Health, Science and Technology, Oxford Brookes University, Oxford OX3 0BP, UK
* Correspondence: dwdi@issas.ac.cn (D.-W.D.); vkriechbaumer@brookes.ac.uk (V.K.)

1. Introduction

The phytohormone auxin, predominantly as indole-3-acetic acid (IAA), functions as a master regulator that orchestrates plant growth, development, and adaptation to environmental cues [1,2]. Its spatiotemporal distribution is fine-tuned by a multi-layered regulatory network encompassing biosynthesis, metabolism, polar transport, and signal transduction, thereby integrating key processes ranging from embryogenesis and organ patterning to stress resilience [3–5]. In the context of escalating global climate change and increasing frequency of abiotic stresses, deciphering how auxin-mediated networks equip plants with phenotypic plasticity and adaptive capacity is not only a fundamental scientific question but also a critical need for sustainable agriculture and ecosystem preservation [2].

In recent years, auxin biology has witnessed a profound conceptual transformation, shifting from the classical nuclear-centric $SCF^{TIR1/AFB}$-Aux/IAA-ARF signaling pathway toward a more complex, decentralized, and multi-tiered regulatory system. Central to this paradigm is the emergence of a dual-channel signaling framework. At the nuclear level, TIR1/AFB receptors utilize their intrinsic adenylate cyclase activity to generate confined cAMP pools, which are essential for the precise transcriptional reprogramming of auxin-responsive genes [6–9]. Concurrently, at the plasma membrane, the ABP1/ABL–TMK module initiates rapid phosphorylation cascades that directly modify pivotal targets such as PIN-formed efflux carriers and H^+-ATPases, thereby coordinating auxin transport and acid-growth responses within minutes [10–16]. This bipartite signaling model not only resolves the long-standing dichotomy between rapid physiological adjustments and delayed transcriptional changes but also illustrates how auxin signaling is temporally and spatially layered to optimize plant adaptation [9,10].

Expanding upon these mechanistic foundations, contemporary research has increasingly focused on the self-organizing properties of the auxin system. Through the integration of functional analyses of self-organizing transport modules with high-resolution structural studies of AUX1/LAX influx and PIN efflux carriers, fundamental biophysical principles governing the spatiotemporal patterning of auxin distribution are being unveiled [11,17–20]. These advances highlight how modularity, feedback loops, and physical constraints collectively shape auxin-mediated development and stress responses.

This Special Issue, titled "Advances in Plant Auxin Biology," assembles a collection of cutting-edge reviews and original research articles that collectively examine the complexity and integrative nature of auxin regulatory networks. Contributions span multiple scales of analysis, from molecular enzymology and real-time ion/metabolite flux measurements to

 https://doi.org/10.3390/plants15050683

whole-plant phenology and polyploid genome regulation, thereby offering a cohesive and multi-dimensional perspective on auxin as a central orchestrator of plant life.

2. Foundational Insights from Comprehensive Reviews

This Special Issue integrates several foundational review articles that collectively refine and advance the theoretical framework of auxin biology. Gao et al. systematically synthesize the current understanding of the auxin regulatory system, offering a detailed elucidation of tryptophan-dependent IAA biosynthesis with a focus on the central IPyA pathway [21]. They further delineate key mechanisms governing IAA inactivation, including GH3-mediated amide conjugation and DAO-catalyzed oxidation. By integrating recent insights into polar auxin transport components—such as the PIN, AUX1/LAX, and ABCB protein families—the authors critically examine the layered complexity of auxin signaling, spanning from canonical TIR1/AFB-mediated transcriptional regulation to rapid non-transcriptional responses. Notably, the review emphasizes the context-dependent role of auxin in abiotic stress adaptation, illustrating how its signaling is often suppressed under drought or salinity, yet can be recruited to drive adaptive morphological responses such as hyponasty under heat stress.

In the context of auxin metabolic regulation, Luo et al. present a focused examination of the IAA–amino acid amide conjugation pathway mediated by the GRETCHEN HAGEN 3 (GH3) family [22]. Their review provides a comprehensive analysis of the enzymatic mechanisms, transcriptional control, and functional divergence of GH3 members across plant species. Building on established evidence, the authors systematically argue that GH3 enzymes extend beyond their classical role as mere "buffers" of IAA homeostasis; instead, they act as pivotal signaling hubs that integrate diverse environmental stress cues. These include drought, temperature variation, pathogen challenge, and nutrient imbalance, through which GH3s dynamically fine-tune active IAA pools. Furthermore, the review underscores that the development of GH3-specific chemical inhibitors represents a crucial experimental strategy for elucidating the functional redundancy and specificity inherent to this gene family.

Ma et al. [23] provide a systematic review of the evolving understanding of IAOx-dependent IAA biosynthesis within Brassicaceae, tracing its conceptual progression from an early linear model to a more complex and debated framework [24,25]. Recent genetic and metabolic evidence has challenged the established linear pathway model involving specific *CYP71A*, *NIT*, and *AMI* gene families [25]. Instead, IAOx is now viewed as a potential metabolic branch point whose flux towards IAA may involve yet-uncharacterized enzymatic routes, operating alongside or independently of its role in defense-related metabolites such as glucosinolates [25]. This ongoing reassessment highlights auxin biosynthesis, particularly via IAOx, as a potentially critical and underexplored interface in the plant's strategic allocation of resources between growth and stress adaptation. By integrating recent molecular insights with persistent knowledge gaps, the authors propose several priority research trajectories, including identifying the enzyme(s) catalyzing the direct conversion of IAOx to IAA; clarifying the biosynthetic origins and signaling roles of intermediates IAN and IAM; elucidating how metabolic channeling and cell-type-specific regulation coordinate growth-defense trade-offs; and examining the evolutionary conservation and adaptive divergence of this pathway beyond Brassicaceae [26].

3. Polar Auxin Transport: The Architect of Form

The translation of hormonal gradients into morphological outcomes is orchestrated by the polar auxin transport (PAT) system. Wenzel et al. present robust experimental support for its conserved morphogenetic function across diverse dicot species. By phar-

 https://doi.org/10.3390/plants15050683

macologically inhibiting PAT in four taxonomically distinct dicots exhibiting varied leaf forms, including pinnate and palmate architectures—the authors establish that PAT is indispensable for establishing discrete auxin maxima, which in turn govern leaf complexity and the elaboration of venation networks [27]. Their findings, corroborated by computational simulations, demonstrate that PAT disruption leads to auxin distribution homogenization, ultimately driving leaf simplification and a convergence of venation toward parallelized patterning. Collectively, this study consolidates the role of PAT as an evolutionarily conserved and fundamental regulator of plant morphological architecture.

4. Context-Specific Regulation: Driving Unique Phenotypes Through Auxin

Auxin also plays a pivotal role in plant adaptation to specific physiological and environmental contexts. Mao et al. investigated the physiological basis of the unique year-round shooting phenomenon in the woody bamboo species *Cephalostachyum pingbianense* [28]. Through comprehensive annual hormone profiling, they demonstrated that sustained high levels of IAA and ABA in dormant rhizome buds, coupled with an elevated IAA/ABA ratio, are robustly correlated with shoot emergence. This indicates that the bamboo's rhythmic growth is governed primarily by internal hormonal homeostasis rather than external seasonal cues, underscoring the significance of auxin homeostasis in shaping adaptive growth strategies.

At the interface of primary and specialized metabolism, Martin et al. revealed that targeted modulation of auxin signaling, via exogenous IAA or the transport inhibitor 2,3,5-Triiodobenzoic Acid (TIBA), can effectively reprogram alkaloid biosynthesis in *Annona emarginata* [29]. TIBA treatment notably enhanced alkaloid accumulation in roots, a response potentially driven by redirected photosynthate allocation. This aligns with earlier studies in other plant systems where auxin signaling has been shown to modulate the expression of key alkaloid biosynthetic genes (e.g., *TDC* in *Catharanthus roseus*) and influence precursor flux through shared pathways such as the shikimate–tryptophan route [30,31]. Thus, while the precise molecular mechanisms in Annona remain to be fully elucidated, the observed metabolic shifts suggest an interplay between auxin physiology and secondary metabolism, highlighting the potential of hormonal intervention as a strategic tool for the targeted production of valuable plant specialized metabolites.

5. Allelic Bias in Polyploids: Reshaping the Auxin Response Network

In polyploid crops, the contribution of homoeologous alleles introduces an additional layer of regulatory complexity. Zhao et al., while investigating embryogenic callus formation in ABB banana, uncovered striking allele-specific dominance within the auxin signaling pathway [32]. Their study revealed that A-genome-derived alleles of key transcription factors and signaling components overwhelmingly dominate transcriptional reprogramming, thereby directing cell fate transitions. This work elucidates how genomic architecture and allelic bias can fundamentally reshape the output of the core auxin response network in polyploid systems.

6. Empowering Auxin Physiology Through Advanced Technologies: NMT as a Key Tool for Real-Time Flux Analysis

Dissecting the rapid physiological functions of auxin necessitates the use of advanced technological platforms. Zhang et al. examine the essential contribution of Non-invasive Micro-test Technology (NMT) to the analysis of plant abiotic stress responses [33]. Capable of real-time, in situ, and high-sensitivity detection of transmembrane ion and molecule fluxes, such as IAA, H^+, Ca^{2+}, and K^+, NMT provides a powerful tool for resolving the

 https://doi.org/10.3390/plants15050683

spatiotemporal dynamics of auxin under stress. The authors emphasize that NMT not only directly measures IAA transport across membranes but also concurrently tracks its coordination with other signaling pathways, including Ca^{2+} signaling and H^+ flux [34,35]. This multimodal capability enables a systems-level understanding of how auxin distribution and activity are regulated during stress adaptation, influencing processes such as stomatal conductance, root system reorganization, and ion homeostasis. Collectively, these insights position NMT as a pivotal methodological bridge that integrates auxin-related molecular mechanisms with organism-level physiology, offering a vital physiological dimension to molecular investigations of auxin-dependent stress responses.

7. Outlook

Over the past decade, auxin research has undergone profound paradigm shifts, transitioning from a singular focus on nuclear receptor mechanisms to the recognition of a sophisticated cell surface co-receptor system. This evolution in perspective extends from earlier models of linear biosynthetic pathways toward an integrated understanding of complex metabolic networks, and from static descriptions of hormone distribution to the exploration of dynamic, self-organizing patterning processes. Collectively, these advances reveal a signaling landscape that is far more intricate, context-dependent, and dynamically regulated than previously appreciated. This refined framework not only redefines our understanding of auxin-mediated development but also highlights the multi-layered nature of hormone signaling in plants. At this new conceptual frontier, the field is poised to address critical challenges in systemic integration, particularly in elucidating how spatially and temporally distinct signaling modules are coordinated across scales.

(i) Elucidating unknown molecular and biochemical mechanisms

Although the core framework of auxin signaling has been largely delineated, critical gaps persist in our understanding of key molecular and biochemical mechanisms. A primary research priority lies in resolving the full-length three-dimensional structures of classical long-loop PIN family proteins (e.g., PIN1), which would enable atomic-level insights into the structural basis of their autoinhibition, kinase-dependent activation, and polar localization. Furthermore, it is essential to comprehensively elucidate the functional network of cAMP produced by TIR1/AFB receptors as a second messenger, including the identification of its direct downstream targets and potential plant-specific cyclic nucleotide effector proteins. Additionally, an integrated approach combining genetics, metabolomics, and other advanced methodologies is urgently needed to uncover the yet-uncharacterized biochemical pathways underlying established phenotypes, such as the precise origin and regulatory nodes of excess auxin synthesis in *superroot 2* (*sur2*) mutants. The elucidation of these molecular and biochemical mechanisms will not only provide a robust foundation for constructing predictive and testable theoretical models but also facilitate the transition of auxin signaling research from qualitative description toward a quantitative systems biology paradigm.

(ii) Deciphering spatiotemporal specificity and evolutionary adaptability of auxin signaling pathways

Auxin signal transduction is not a fixed, rigid pathway but exhibits remarkable context dependency and evolutionary plasticity. For instance, the TMK signaling module performs antagonistic functions in roots and hypocotyls, while the IAOx biosynthetic pathway has evolved a new role in defense responses within Brassicaceae. Future research should systematically map the activity dynamics and interaction networks of various signaling modules across different cell types, developmental stages, and environmental conditions with higher spatiotemporal resolution. Concurrently, moving beyond model plant studies,

 https://doi.org/10.3390/plants15050683

comparative genomics and evolutionary developmental biology approaches should be employed to trace the innovation and evolution of core components involved in auxin perception, signal transduction, and transport during the adaptive radiation of land plants. This will help address a fundamental scientific question: how can a relatively conserved hormonal system be "reprogrammed" through regulatory mechanisms to support the vast diversity of plant morphogenesis?

Collectively, recent findings reinforce the view that auxin-mediated development emerges from self-organizing systems integrating local biosynthesis, directional transport, and dynamic signaling feedback. The interplay between nuclear and non-nuclear auxin signaling pathways, coupled with interactions with other hormonal networks, enables plants to generate robust yet flexible developmental patterns. Future research aimed at linking molecular mechanisms to emergent properties across spatial and temporal scales will be crucial for fully understanding how auxin orchestrates plant form and function. In-depth exploration of these questions will not only reveal fundamental principles of plant life but also provide important theoretical insights and technical inspiration for sustainable agricultural development and biomaterial innovation.

Author Contributions: The authors confirm contributions to the paper as follows: D.-W.D. and V.K.: draft manuscript preparation; D.-W.D. and V.K.: manuscript revision. D.-W.D. and V.K.: funding acquisition. All authors have read and agreed to the published version of the manuscript.

Funding: This work was supported by grants from the National Natural Science Foundation of China (32573134), Natural Science Foundation of Jiangsu Province (No. BK20240214), and the Field Frontier Program of the Institute of Soil Science (ISSAS2412). V.K. was supported by a BBSRC responsive mode grant (BB/X006417/1).

Data Availability Statement: No new data were created or analyzed in this study. Data sharing is not applicable to this article.

Acknowledgments: The Guest Editors extend their sincere gratitude to all the authors and reviewers for their invaluable contributions to this Special Issue. We also wish to thank the Plants Editorial Office for their continuous professional support throughout the entire publication process.

Conflicts of Interest: The authors declare no conflicts of interest.

References

1. Di, D.-W.; Zhang, C.; Luo, P.; An, C.-W.; Guo, G.-Q. The biosynthesis of auxin: How many paths truly lead to IAA? *Plant Growth Regul.* **2016**, *78*, 275–285. [CrossRef]
2. Vanneste, S.; Pei, Y.; Friml, J. Mechanisms of auxin action in plant growth and development. *Nat. Rev. Mol. Cell Biol.* **2025**, *26*, 648–666. [CrossRef] [PubMed]
3. Luo, P.; Di, D.-W. Precise regulation of the TAA1/TAR-YUCCA auxin biosynthesis pathway in plants. *IJMS* **2023**, *24*, 8514. [CrossRef]
4. Kriechbaumer, V.; Botchway, S.W.; Hawes, C. Localization and interactions between Arabidopsis auxin biosynthetic enzymes in the TAA/YUC-dependent pathway. *J. Exp. Bot.* **2016**, *67*, 4195–4207. [CrossRef] [PubMed]
5. Blakeslee, J.J.; Spatola Rossi, T.; Kriechbaumer, V. Auxin biosynthesis: Spatial regulation and adaptation to stress. *J. Exp. Bot.* **2019**, *70*, 5041–5049. [CrossRef]
6. Luo, P.; Luan, C.-S.; Li, T.-T.; Kriechbaumer, V.; Di, D.-W. 3′,5′-cAMP in plants: An integrated view of homeostasis, effectors, and physiological functions. *J. Exp. Bot.* **2025**, eraf517. [CrossRef]
7. Qi, L.; Kwiatkowski, M.; Chen, H.; Hoermayer, L.; Sinclair, S.; Zou, M.; Del Genio, C.I.; Kubeš, M.F.; Napier, R.; Jaworski, K.; et al. Adenylate cyclase activity of TIR1/AFB auxin receptors in plants. *Nature* **2022**, *611*, 133–138. [CrossRef]
8. Chen, H.; Qi, L.; Zou, M.; Lu, M.; Kwiatkowski, M.; Pei, Y.; Jaworski, K.; Friml, J. TIR1-produced cAMP as a second messenger in transcriptional auxin signaling. *Nature* **2025**, *640*, 1011–1016. [CrossRef]
9. Wu, L.; Kriechbaumer, V.; Di, D.-W. Emerging roles of cAMP: A transcriptional master regulator in the canonical TIR1/AFB-mediated auxin signaling. *Plant Horm.* **2025**, *1*, e018. [CrossRef]

https://doi.org/10.3390/plants15050683

10. Friml, J.; Gallei, M.; Gelová, Z.; Johnson, A.; Mazur, E.; Monzer, A.; Rodriguez, L.; Roosjen, M.; Verstraeten, I.; Živanović, B.D.; et al. ABP1-TMK auxin perception for global phosphorylation and auxin canalization. *Nature* **2022**, *609*, 575–581. [CrossRef]
11. Rodriguez, L.; Fiedler, L.; Zou, M.; Giannini, C.; Monzer, A.; Vladimirtsev, D.; Randuch, M.; Yu, Y.; Gelová, Z.; Verstraeten, I.; et al. ABP1/ABL3-TMK1 cell-surface auxin signaling targets PIN2-mediated auxin fluxes for root gravitropism. *Cell* **2025**, *188*, 6138–6150.e17. [CrossRef] [PubMed]
12. Li, L.; Verstraeten, I.; Roosjen, M.; Takahashi, K.; Rodriguez, L.; Merrin, J.; Chen, J.; Shabala, L.; Smet, W.; Ren, H.; et al. Cell surface and intracellular auxin signalling for H^+ fluxes in root growth. *Nature* **2021**, *599*, 273–277. [CrossRef]
13. Lin, W.; Zhou, X.; Tang, W.; Takahashi, K.; Pan, X.; Dai, J.; Ren, H.; Zhu, X.; Pan, S.; Zheng, H.; et al. TMK-based cell-surface auxin signalling activates cell-wall acidification. *Nature* **2021**, *599*, 278–282. [CrossRef]
14. Wang, J.-L.; Wang, M.; Zhang, L.; Li, Y.-X.; Li, J.-J.; Li, Y.-Y.; Pu, Z.-X.; Li, D.-Y.; Liu, X.-N.; Guo, W.; et al. WAV E3 ubiquitin ligases mediate degradation of IAA32/34 in the TMK1-mediated auxin signaling pathway during apical hook development. *Proc. Natl. Acad. Sci. USA* **2024**, *121*, e2314353121. [CrossRef]
15. Yu, Y.; Tang, W.; Lin, W.; Li, W.; Zhou, X.; Li, Y.; Chen, R.; Zheng, R.; Qin, G.; Cao, W.; et al. ABLs and TMKs are co-receptors for extracellular auxin. *Cell* **2023**, *186*, 5457–5471.e17. [CrossRef]
16. Wang, J.; Jin, D.; Deng, Z.; Zheng, L.; Guo, P.; Ji, Y.; Song, Z.; Zeng, H.Y.; Kinoshita, T.; Liao, Z.; et al. The apoplastic pH is a key determinant in the hypocotyl growth response to auxin dosage and light. *Nat. Plants* **2025**, *11*, 279–294. [CrossRef]
17. Yang, Z.; Wei, H.; Gan, Y.; Liu, H.; Cao, Y.; An, H.; Que, X.; Gao, Y.; Zhu, L.; Tan, S.; et al. Structural insights into auxin influx mediated by the Arabidopsis AUX1. *Cell* **2025**, *188*, 3960–3973.e15. [CrossRef]
18. Ung, K.L.; Winkler, M.; Schulz, L.; Kolb, M.; Janacek, D.P.; Dedic, E.; Stokes, D.L.; Hammes, U.Z.; Pedersen, B.P. Structures and mechanism of the plant PIN-FORMED auxin transporter. *Nature* **2022**, *609*, 605–610. [CrossRef] [PubMed]
19. Pérez-Henríquez, P.; Nagawa, S.; Liu, Z.; Pan, X.; Michniewicz, M.; Tang, W.; Rasmussen, C.; Cui, X.; Van Norman, J.; Strader, L.; et al. PIN2-mediated self-organizing transient auxin flow contributes to auxin maxima at the tip of Arabidopsis cotyledons. *Nat. Commun.* **2025**, *16*, 1380. [CrossRef] [PubMed]
20. Huang, R.; Wang, J.; Chang, M.; Tang, W.; Yu, Y.; Zhang, Y.; Peng, Y.; Wang, Y.; Guo, Y.; Lu, T.; et al. TMK-PIN1 drives a short self-organizing circuit for auxin export and signaling in Arabidopsis. *Dev. Cell* **2025**, *61*, 73–84.E6. [CrossRef]
21. Gao, J.; Zhuang, S.; Zhang, W. Advances in plant auxin biology: Synthesis, metabolism, signaling, interaction with other hormones, and roles under abiotic stress. *Plants* **2024**, *13*, 2523. [CrossRef]
22. Luo, P.; Li, T.-T.; Shi, W.-M.; Ma, Q.; Di, D.-W. The Roles of GRETCHEN HAGEN3 (GH3)-dependent auxin conjugation in the regulation of plant development and stress adaptation. *Plants* **2023**, *12*, 4111. [CrossRef] [PubMed]
23. Ma, M.-K.; Kriechbaumer, V.; Di, D.-W. The IAOx-Dependent IAA Biosynthesis Pathway: Acquired Insights, Paradigm Shifts, and Unresolved Questions. *Plants* **2026**, *15*, 436. [CrossRef] [PubMed]
24. Sugawara, S.; Hishiyama, S.; Jikumaru, Y.; Hanada, A.; Nishimura, T.; Koshiba, T.; Zhao, Y.; Kamiya, Y.; Kasahara, H. Biochemical analyses of indole-3-acetaldoxime-dependent auxin biosynthesis in *Arabidopsis*. *Proc. Natl. Acad. Sci. USA* **2009**, *106*, 5430–5435. [CrossRef]
25. Fenech, M.; Brumos, J.; Pěnčík, A.; Edwards, B.; Belcapo, S.; DeLacey, J.; Patel, A.; Kater, M.; Li, X.; Ljung, K.; et al. The *CYP71A*, *NIT*, *AMI*, and *IAMH* gene families are dispensable for indole-3-acetaldoxime-mediated auxin biosynthesis in *Arabidopsis*. *Plant Cell* **2025**, *37*, koaf242. [CrossRef] [PubMed]
26. Perez, V.C.; Dai, R.; Bai, B.; Tomiczek, B.; Askey, B.C.; Zhang, Y.; Garret, M.R.; Ding, Y.; Grenning, A.; Block, A.K.; et al. Aldoximes are precursors of auxins in Arabidopsis and maize. *New Phytol.* **2021**, *231*, 1449–1461. [CrossRef]
27. Wenzel, C.L.; Holloway, D.M.; Mattsson, J. The effects of auxin transport inhibition on the formation of various leaf and vein patterns. *Plants* **2024**, *13*, 2566. [CrossRef]
28. Mao, W.; Bao, C.; Cheng, Q.; Liang, N.; Wang, L.; Yang, H. All-Year High IAA and ABA contents in rhizome buds may contribute to natural four-season shooting in woody bamboo *Cephalostachyum pingbianense*. *Plants* **2024**, *13*, 410. [CrossRef]
29. Martin, B.C.; De-la-Cruz-Chacón, I.; Mimi, C.O.; Boaro, C.S.F.; Campos, F.G.; Moreira-Coneglian, I.R.; Ferreira, G. Impact of external sources of indole acetic acid and 2,3,5-triiodobenzoic acid on alkaloid production and their relationships with primary metabolism and antioxidant activity in *Annona emarginata* (Schltdl.) H. Rainer. *Plants* **2024**, *13*, 2637. [CrossRef] [PubMed]
30. Goddijn, O.J.M.; de Kam, R.J.; Zanetti, A.; Schilperoort, R.A.; Hoge, J.H.C. Auxin rapidly down-regulates transcription of the tryptophan decarboxylase gene from *Catharanthus roseus*. *Plant Mol. Biol.* **1992**, *18*, 1113–1120. [CrossRef]
31. Pasquali, G.; Goddijn, O.J.M.; de Waal, A.; Verpoorte, R.; Schilperoort, R.A.; Hoge, J.H.C.; Memelink, J. Coordinated regulation of two indole alkaloid biosynthetic genes from *Catharanthus roseus* by auxin and elicitors. *Plant Mol. Biol.* **1992**, *18*, 1121–1131. [CrossRef]
32. Zhao, X.; Zhuang, Y.; Xie, W.; Yang, Y.; Pu, J.; Fan, Z.; Chen, Y.; Lin, Y.; Lai, Z. Allelic expression dynamics of regulatory factors during embryogenic callus induction in ABB banana (*Musa* Spp. Cv. Bengal, ABB Group). *Plants* **2025**, *14*, 761. [CrossRef] [PubMed]

 https://doi.org/10.3390/plants15050683

33. Zhang, T.; Yin, P.; Yang, X.; Liu, Y.; Xu, R. Non-invasive micro-test technology in plant physiology under abiotic stress: From mechanism to application. *Plants* **2025**, *14*, 1932. [CrossRef] [PubMed]
34. Di, D.-W.; Wu, J.; Ma, M.; Li, G.; Wang, M.; Kronzucker, H.J.; Shi, W. PIN5 is involved in regulating NH_4^+ efflux and primary root growth under high-ammonium stress via mediating intracellular auxin transport. *Plant Soil* **2024**, *505*, 25–40. [CrossRef]
35. Di, D.; Sun, L.; Wang, M.; Wu, J.; Kronzucker, H.J.; Fang, S.; Chu, J.; Shi, W.; Li, G. WRKY46 promotes ammonium tolerance in Arabidopsis by repressing NUDX9 and indole-3-acetic acid-conjugating genes and by inhibiting ammonium efflux in the root elongation zone. *New Phytol.* **2021**, *232*, 190–207. [CrossRef]

https://doi.org/10.3390/plants15050683

plants

MDPI

Review

The Roles of GRETCHEN HAGEN3 (GH3)-Dependent Auxin Conjugation in the Regulation of Plant Development and Stress Adaptation

Pan Luo [1,†], Ting-Ting Li [2,3,†], Wei-Ming Shi [2,3], Qi Ma [1,*] and Dong-Wei Di [2,3,*]

1 College of Life Science and Technology, Gansu Agricultural University, Lanzhou 730070, China; luopan@gsau.edu.cn
2 State Key Laboratory of Soil and Sustainable Agriculture, Institute of Soil Science, Chinese Academy of Sciences, Nanjing 210008, China; litingting@issas.ac.cn (T.-T.L.); wmshi@issas.ac.cn (W.-M.S.)
3 University of Chinese Academy of Sciences, Beijing 100049, China
* Correspondence: maq@gsau.edu.cn (Q.M.); dwdi@issas.ac.cn (D.-W.D.)
† These authors contribute equally to this work.

Abstract: The precise control of free auxin (indole-3-acetic acid, IAA) gradient, which is orchestrated by biosynthesis, conjugation, degradation, hydrolyzation, and transport, is critical for all aspects of plant growth and development. Of these, the GRETCHEN HAGEN 3 (GH3) acyl acid amido synthetase family, pivotal in conjugating IAA with amino acids, has garnered significant interest. Recent advances in understanding GH3-dependent IAA conjugation have positioned GH3 functional elucidation as a hot topic of research. This review aims to consolidate and discuss recent findings on (i) the enzymatic mechanisms driving GH3 activity, (ii) the influence of chemical inhibitor on GH3 function, and (iii) the transcriptional regulation of GH3 and its impact on plant development and stress response. Additionally, we explore the distinct biological functions attributed to IAA-amino acid conjugates.

Keywords: auxin; IAA; GH3; chemical inhibitor; transcriptional regulation

1. Introduction

Auxins, a group of phytohormones, are integral to the regulation of plant development and stress responses [1,2]. Previous research has identified three naturally occurring auxins: indole-3-acetic acid (IAA), phenylacetic acid (PAA), and 4-chloro-indole-3-acetic acid (4-Cl-IAA), with IAA being the most prevalent and significant in plants [1]. The meticulous modulation of IAA levels, governed by biosynthesis, transport, and inactivation, is essential for normal plant growth, development, and adaptation to both biotic and abiotic environmental stresses [1,3,4].

Four primary pathways have been documented for IAA inactivation in plants: (i) IAA CARBOXYL METHYLTRANSFERASE1 (IAMT1) converts IAA to methyl IAA; (ii) UDP-glucosyltransferase (UGTs) generate ester-linked IAA conjugates; (iii) GRETCHEN HAGEN 3 (GH3) acyl amido synthetases facilitate the formation of amide-linked IAA conjugates; and (iv) IAA oxidation carried out by DIOXYGENASE FOR AUXIN OXIDATION (DAO) [5–10]. Methyl IAA and ester-linked IAA, both subject to reconversion into IAA by specific hydrolases, are predominantly regarded as forms of IAA storage [11]. Contrastingly, the reversibility of amide-linked conjugates varies based on the amino acid involved. Most, such as IAA-alanine (IAA-Ala), IAA-phenylalanine (IAA-Phe), and IAA-leucine (IAA-Leu), revert to free IAA, while others like IAA-glutamate (IAA-Glu) and IAA-aspartate (IAA-Asp) are directly degraded [5,11–13]. However, recent findings suggest that IAA-Glu and IAA-Asp are also hydrolyzed to free IAA by IAA-LEU-RESISTANT1 (ILR1), and that DAO-dependent oxidation also occurs by oxidizing IAA-Asp and IAA-Glu to oxIAA-Asp and oxIAA-Glu, which in turn are hydrolyzed to oxIAA in the presence of the hydrolase ILR1, rather than by direct IAA oxidation to oxIAA. This suggests that GH3-dependent IAA conjugation may be a key node in IAA storage and IAA oxidative degradation [6,14].

The first GH3 gene was identified from *Glycine max* as a rapid early auxin-responsive gene. Subsequent research has established the widespread distribution of the GH3 gene family across the plant kingdom, encompassing species from *Arabidopsis thaliana* to *Oryza sativa*, *Zea mays*, *Triticum aestivum*, and even non-vascular plants like *Physcomitrella patens* and *Marchantia polymorpha* [8,11,15–19]. Based on sequence homology and substrate specificity, the GH3 family in Arabidopsis is categorized into three distinct groups: Group I, II, and III. Group I GH3 genes are known to encode enzymes that synthesize amides from jasmonic acid (JA) or salicylic acid (SA). Group II GH3 enzymes function as IAA-amido synthetases, and Group III has been shown to catalyze the conjugation between amino acids and 4-substituted benzoates or indole-3-butyric acid (IBA) [8,20]. Emerging evidence underscores the role of GH3-mediated IAA conjugation not only in modulating free IAA availability but also in its potential as a signaling molecule or inhibitor, impacting plant growth and development [21–24]. This review will delve into the biochemical mechanisms of GH3-mediated IAA conjugation and its transcriptional regulation.

2. The Catalytic Mechanisms and Substrate Specificity of GH3 Acyl Acid Amido Synthetase Enzyme

Chen et al. firstly utilized a combination of initial velocity and product inhibition analyses, alongside mass spectrometry, to delineate the kinetic and chemical mechanisms governing OsGH3.8 activity [25]. They discovered that the conjugation of IAA with Asp operates via a 'Bi Uni Uni Bi Ping Pong' mechanism, as depicted in Figure 1A. The process initiates with the binding of IAA and ATP, in the presence of Mg^{2+}, to the unoccupied enzyme. This interaction results in the formation of an adenylated IAA intermediate (IAA-AMP) and the concurrent release of pyrophosphate (PPi). Following this, Asp attaches to the enzyme•IAA•AMP complex, leading to the displacement of AMP and the establishment of an amide linkage between IAA and Asp. The final reaction products, IAA-Asp and AMP, are then released from the OsGH3.8 enzyme's active site [25].

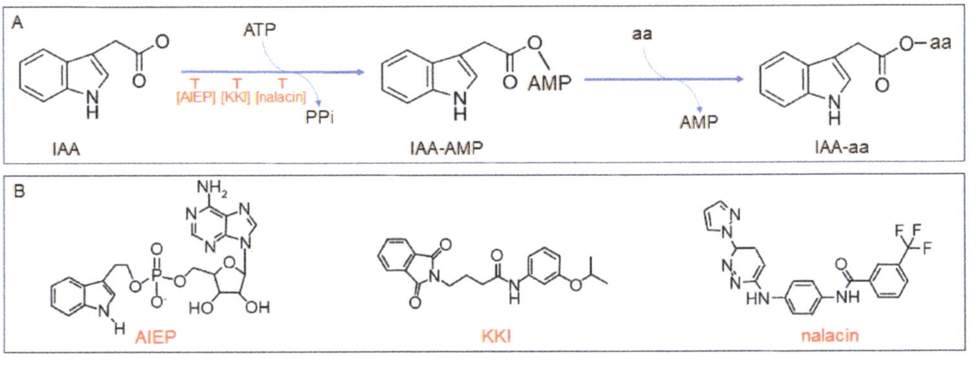

Figure 1. Catalytic reaction and inhibitor structures of Group II GH3 amido synthetase. (**A**) Schematic representation of the total reaction mediated by Group II GH3 amido synthetases. (**B**) Chemical structures of inhibitors targeting Group II GH3 amido synthetases: AIEP (adenosine-59-[2-(1H-indol-3-yl) ethyl] phosphate), KKI (kakeimide), and nalacin (N-[4-[[6-(1H-pyrazol-1-yl)-3-pyridazinyl] amino] phenyl]-3-(trifluoromethyl)benzamide).

Structural analyses of the GH3 enzyme in Arabidopsis, grape, and rice have illuminated that both monocotyledons and dicotyledons employ a similar mechanism for AMP and IAA binding [26]. The GH3 enzyme exhibits distinct acyl acid binding preferences, with specific residues within its active site conferring selectivity for particular substrates [26]. In OsGH3.8, the amino acids Arg130 and Leu137 play a crucial role in substrate specificity. The mutation of Arg130 to Leu (Arg130-Leu) shifts the enzyme's substrate preference from IAA to benzoate/SA, while an Arg130-Thr substitution favors JA over IAA. Similarly, Leu137-Ser mutation leads to a benzoate/SA preference, and the replacement of Leu137 with Arg/Ile induces a preference for JA [25,26]. GH3 proteins also exhibit amino acid specificity; for instance, the carboxylate group of Asp is a determinant for the active site's specificity in OsGH3.8 [25,26]. Ser341 participates in adenylate formation by forming hydrogen bonds with phosphate groups [26]. Moreover, Mg^{2+} is essential for the enzyme's maximum activity, aiding in AMP orientation, with Glu342 being critical for Mg^{2+} coordination [25–27]. Additionally, seven residues identified through the sequence comparison of acyl-substrate binding sites (Arg130, Leu137, Valine 174 (Val174), Leu175, Methionine 337 (Met337), Alanine 339 (Ala339), and Tyrosine 344 (Tyr344)) are thought to be involved in the substrate-specific selection of IAA [26].

It has been observed that the residues involved in the AMP binding site of acyl adenylate cleavage enzymes exhibit a high degree of conservation across the GH3 protein family. In contrast, the residues that interact with acyl substrates show variability, accommodating the binding of diverse substrates [25,26,28,29]. This variation in amino acid residues leads to the formation of different binding pockets, each tailored for specific substrates [29]. This may account for the fact that auxin function at all stages of plant growth and development, a process that offers great flexibility in regulating auxin action is necessary. The existence of a complex auxin conjugation system, as evidenced by these variations in GH3 proteins, is likely a strategic evolutionary development to facilitate this flexibility [28].

3. The Modulation of IAA Homeostasis by Small Chemical Molecules via the Inhibition of GH3 Enzyme Activity

The functional redundancy of class II GH3 enzymes in plants presents a challenge to traditional genetic approaches when exploring their biological roles [30]. To overcome this obstacle, small molecule inhibitors have emerged as a powerful alternative. These inhibitors can be applied to any plant tissue at any developmental stage, given in appropriate concentrations. Their utility lies in the ability to bypass gene redundancy and the potential detrimental effects of lethal mutations often associated with simultaneous multi-gene mutations [31]. To date, three potent inhibitors have been identified that modulate GH3-dependent IAA conjugation: adenosine-59-[2-(1H-indol-3-yl) ethyl] phosphate (AIEP), kakeimide (KKI), and N-[4-[[6-(1H-pyrazol-1-yl)-3-pyridazinyl]amino] phenyl]-3-(trifluoromethyl) benzamide (nalacin), as represented in Figure 1B [30–32].

3.1. AIEP, the First Chemical Inhibitor of Auxin Conjugation

The GH3-dependent IAA conjugation initiates when the unbound GH3 enzyme interacts with ATP and IAA, leading to the formation of IAA•AMP. Böttcher et al. engineered and synthesized a stable analogue of IAA•AMP, named AIEP. This molecule competes with ATP and IAA for the binding sites on the GH3 enzyme at the onset of catalysis [30]. The competitive inhibitory effect of AIEP on ATP and IAA binding was validated through substrate velocity experiments involving VvGH3.1 and VvGH3.6 from grape. However, the study did not extend to phenotypic examinations to assess the broader biological impacts of this inhibition (Figure 1).

3.2. KKI, a Specific Inhibitor of IAA-Conjugating GH3 Enzymes

In the quest to identify inhibitors of IAA-conjugating GH3 enzymes, researchers leveraged Arabidopsis *AtGH3.6* overexpression plants as a biological assay system. They screened a synthetic chemical library comprising 10,000 compounds for agents capable of reverting the altered root hair growth phenotype of AtGH3.6-overexpressed lines. This led to the initial identification of compound '1', followed by the synthesis of 25 derivatives of '1'. Among these, kakeimide (KKI) emerged as a highly potent inhibitor. (Figure 1B). KKI functions by directly interacting with the IAA binding site within the GH3•ATP complex, forming a stable GH3•ATP•KKI ternary complex that impedes the synthesis of IAA-amino acid conjugates [32]. Validation experiments demonstrated KKI's effectiveness in targeting the IAA binding sites of various GH3 enzymes, notably VvGH3.1, AtGH3.5, and OsGH3.8, while sparing the IBA binding site of AtGH3.15. This specificity, coupled with KKI's lack of interference in jasmonic acid (JA) homeostasis, underscores its role as a specific inhibitor of IAA-conjugating GH3 enzymes [32].

3.3. Nalacin, a Potent Inhibitor Targeting Group II GH3 Enzymes

Nalacin was identified from a chemical screen by observing the auxin-related root phenotypes in the Arabidopsis wild-type Col-0 (Figure 1B). Subsequent studies have shown that nalacin competitively inhibits substrate acceptance by AtGH3.6 and AtGH3.11 through trifluoromethyl phenyl occupancy of the IAA binding site of AtGH3s, suggesting that nalacin also functions in the first step of the 'Bi Uni Uni Bi Ping Pong' reaction of GH3 enzymes. Unlike KKI, which selectively inhibits only class II GH3 enzymes, nalacin also impedes the formation of JA amino acid conjugates mediated by AtGH3.11, albeit through a distinct binding mode [31]. Consequently, there is potential for further chemical modifications of nalacin to enhance its selective inhibition of different GH3 enzyme members.

4. The Transcriptional Control of GH3 Enzymes in Plant Growth, Development, and Stress Adaptation

Research into GH3-dependent IAA conjugation has revealed its critical role in the modulation of plant growth, development, and stress responses across a variety of species, with extensive studies conducted particularly in maize, wheat, rice, and Arabidopsis. The forthcoming sections will focus on the advances in understanding GH3-dependent IAA conjugation in Arabidopsis and rice. Additionally, progress in other species, reflecting the broader impact and relevance of GH3 enzymes in plant biology, has been systematically compiled in Table 1.

Table 1. Compilation of Group II GH3 genes across various species and their identified biological roles.

Species	Members	TFs	Biological Process	Ref.
Physcomitrella patens	*PpGH3.1* *PpGH3.2*		High temperature and salt tolerance	[33,34]
Stylosanthes guianensis	*SgGH3.1*		Chilling and cold tolerance	[35]
Dianthus caryophyllus	*DcGH3.1*		Adventitious root development	[36]
Brassica oleracea	*BoGH3.12*		Cadmium tolerance	[37]
Pisum sativum	*PsGH3.5*		Seedlings development	[38]
Cucumis sativus	*CsGH3.5*		Adventitious root formation	[39]
Capsicum chinense	*CcGH3*		Fruit ripening	[40]
Solanum lycopersicum	*SlGH3.8*	YABBY2b	Plant height	[41]
	SlGH3.2		Fruit ripening	[42]
	SlGH3.15		Lateral root development; gravitropism	[43]
Coffea canephora	*CcGH3.1* *CcGH3.6* *CcGH3.17*		Somatic embryogenesis	[44,45]
Zea mays	*ZmGH3.2*	DREB2A	Seed aging tolerance	[46]
Vitis vinifera	*VvGH3.1* *VvGH3.2*		Berry ripening	[30] [21]
	VvGH3.6		Tissue auxin homeostasis	[28,30]
Citrus sinensis	*CsGH3.1* *CsGH3.1L*		Susceptibility to pathogen	[47]
Malus sieversii	*MsGH3.5*	RR1a	Shoot and root development	[48]
Malus domestica	*MdGH3-2*	bHLH3	Leaf shape	[49]

Table 1. *Cont.*

Species	Members	TFs	Biological Process	Ref.
Castanea sativa	*CsGH3.1*		Adventitious root development	[50]
Carya cathayensis	*CcGH3*		Grafting	[51]
Picea abies	*PaGH3.gII.8* *PaGH3.gII.9* *PaGH3.17*		Tissue auxin homeostasis	[52]
Betula platyphylla	*BpGH3.3* *BpGH3.5a* *BpGH3.5b* *BpGH3.9*		Tissue auxin homeostasis	[53]

4.1. GH3-Dependent IAA Conjugation Is Involved in Regulating Multiple Developmental Processes

In Arabidopsis, eight Group II *GH3* genes are involved in catalyzing IAA conjugation to amino acids. Due to redundant gene functions, mutations in single genes result in only subtle phenotypic changes and modified sensitivity to exogenous IAA [54]. In contrast, mutants with overexpressed *GH3* genes, obtained through activation tag insertion, provide a more discernible phenotype for study. For instance, *AtGH3.2* and *AtGH3.6* were identified through the screening of their overexpressed mutants, *ydk1-D* and *dfl1-D*, respectively [54,55]. Interestingly, however, despite all overexpressing genes being closely related to *GH3* Group II family members, they still showed inconsistent phenotypes. Under various light conditions, the *dfl1-D* mutant displayed shortened hypocotyls exclusively in light, while the *ydk1-D* mutant showed this phenotype under both light and dark conditions. Additionally, the *ydk1-D* mutant had a shorter primary root but did not exhibit significant difference in susceptibility to auxin-mediated root growth inhibition. In contrast, the *dfl1-D* mutant was resistant to IAA-mediated root growth inhibition and did not present a short-root phenotype compared with the wild type [54,55]. Furthermore, several *Group II GH3* genes are transcriptionally induced by IAA, whereas *AtGH3.9* is repressed in response to exogenous IAA application [56]. Altogether, these data indicate that, while Group II GH3 members share some commonalities and function in a similar pathway by regulating free IAA conjugation, they each play distinct roles in plant development, which may be due to differences in their tissue specificity and/or hormonal regulation (e.g., IAA, JA, etc.).

As early auxin response genes, Group II *GH3* genes play a significant role downstream of auxin response factors (ARFs), which are key elements in the auxin signaling pathway [57]. Research has elucidated the involvement of ARF6, AtARF7, AtARF8, and AtARF17 in the transcriptional regulation of several *AtGH3* genes. AtARF7 and AtARF8 are known to positively regulate the transcription of *AtGH3.2*, *AtGH3.5*, and *AtGH3.6*, influencing hypocotyl elongation under different light conditions [54,58]. In contrast, AtARF17 has a negative regulatory role on *AtGH3.5* and *AtGH3.6* transcription, which is essential for proper plant development. The microRNA AtmiR160 directly targets *AtARF17* mRNA, which is crucial for normal leaf and root growth [59]. In terms of adventitious root development, AtARF6 and AtARF8 act as positive regulators, while AtARF17 functions as a negative regulator. They co-regulate the transcription of *AtGH3.3*, *AtGH3.5*, and *AtGH3.6*, impacting JA conjugation with amino acids, but not IAA conjugation [60]. Additionally, AtARF7 actives another auxin-induced transcription factor (TF), AtMYB77, which subsequently upregulates *AtGH3.2* and *AtGH3.3* to control root development [61]. AtMYB30, another MYB TF, directly binds to the promoters of *AtGH3.2* and *AtGH3.3*, repressing their transcription to foster root elongation [62]. WRINKLED1 (WRI1), yet another TF, binds to the *AtGH3.3* promoter. Although *AtGH3.3* transcription increases in the *Atwri1-1* mutant, AtWRI1 does not appear to repress *AtGH3.3* directly, suggesting it may function as a transcriptional co-factor [63,64]. Subsequent research identified that AtTCP20, a TF that interacts with AtWRI1, binds to the *AtGH3.3* promoter, and activates its transcription [63]. Electrophoretic mobility shift assays have shown that the addition of AtWRI1 decreases the binding activity of AtTCP20 to the *AtGH3.3* promoter in a dose-dependent manner, indicating that AtWRI1 regulates *AtGH3.3* transcription by antagonizing AtTCP20's binding [63]. Several basic leucine zipper (bZIP) transcription factors, including AtbZIP2, 11, 44, 53, and 63, also directly bind to the *AtGH3.3* promoter and enhance its transcription [65]. Furthermore, cytokinin influences the root meristem by maintaining IAA concentration, with AtARR1 directly activating *AtGH3.17* transcription [66]. There are also indications of TFs AtHLS1, AtURO, and AtSTY1 being involved in modulating GH3-dependent IAA conjugation in plant developmental processes, although direct regulatory evidence is yet to be established (Figure 2) [67–69].

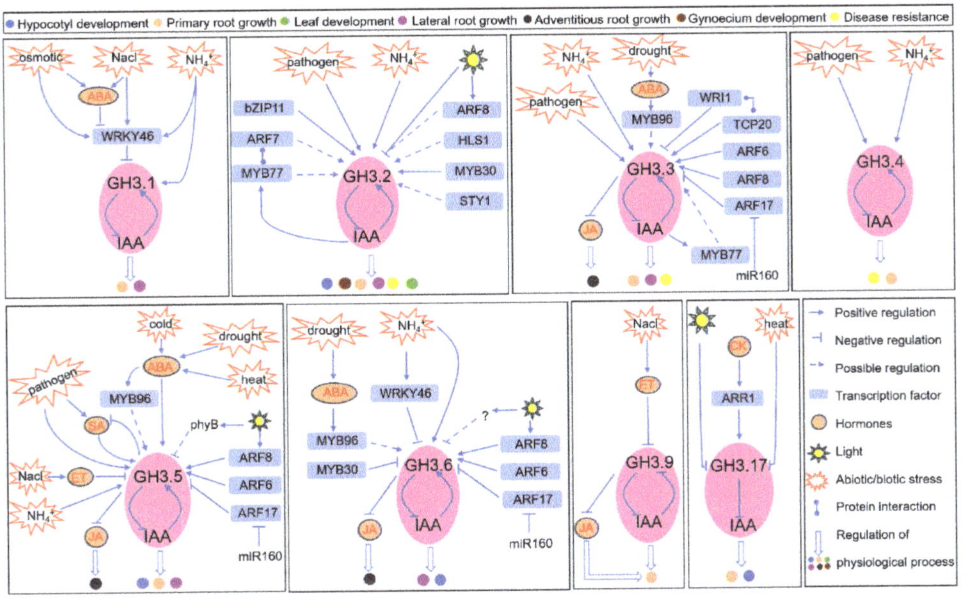

Figure 2. Regulatory network of Arabidopsis GH3-dependent IAA conjugation in abiotic and biotic stress responses. Abbreviation: NH_4^+, ammonium; WRKY46, WRKY DNA-BINDING PROTEIN46; bZIP11, basic leucine Zipper11; ARF6/7/8/17, auxin response factor6/7/8/17; MYB30/77/96, myeloblastosis30/77/96; WRI1, WRINKLED1; TCP20, TEOSINTE BRANCHED 1; CYCLOIDEA, PCF (TCP)-DOMAIN FAMILY PROTEIN 20; ARR1, Arabidopsis response regulator1; miR160, microRNA; IAA, indole-3-acetic acid; ABA, abscisic acid; JA, jasmonic acid; ET, ethylene; SA, salicylic acid; CK, cytokinin.

In rice, the Group II GH3 family, which includes *OsGH3.1*, *OsGH3.2*, *OsGH3.5*, *OsGH3.8*, and *OsGH3.13*, plays a significant role in various developmental aspects such as shoot height, leaf angle, floret fertility, and tiller number. These effects are primarily mediated through the modulation of IAA conjugation [70–74]. Recent findings highlight several TFs that directly activate the transcription of *OsGH3* genes, leading to decreased levels of free IAA and consequent impacts on rice morphology. For example, OsbZIP49 has been shown to upregulate *OsGH3.2* and *OsGH3.13* by binding to TGACG motifs in their promoters. This upregulation results in reduced free IAA concentrations, affecting cell elongation mediated by expansins and thus impacting shoot gravitropism [75]. Furthermore, *OsGH3.2* is regulated by OsARF8, a downstream target of OsmiR167, indicating a critical role for the OsmiR167-OsARF8-OsGH3.2 pathway in cellular auxin homeostasis, particularly in response to exogenous auxin [76]. Another ARF, OsARF19, has been found to reduce free IAA levels by activating *OsGH3.5*, thereby influencing leaf angulation. Intriguingly, the transcription of *OsARF19* itself is induced by IAA and brassinolide (BR), suggesting the

OsARF19-OsGH3.5 module's involvement in integrating IAA-BR signals [73]. Additionally, OsSPL7, a target of OsmiR156f, directly activates *OsGH3.8*, affecting tiller number and shoot height [70]. The involvement of OsMADS1 and OsMADS6 in binding to the *OsGH3.8* promoter and regulating floret fertility has also been documented [77]. Collectively, these findings demonstrate that *Group II GH3* genes, together with their upstream TFs, form a complex regulatory network that integrates light, miRNA, and hormonal signals to control plant growth and development (Figure 3) [78].

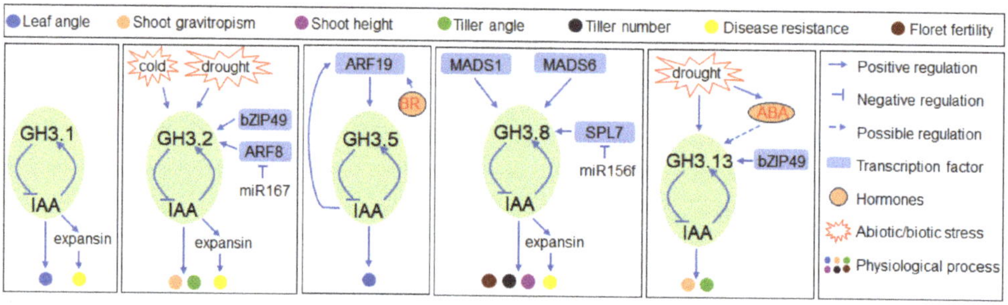

Figure 3. Role of rice GH3-dependent IAA conjugation in response to abiotic and biotic stresses and its transcriptional regulation. Abbreviation: bZIP49, basic leucine Zipper49; ARF8/19, auxin response factor; MADS1/6, MADS-domain transcription factor1/6; SPL7, SQUAMOSA PROMOTER BINDING PROTEIN-LIKE 7; miR156f/167, microRNA156f/167; IAA, indole-3-acetic acidl; BR, brassinolide; ABA, abscisic acid.

4.2. The Integration of Hormonal Signals in GH3-Dependent IAA Conjugation's Responses to Abiotic Stresses

4.2.1. Drought Stress

In Arabidopsis, *AtGH3.5* has been observed to respond rapidly to drought conditions, with the *wes1-D* mutant, which overexpresses *AtGH3.5*, exhibiting enhanced drought resistance [79]. A subsequent study revealed that AtMYB96 modulates the expression of several *AtGH3* genes, including *AtGH3.3*, *AtGH3.5*, and *AtGH3.6*, under drought stress through an abscisic acid (ABA)-dependent pathway [80]. This finding underscores the importance of ABA signaling in modulating *GH3* gene expression during drought response. Additionally, the *gh3oct* mutant, with knockouts of all *Group II GH3* genes (*GH3.1,2,3,4,5,6,9,17*), exhibits increased drought tolerance, further highlighting the role of these genes in drought response mechanisms [81]. In rice, the activation of *OsGH3.13* has been linked to a reduction in free IAA levels, leading to a structural adaptation in the leaves, such as thicker blades, which enhance drought tolerance by minimizing water loss [74]. However, the response to drought stress in rice is complex, as evidenced by the contrasting effects observed with *OsGH3.2*. While *OsGH3.2* is also upregulated in response to drought, its overexpression leads to decreased drought tolerance. This discrepancy is attributed to the inhibition of carotenoid biosynthesis by overexpressed *OsGH3.2*, which consequently reduces ABA synthesis. This is in stark contrast to the increased ABA levels seen in lines overexpressing *OsGH3.13* [72]. These findings collectively indicate that, while *GH3* genes are integral to stress responses through IAA homeostasis regulation, the distinct spatial–temporal expression patterns and secondary growth effects can result in varying stress sensitivities [72,74].

4.2.2. Temperature (Heat/Cold/Freezing) Stress

In Arabidopsis, the transcription of *AtGH3.5* is notably responsive to temperature extremes, showing increased levels under both low (4 °C) and high (37 °C) temperature conditions. The *wes1-D* mutant, characterized by the overexpression of *AtGH3.5*, shows increased survival after exposure to freezing temperatures (−7 °C). This suggests a broad regulatory role for *AtGH3.5* across a spectrum of temperature stresses [82]. In rice, the overexpression of *OsGH3.2* leads to a reduction in free IAA levels, thereby activating cold-responsive genes and enhancing the plant's ability to scavenge reactive oxygen species (ROS). Consequently, this confers increased resistance to cold stress [72].

4.2.3. Salt and Osmotic Stress

All root-expressed Group II *GH3* genes in Arabidopsis are upregulated following treatment with NaCl at concentrations of 75 mM and 150 mM. The *Atgh3oct* mutant, with combined knockouts of all Group II *GH3* genes, exhibits greater resilience to NaCl stress compared to the wild type [81]. This enhanced tolerance also extends to sorbitol and mannitol exposure, suggesting that Group II GH3s may confer broad osmotic stress resistance, inclusive of salinity stress. Further investigation reveals that NaCl treatment increases *AtACS2* transcription, leading to the accumulation of the ethylene precursor ACC, which in turn downregulates *AtGH3.5* and *AtGH3.9* transcription, maintaining free IAA levels and primary root growth [83]. Additionally, both osmotic and salt stresses are known to stimulate the accumulation of ABA, which reduces free IAA content, inhibiting lateral root (LR) development. At the same time, these stresses activate AtWRKY46, a transcription factor that suppresses *AtGH3.1* transcription, thus maintaining free IAA levels and LR development. Additionally, ABA inhibits *AtWRKY46* transcription. This antagonistic interaction between ABA and AtWRKY46 fine-tunes free IAA levels and LR development, allowing for improved adaptation to osmotic and salt stresses [84].

4.2.4. Ammonium (NH_4^+) Stress

NH_4^+ serves as a vital nitrogen source for plants, but when available in excess, it can be detrimental to growth [85,86]. Prior research indicates that high NH_4^+ levels lead to a reduction in free IAA [87–89]. In Arabidopsis, elevated NH_4^+ conditions trigger the induction of nearly all Group II *GH3* genes, which in turn accelerates the conjugation of free IAA to amino acids [87]. Concurrently, high NH_4^+ levels also enhance the transcription of *AtWRKY46*, a transcription factor that binds to the promoters of *AtGH3.1* and *AtGH3.6*, repressing their transcription. This response serves to maintain free IAA levels and support primary root growth under high NH_4^+stress conditions. The overexpression of *AtWRKY46* improves NH_4^+ tolerance, suggesting its critical role in modulating primary root development during NH_4^+ stress [90].

4.2.5. Pathogen Stress

In response to pathogen attacks, plants activate a comprehensive defense strategy: (1) they initiate a hypersensitive response leading to rapid programmed cell death at the infection site alongside other defense responses; (2) they activate systemic-acquired resistance (SAR) in distal tissues; and (3) they activate basal immunity to limit pathogen growth [91]. GH3-dependent IAA conjugation is intricately involved in these plant defense mechanisms. In Arabidopsis, the pathogens *B. cinerea* and *P. syringae pv tomato* (*Pst*) DC3000 significantly upregulate *AtGH3.2* and *AtGH3.3* transcription. Loss-of-function mutations in *AtGH3.2* or *AtGH3.4* enhance resistance to both pathogens, suggesting that *AtGH3.2*, *AtGH3.3*, and *AtGH3.4* may negatively influence the plant's response to *B. cinerea* and *Pst* DC3000 [92]. The overexpression of mutant *gh3.5-1D* shows a compromised hypersensitive response but retains normal SAR and basal immunity, whereas the *Atgh3.5* mutant exhibits a defective SAR response yet maintains a typical hypersensitive response and basal immunity. In contrast, the *dfl1-D* mutant displays altered hypersensitive and basal immune responses [91].

In rice, IAA has been linked to increased susceptibility to various pathogens, partly due to IAA-induced expansins that relax the cell wall, a plant's primary defense barrier [93]. Overexpressing certain Group II *GH3* genes, such as *OsGH3.1*, *OsGH3.2*, and *OsGH3.8*, bolsters resistance against bacterial and fungal pathogens. This resistance is attributed to the suppression of pathogen-induced IAA accumulation, leading to the downregulation of expansin expression and subsequent stabilization of the cell wall [71,93,94].

The varied effects of Group II *GH3* genes on pathogen stress between rice and Arabidopsis could be attributed to multiple factors. In rice, OsGH3s appear to primarily contribute to forming a robust cell wall barrier, which acts as the first line of defense against pathogen entry, without further invoking a hypersensitive response or SAR [39,94]. In contrast, the expression of *AtGH3* in Arabidopsis is highly specific to tissue type and developmental stage, and unspecific overexpression could lead to unintended consequences, such as disruptions in the levels of IAA and salicylic acid (SA), which are critical for the plant's defense response [91,92]. Additionally, the GH3-catalyzed IAA-Asp conjugate is hypothesized to act as a susceptibility signal [23]. While GH3 activity reduces free IAA levels, it simultaneously results in the accumulation of IAA-Asp. The specific distribution and balance of IAA and IAA-Asp within the plant may significantly influence the outcome of GH3's role in pathogen response [92].

5. Atypical Roles of Group II GH3 and IAA-Amino Acids

The exploration of Group II GH3 enzymes as IAA-acyl acid amino synthetases in 2002 marked a significant advancement in our understanding of these enzymes [11]. In addition to the determination of their three-dimensional structures and the key amino acid sites for enzyme activity, research has expanded their known functions to include the conjugation of other auxins like PAA and IBA with amino acids, IAA conjugation with proteins, and the metabolism of auxinic herbicides [95]. Furthermore, AtGH3.15, previously an undefined member of Group III, is now recognized as an IBA-conjugating acyl acid amido synthetase. IBA serves as a precursor to IAA, which is converted through the peroxisomal β-oxidation pathway [96]. Intriguingly, AtGH3.13, AtGH3.14, and AtGH3.16—other members of Group III—possess acyl acid binding sites similar to AtGH3.15, suggesting their possible involvement in IAA homeostasis [20,95]. Here, we will discuss the atypical roles of Group II GH3 and IAA-amino acids.

5.1. The Roles of Group II GH3 beyond the Catalyzation of IAA-Amino Acid Conjugate Formation

The capacity of Group II GH3 enzymes extends beyond the synthesis of IAA-amino acid conjugates. Research using recombinant GH3 IAA-amino acid synthetase from pea has revealed the enzyme's ability to conjugate IAA not only to aspartate but also to proteins in immature seeds. The proposition that IAA conjugation to proteins may serve a regulatory function acting as a prosthetic group and influencing protein activity via posttranslational modifications is a compelling avenue for further exploration [97,98]. Besides proteins, Group II GH3 enzymes also facilitate the conjugation of PAA to amino acids. PAA, another natural but less active auxin than IAA, exhibits unique distribution and transport characteristics, implying a role in sustaining the auxin equilibrium necessary for plant cellular processes [99–101]. Notably, PAA can stimulate GH3-dependent IAA conjugation, while high IAA levels can suppress PAA biosynthesis, underscoring the importance of Group II GH3 enzymes in balancing IAA/PAA ratios [99]. Moreover, this enzyme group, along with AtGH3.17, has been implicated in the detoxification of the auxinic herbicide 2,4-DB, suggesting their potential application in herbicide resistance strategies [102].

5.2. The Specialized Functions of IAA-Amino Acid Conjugates beyond Their Role as Auxin Stock

Previous studies found that the exogenous addition of IAA-aa and IAA both rapidly increased content-free IAA levels and exhibited similar high growth factor phenotypes, suggesting that IAA-aa is a storage form of IAA [103]. Subsequent studies found that these IAA-aa, IAA-Leu, and IAA-Ala could be reversibly converted to free IAA by the action of the hydrolases, IAA-LEUCINE RESIS TANT1 (ILR1), ILR1-LIKE proteins (ILLs), and IAA-ALANINE RESISTANT3 (IAR3) [5,104,105]. IAA-Glu and IAA-Asp, once considered only as degradation intermediates, are now recognized as reversible storage forms [14]. Beyond storage, IAA-aa have been identified as possessing unique biological functions. IAA-Trp, for instance, acts as a 'super inactivator' by not only consuming free IAA for its synthesis but also antagonizing the activity of residual IAA, with IAA-Trp significantly mitigating root inhibition effects caused by IAA [24]. IAA-Asp has been reported to have more diverse roles: (1) correcting the temperature sensitivity of henbane (*Hyoscyamus muticus*) XIIB2 (temperature-sensitive variant) cells [22]; (2) IAA-Asp directly and specifically enhance the pea (*Pisum sativum*) responses to abiotic stress by increasing the antioxidant enzyme activity and then reducing the H_2O_2 concentration [23]; (3) IAA-Asp as a ripening signal in grapes (*Vitis vinifera*) can be perceived at a certain stage of fruit development; however, the mechanism of sensing remains unknown [21]; and (4) IAA-Asp promotes pathogen development in plants by regulating the transcription of virulence genes [92]. These insights suggest that IAA-amino acids are not just byproducts of GH3 activity but are biologically active molecules with specific roles.

6. Concluding Remarks

Our review underscores that Group II GH3 enzymes and IAA-amino acids are integral components of plant biology, with roles extending beyond the simple conversion of free IAA to its stored form. The functional diversity of these enzymes and their products in plant growth, development, and stress adaptation is a testament to the complexity of plant hormonal regulation. Each *GH3* gene exhibits specific spatial and temporal expression patterns, contributing to a nuanced regulatory network. These findings challenge the traditional view of IAA-amino acids as mere storage forms and highlight the need to consider their biological activity in understanding plant physiology and development.

Author Contributions: P.L., T.-T.L. and D.-W.D. drafted and wrote this review; D.-W.D., Q.M. and W.-M.S. edited the manuscript. All authors have read and agreed to the published version of the manuscript.

Funding: This work was supported by grants from the National Natural Science Foundation of China (Grant No. 32030099) and Special Project on people's livelihood of Gansu Province (Grant No. 21CX6NA082).

Data Availability Statement: Not applicable.

Conflicts of Interest: The authors declare no conflict of interest.

References

1. Di, D.W.; Zhang, C.; Luo, P.; An, C.W.; Guo, G.Q. The biosynthesis of auxin: How many paths truly lead to IAA? *Plant Growth Regul.* **2016**, *78*, 275–285. [CrossRef]
2. Fiedler, L.; Friml, J. Rapid auxin signaling: Unknowns old and new. *Curr. Opin. Plant Biol.* **2023**, *75*, 102443. [CrossRef] [PubMed]
3. Jing, H.; Wilkinson, E.G.; Sageman-Furnas, K.; Strader, L.C. Auxin and abiotic stress responses. *J. Exp. Bot.* **2023**, *74*, 7000–7014. [CrossRef] [PubMed]
4. Luo, P.; Di, D.W. Precise regulation of the TAA1/TAR-YUCCA auxin biosynthesis pathway in plants. *Int. J. Mol. Sci.* **2023**, *24*, 8514. [CrossRef] [PubMed]
5. LeClere, S.; Tellez, R.; Rampey, R.A.; Matsuda, S.P.T.; Bartel, B. Characterization of a family of IAA-amino acid conjugate hydrolases from Arabidopsis. *J. Biol. Chem.* **2002**, *277*, 20446–20452. [CrossRef]
6. Mellor, N.; Band, L.R.; Pencík, A.; Novák, O.; Rashed, A.; Holman, T.; Wilson, M.H.; Voss, U.; Bishopp, A.; King, J.R.; et al. Dynamic regulation of auxin oxidase and conjugating enzymes and modulates auxin homeostasis. *Proc. Natl. Acad. Sci. USA* **2016**, *113*, 11022–11027. [CrossRef]
7. Porco, S.; Pencík, A.; Rashed, A.; Voss, U.; Casanova-Sáez, R.; Bishopp, A.; Golebiowska, A.; Bhosale, R.; Swarup, R.; Swarup, K.; et al. Dioxygenase-encoding gene controls IAA oxidation and homeostasis in Arabidopsis. *Proc. Natl. Acad. Sci. USA* **2016**, *113*, 11016–11021. [CrossRef]
8. Staswick, P.E.; Serban, B.; Rowe, M.; Tiryaki, I.; Maldonado, M.T.; Maldonado, M.C.; Suza, W. Characterization of an Arabidopsis enzyme family that conjugates amino acids to indole-3-acetic acid. *Plant Cell* **2005**, *17*, 616–627. [CrossRef]
9. Yang, Y.; Xu, R.; Ma, C.J.; Vlot, A.C.; Klessig, D.F.; Pichersky, E. Inactive methyl indole-3-acetic acid ester can be hydrolyzed and activated by several esterases belonging to the AMES esterase family of Arabidopsis. *Plant Physiol.* **2008**, *147*, 1034–1045. [CrossRef]
10. Zhang, J.; Lin, J.E.; Harris, C.; Pereira, F.C.M.; Wu, F.; Blakeslee, J.J.; Peer, W.A. DAO1 catalyzes temporal and tissue-specific oxidative inactivation of auxin in. *Proc. Natl. Acad. Sci. USA* **2016**, *113*, 11010–11015. [CrossRef] [PubMed]

11. Casanova-Saez, R.; Mateo-Bonmati, E.; Ljung, K. Auxin Metabolism in Plants. *CSH Perspect. Biol.* **2021**, *13*, a039867. [CrossRef] [PubMed]
12. Ostin, A.; Kowalyczk, M.; Bhalerao, R.P.; Sandberg, G. Metabolism of indole-3-acetic acid in Arabidopsis. *Plant Physiol.* **1998**, *118*, 285–296. [CrossRef] [PubMed]
13. Rampey, R.A.; LeClere, S.; Kowalczyk, M.; Ljung, K.; Sandberg, G.; Bartel, B. A family of auxin-conjugate hydrolases that contributes to free indole-3-acetic acid levels during Arabidopsis germination. *Plant Physiol.* **2004**, *135*, 978–988. [CrossRef]
14. Hayashi, K.I.; Arai, K.; Aoi, Y.; Tanaka, Y.; Hira, H.; Guo, R.; Hu, Y.; Ge, C.; Zhao, Y.; Kasahara, H.; et al. The main oxidative inactivation pathway of the plant hormone auxin. *Nat. Commun.* **2021**, *12*, 6752. [CrossRef] [PubMed]
15. Feng, S.; Yue, R.; Tao, S.; Yang, Y.; Zhang, L.; Xu, M.; Wang, H.; Shen, C. Genome-wide identification, expression analysis of auxin-responsive *GH3* family genes in maize (*Zea mays* L.) under abiotic stresses. *J. Integr. Plant Biol.* **2015**, *57*, 783–795. [CrossRef]
16. Hagen, G.; Kleinschmidt, A.; Guilfoyle, T. Auxin-regulated gene expression in intact soybean hypocotyl and excised hypocotyl sections. *Planta* **1984**, *162*, 147–153. [CrossRef] [PubMed]
17. Jain, M.; Kaur, N.; Tyagi, A.K.; Khurana, J.P. The auxin-responsive *GH3* gene family in rice (*Oryza sativa*). *Funct. Integr. Genom.* **2006**, *6*, 36–46. [CrossRef]
18. Jiang, W.; Yin, J.; Zhang, H.; He, Y.; Shuai, S.; Chen, S.; Cao, S.; Li, W.; Ma, D.; Chen, H. Genome-wide identification, characterization analysis and expression profiling of auxin-responsive GH3 family genes in wheat (*Triticum aestivum* L.). *Mol. Biol. Rep.* **2020**, *47*, 3885–3907. [CrossRef]
19. Ludwig-Muller, J.; Julke, S.; Bierfreund, N.M.; Decker, E.L.; Reski, R. Moss (*Physcomitrella patens*) GH3 proteins act in auxin homeostasis. *New Phytol.* **2009**, *181*, 323–338. [CrossRef]
20. Sherp, A.M.; Westfall, C.S.; Alvarez, S.; Jez, J.M. Arabidopsis thaliana GH3.15 acyl acid amido synthetase has a highly specific substrate preference for the auxin precursor indole-3-butyric acid. *J. Biol. Chem.* **2018**, *293*, 4277–4288. [CrossRef]
21. Böttcher, C.; Burbidge, C.A.; Boss, P.K.; Davies, C. Interactions between ethylene and auxin are crucial to the control of grape (*Vitis vinifera* L.) berry ripening. *BMC Plant Biol.* **2013**, *13*, 222. [CrossRef]
22. Oetiker, J.H.; Aeschbacher, G. Temperature-sensitive plant cells with shunted indole-3-acetic acid conjugation. *Plant Physiol.* **1997**, *114*, 1385–1395. [CrossRef] [PubMed]
23. Ostrowski, M.; Ciarkowska, A.; Jakubowska, A. The auxin conjugate indole-3-acetyl-aspartate affects responses to cadmium and salt stress in *Pisum sativum* L. *J. Plant Physiol.* **2016**, *191*, 63–72. [CrossRef] [PubMed]
24. Staswick, P.E. The tryptophan conjugates of jasmonic and indole-3-acetic acids are endogenous auxin inhibitors. *Plant Physiol.* **2009**, *150*, 1310–1321. [CrossRef]
25. Chen, Q.; Westfall, C.S.; Hicks, L.M.; Wang, S.; Jez, J.M. Kinetic basis for the conjugation of auxin by a GH3 family indole-acetic acid-amido synthetase. *J. Biol. Chem.* **2010**, *285*, 29780–29786. [CrossRef]
26. Xu, G.L.; Zhang, Y.K.; Li, M.Y.; Jiao, X.; Zhou, L.; Ming, Z.H. Crystal structure of the acyl acid amido synthetase GH3-8 from *Oryza sativa*. *Biochem. Biophys. Res. Commun.* **2021**, *534*, 266–271. [CrossRef] [PubMed]
27. Takase, T.; Nakazawa, M.; Ishikawa, A.; Manabe, K.; Matsui, M. *DFL2*, a new member of the Arabidopsis *GH3* gene family, is involved in red light-specific hypocotyl elongation. *Plant Cell Physiol.* **2003**, *44*, 1071–1080. [CrossRef]
28. Peat, T.S.; Bottcher, C.; Newman, J.; Lucent, D.; Cowieson, N.; Davies, C. Crystal structure of an indole-3-acetic acid amido synthetase from grapevine involved in auxin homeostasis. *Plant Cell* **2012**, *24*, 4525–4538. [CrossRef]
29. Westfall, C.S.; Zubieta, C.; Herrmann, J.; Kapp, U.; Nanao, M.H.; Jez, J.M. Structural basis for prereceptor modulation of plant hormones by GH3 proteins. *Science* **2012**, *336*, 1708–1711. [CrossRef]
30. Bottcher, C.; Dennis, E.G.; Booker, G.W.; Polyak, S.W.; Boss, P.K.; Davies, C. A novel tool for studying auxin-metabolism: The inhibition of grapevine indole-3-acetic acid-amido synthetases by a reaction intermediate analogue. *PLoS ONE* **2012**, *7*, e37632. [CrossRef]
31. Xie, Y.; Zhu, Y.; Wang, N.; Luo, M.; Ota, T.; Guo, R.; Takahashi, I.; Yu, Z.; Aizezi, Y.; Zhang, L.; et al. Chemical genetic screening identifies nalacin as an inhibitor of GH3 amido synthetase for auxin conjugation. *Proc. Natl. Acad. Sci. USA* **2022**, *119*, e2209256119. [CrossRef]
32. Fukui, K.; Arai, K.; Tanaka, Y.; Aoi, Y.; Kukshal, V.; Jez, J.M.; Kubes, M.F.; Napier, R.; Zhao, Y.; Kasahara, H.; et al. Chemical inhibition of the auxin inactivation pathway uncovers the roles of metabolic turnover in auxin homeostasis. *Proc. Natl. Acad. Sci. USA* **2022**, *119*, e2206869119. [CrossRef] [PubMed]
33. Koochak, H.; Ludwig-Muller, J. *Physcomitrium patens* mutants in auxin conjugating GH3 proteins show salt stress tolerance but auxin homeostasis is not involved in regulation of oxidative stress factors. *Plants* **2021**, *10*, 10071398. [CrossRef] [PubMed]
34. Mittag, J.; Gabrielyan, A.; Ludwig-Müller, J. Knockout of *GH3* genes in the moss *Physcomitrella patens* leads to increased IAA levels at elevated temperature and in darkness. *Plant Physiol. Biochem.* **2015**, *97*, 339–349. [CrossRef] [PubMed]
35. Jiang, M.; Ma, L.L.; Huang, H.A.; Ke, S.W.; Gui, C.S.; Ning, X.Y.; Zhang, X.Q.; Zhong, T.X.; Xie, X.M.; Chen, S. Overexpression of *SgGH3.1* from fine-stem stylo (*Stylosanthes guianensis* var. *intermedia*) enhances chilling and cold tolerance in *Arabidopsis thaliana*. *Genes* **2021**, *12*, 1367. [CrossRef] [PubMed]
36. Cano, A.; Sanchez-Garcia, A.B.; Albacete, A.; Gonzalez-Bayon, R.; Justamante, M.S.; Ibanez, S.; Acosta, M.; Perez-Perez, J.M. Enhanced conjugation of auxin by GH3 enzymes leads to poor adventitious rooting in carnation stem cuttings. *Front. Plant Sci.* **2018**, *9*, 566. [CrossRef] [PubMed]

37. Jiang, J.; Wang, Z.; Kong, X.; Chen, Y.; Li, J. Exogenous tryptophan application improves cadmium tolerance and inhibits cadmium upward transport in broccoli (*Brassica oleracea* var. *italica*). *Front. Plant Sci.* **2022**, *13*, 969675. [CrossRef]
38. Ostrowski, M.; Świdziński, M.; Ciarkowska, A.; Jakubowska, A. IAA-amido synthetase activity and GH3 expression during development of pea seedlings. *Acta Physiol. Plant* **2014**, *36*, 3029–3037. [CrossRef]
39. Dong, C.J.; Liu, X.Y.; Xie, L.L.; Wang, L.L.; Shang, Q.M. Salicylic acid regulates adventitious root formation via competitive inhibition of the auxin conjugation enzyme CsGH3.5 in cucumber hypocotyls. *Planta* **2020**, *252*, 75. [CrossRef]
40. Liu, K.D.; Kang, B.C.; Jiang, H.; Moore, S.L.; Li, H.X.; Watkins, C.B.; Setter, T.L.; Jahn, M.M. A GH3-like gene, CcGH3, isolated from *Capsicum chinense* L. fruit is regulated by auxin and ethylene. *Plant Mol. Biol.* **2005**, *58*, 447–464. [CrossRef]
41. Sun, M.H.; Li, H.; Li, Y.B.; Xiang, H.Z.; Liu, Y.D.; He, Y.; Qi, M.F.; Li, T.L. Tomato YABBY2b controls plant height through regulating indole-3-acetic acid-amido synthetase (GH3.8) expression. *Plant Sci.* **2020**, *297*, 110530. [CrossRef]
42. Sravankumar, T.; Akash; Naik, N.; Kumar, R. A ripening-induced *SlGH3-2* gene regulates fruit ripening via adjusting auxin-ethylene levels in tomato (*Solanum lycopersicum* L.). *Plant Mol. Biol.* **2018**, *98*, 455–469. [CrossRef]
43. Ai, G.; Huang, R.; Zhang, D.; Li, M.; Li, G.; Li, W.; Ahiakpa, J.K.; Wang, Y.; Hong, Z.; Zhang, J. *SlGH3.15*, a member of the GH3 gene family, regulates lateral root development and gravitropism response by modulating auxin homeostasis in tomato. *Plant Sci.* **2023**, *330*, 111638. [CrossRef]
44. Ayil-Gutierrez, B.; Galaz-Avalos, R.; Pena-Cabrera, E.; Loyola-Vargas, V. Dynamics of the concentration of IAA and some of its conjugates during the induction of somatic embryogenesis in *Coffea canephora*. *Plant Signal. Behav.* **2013**, *8*, e26998. [CrossRef]
45. Mendez-Hernandez, H.A.; Quintana-Escobar, A.O.; Uc-Chuc, M.A.; Loyola-Vargas, V.M. Genome-wide analysis, modeling, and identification of amino acid binding motifs suggest the involvement of *GH3* genes during somatic embryogenesis of *Coffea canephora*. *Plants* **2021**, *10*, 2034. [CrossRef]
46. Han, Q.; Chen, K.; Yan, D.; Hao, G.; Qi, J.; Wang, C.; Dirk, L.M.A.; Bruce Downie, A.; Gong, J.; Wang, J.; et al. ZmDREB2A regulates *ZmGH3.2* and *ZmRAFS*, shifting metabolism towards seed aging tolerance over seedling growth. *Plant J.* **2020**, *104*, 268–282. [CrossRef]
47. Zou, X.; Long, J.; Zhao, K.; Peng, A.; Chen, M.; Long, Q.; He, Y.; Chen, S. Overexpressing *GH3.1* and *GH3.1L* reduces susceptibility to *Xanthomonas citri* subsp. citri by repressing auxin signaling in citrus (*Citrus sinensis* Osbeck). *PLoS ONE* **2019**, *14*, e0220017. [CrossRef]
48. Zhao, D.; Wang, Y.; Feng, C.; Wei, Y.; Peng, X.; Guo, X.; Guo, X.; Zhai, Z.; Li, J.; Shen, X.; et al. Overexpression of *MsGH3.5* inhibits shoot and root development through the auxin and cytokinin pathways in apple plants. *Plant J.* **2020**, *103*, 166–183. [CrossRef] [PubMed]
49. Hu, D.G.; Wang, N.; Wang, D.H.; Cheng, L.; Wang, Y.X.; Zhao, Y.W.; Ding, J.Y.; Gu, K.D.; Xiao, X.; Hao, Y.J. A basic/helix-loop-helix transcription factor controls leaf shape by regulating auxin signaling in apple. *New Phytol.* **2020**, *228*, 1897–1913. [CrossRef] [PubMed]
50. Vielba, J.M.; VERico, S.; Covelo, P.; Vidal, N.; Sánchez, C. Expression of a GH3 gene during adventitious rooting in Chestnut. In Proceedings of the 4th International Conference of the IUFRO Unit 2.09.02 on "Development and Application of Vegetative Propagation Technologies in Plantation Forestry to Cope with a Changing Climate and Environment", La Plata, Argentina, 19–23 September 2016.
51. Xu, D.; Yang, Y.; Tao, S.; Wang, Y.; Yuan, H.; Sharma, A.; Wang, X.; Shen, C.; Yan, D.; Zheng, B. Identification and expression analysis of auxin-responsive *GH3* family genes in Chinese hickory (*Carya cathayensis*) during grafting. *Mol. Biol. Rep.* **2020**, *47*, 4495–4506. [CrossRef] [PubMed]
52. Brunoni, F.; Collani, S.; Casanova-Saez, R.; Simura, J.; Karady, M.; Schmid, M.; Ljung, K.; Bellini, C. Conifers exhibit a characteristic inactivation of auxin to maintain tissue homeostasis. *New Phytol.* **2020**, *226*, 1753–1765. [CrossRef]
53. Zhang, R.S.; Wang, Y.C.; Wang, C.; Wei, Z.G.; Xia, D.; Wang, Y.F.; Liu, G.F.; Yang, C.P. Time-course analysis of levels of indole-3-acetic acid and expression of auxin-responsive *GH3* genes in *Betula platyphylla*. *Plant Mol. Biol. Rep.* **2011**, *29*, 898–905. [CrossRef]
54. Takase, T.; Nakazawa, M.; Ishikawa, A.; Kawashima, M.; Ichikawa, T.; Takahashi, N.; Shimada, H.; Manabe, K.; Matsui, M. *ydk1-D*, an auxin-responsive GH3 mutant that is involved in hypocotyl and root elongation. *Plant J.* **2004**, *37*, 471–483. [CrossRef]
55. Nakazawa, M.; Yabe, N.; Ichikawa, T.; Yamamoto, Y.Y.; Yoshizumi, T.; Hasunuma, K.; Matsui, M. *DFL2*, an auxin-responsive *GH3* gene homologue, negatively regulates shoot cell elongation and lateral root formation, and positively regulates the light response of hypocotyl length. *Plant J.* **2001**, *25*, 213–221.
56. Khan, S.; Stone, J.M. *Arabidopsis thaliana* GH3.9 in auxin and jasmonate cross talk. *Plant Signal Behav.* **2007**, *2*, 483–485. [CrossRef]
57. Di, D.W.; Zhang, C.G.; Guo, G.Q. Involvement of secondary messengers and small organic molecules in auxin perception and signaling. *Plant Cell Rep.* **2015**, *34*, 895–904. [CrossRef]
58. Tian, C.; Muto, H.; Higuchi, K.; Matamura, T.; Tatematsu, K.; Koshiba, T.; Yamamoto, K.T. Disruption and overexpression of auxin response factor 8 gene of Arabidopsis affect hypocotyl elongation and root growth habit, indicating its possible involvement in auxin homeostasis in light condition. *Plant J.* **2004**, *40*, 333–343. [CrossRef] [PubMed]
59. Mallory, A.C.; Bartel, D.P.; Bartel, B. MicroRNA-directed regulation of Arabidopsis is essential for proper development and modulates expression of early auxin response genes. *Plant Cell* **2005**, *17*, 1360–1375. [CrossRef] [PubMed]

60. Gutierrez, L.; Mongelard, G.; Flokova, K.; Pacurar, D.I.; Novak, O.; Staswick, P.; Kowalczyk, M.; Pacurar, M.; Demailly, H.; Geiss, G.; et al. Auxin controls Arabidopsis adventitious root initiation by regulating jasmonic acid homeostasis. *Plant Cell* **2012**, *24*, 2515–2527. [CrossRef] [PubMed]
61. Shin, R.; Burch, A.Y.; Huppert, K.A.; Tiwari, S.B.; Murphy, A.S.; Guilfoyle, T.J.; Schachtman, D.P. The Arabidopsis transcription factor MYB77 modulates auxin signal transduction. *Plant Cell* **2007**, *19*, 2440–2453. [CrossRef] [PubMed]
62. Zhao, C.Y.; Xue, H.W. PI4Kγ2 interacts with E3 ligase MIEL1 to regulate auxin metabolism and root development. *Plant Physiol.* **2020**, *184*, 933–944. [CrossRef] [PubMed]
63. Kong, Q.; Low, P.M.; Lim, A.R.Q.; Yang, Y.; Yuan, L.; Ma, W.; Kong, Q.; Low, P.M.; Lim, A.R.Q.; Yang, Y.; et al. Functional antagonism of WRI1 and TCP20 modulates *GH3.3* expression to maintain auxin homeostasis in roots. *Plants* **2022**, *11*, 454. [CrossRef] [PubMed]
64. Kong, Q.; Ma, W.; Yang, H.; Ma, G.; Mantyla, J.J.; Benning, C. The Arabidopsis WRINKLED1 transcription factor affects auxin homeostasis in roots. *J. Exp. Bot.* **2017**, *68*, 4627–4634. [CrossRef] [PubMed]
65. Weiste, C.; Dröge-Laser, W. The *Arabidopsis* transcription factor bZIP11 activates auxin-mediated transcription by recruiting the histone acetylation machinery. *Nat. Commun.* **2014**, *5*, 3883. [CrossRef] [PubMed]
66. Di Mambro, R.; De Ruvo, M.; Pacifici, E.; Salvi, E.; Sozzani, R.; Benfey, P.N.; Busch, W.; Novak, O.; Ljung, K.; Di Paola, L.; et al. Auxin minimum triggers the developmental switch from cell division to cell differentiation in the Arabidopsis root. *Proc. Natl. Acad. Sci. USA* **2017**, *114*, E7641–E7649. [CrossRef] [PubMed]
67. Ohto, M.A.; Hayashi, S.; Sawa, S.; Hashimoto-Ohta, A.; Nakamura, K. Involvement of HLS1 in sugar and auxin signaling in Arabidopsis leaves. *Plant Cell Physiol.* **2006**, *47*, 1603–1611. [CrossRef] [PubMed]
68. Sohlberg, J.J.; Myrenas, M.; Kuusk, S.; Lagercrantz, U.; Kowalczyk, M.; Sandberg, G.; Sundberg, E. STY1 regulates auxin homeostasis and affects apical-basal patterning of the Arabidopsis gynoecium. *Plant J.* **2006**, *47*, 112–123. [CrossRef]
69. Sun, Y.; Yang, Y.; Yuan, Z.; Ludwig-Müller, J.; Yu, C.; Xu, Y.F.; Shao, X.H.; Li, X.F.; Decker, E.L.; Reski, R.; et al. Overexpression of the Arabidopsis gene *UPRIGHT ROSETTE* reveals a homeostatic control for indole-3-acetic acid. *Plant Physiol.* **2010**, *153*, 1311–1320. [CrossRef]
70. Dai, Z.; Wang, J.; Yang, X.; Lu, H.; Miao, X.; Shi, Z. Modulation of plant architecture by the miR156f-OsSPL7-OsGH3.8 pathway in rice. *J. Exp. Bot.* **2018**, *69*, 5117–5130. [CrossRef]
71. Domingo, C.; Andrés, F.; Tharreau, D.; Iglesias, D.J.; Talón, M. Constitutive Expression of OsGH3.1 Reduces Auxin Content and Enhances Defense Response and Resistance to a Fungal Pathogen in Rice. *Mol. Plant-Microbe Interact.* **2009**, *22*, 201–210. [CrossRef]
72. Du, H.; Wu, N.; Fu, J.; Wang, S.P.; Li, X.H.; Xiao, J.H.; Xiong, L.Z. A GH3 family member, OsGH3-2, modulates auxin and abscisic acid levels and differentially affects drought and cold tolerance in rice. *J. Exp. Bot.* **2012**, *63*, 6467–6480. [CrossRef] [PubMed]
73. Zhang, S.N.; Wang, S.K.; Xu, Y.X.; Yu, C.L.; Shen, C.J.; Qian, Q.; Geisler, M.; Jiang, D.A.; Qi, Y.H. The auxin response factor, OsARF19, controls rice leaf angles through positively regulating OsGH3-5 and OsBRI1. *Plant Cell Environ.* **2015**, *38*, 638–654. [CrossRef] [PubMed]
74. Zhang, S.W.; Li, C.H.; Cao, J.; Zhang, Y.C.; Zhang, S.Q.; Xia, Y.F.; Sun, D.Y.; Sun, Y. Altered architecture and enhanced drought tolerance in rice via the down-regulation of indole-3-acetic acid by TLD1/OsGH3.13 activation. *Plant Physiol.* **2009**, *151*, 1889–1901. [CrossRef] [PubMed]
75. Ding, C.H.; Lin, X.H.; Zuo, Y.; Yu, Z.L.; Baerson, S.R.; Pan, Z.Q.; Zeng, R.S.; Song, Y.Y. Transcription factor OsbZIP49 controls tiller angle and plant architecture through the induction of indole-3-acetic acid-amido synthetases in rice. *Plant J.* **2021**, *108*, 1346–1364. [CrossRef]
76. Yang, J.H.; Han, S.J.; Yoon, E.K.; Lee, W.S. Evidence of an auxin signal pathway, microRNA167-ARF8-GH3, and its response to exogenous auxin in cultured rice cells. *Nucleic Acids Res.* **2006**, *34*, 1892–1899. [CrossRef]
77. Yadav, S.R.; Khanday, I.; Majhi, B.B.; Veluthambi, K.; Vijayraghavan, U. Auxin-responsive OsMGH3, a common downstream target of OsMADS1 and OsMADS6, controls rice floret fertility. *Plant Cell Physiol.* **2011**, *52*, 2123–2135. [CrossRef]
78. Luo, P.; Di, D.; Wu, L.; Yang, J.; Lu, Y.; Shi, W. MicroRNAs are involved in regulating plant development and stress response through fine-tuning of TIR1/AFB-dependent auxin signaling. *Int. J. Mol. Sci.* **2022**, *23*, 510. [CrossRef]
79. Park, J.E.; Park, J.Y.; Kim, Y.S.; Staswick, P.E.; Jeon, J.; Yun, J.; Kim, S.Y.; Kim, J.; Lee, Y.H.; Park, C.M. GH3-mediated auxin homeostasis links growth regulation with stress adaptation response in Arabidopsis. *J. Biol. Chem.* **2007**, *282*, 10036–10046. [CrossRef] [PubMed]
80. Seo, P.J.; Xiang, F.; Qiao, M.; Park, J.Y.; Lee, Y.N.; Kim, S.G.; Lee, Y.H.; Park, W.J.; Park, C.M. The MYB96 transcription factor mediates abscisic acid signaling during drought stress response in Arabidopsis. *Plant Physiol.* **2009**, *151*, 275–289. [CrossRef]
81. Casanova-Saez, R.; Mateo-Bonmati, E.; Simura, J.; Pencik, A.; Novak, O.; Staswick, P.; Ljung, K. Inactivation of the entire Arabidopsis group II GH3s confers tolerance to salinity and water deficit. *New Phytol.* **2022**, *235*, 263–275. [CrossRef] [PubMed]
82. Park, J.E.; Seo, P.J.; Lee, A.K.; Jung, J.H.; Kim, Y.S.; Park, C.M. An Arabidopsis GH3 gene, encoding an auxin-conjugating enzyme, mediates phytochrome B-regulated light signals in hypocotyl growth. *Plant Cell Physiol.* **2007**, *48*, 1236–1241. [CrossRef]
83. Han, S.; Jia, M.Z.; Yang, J.F.; Jiang, J. The integration of ACS2-generated ACC with GH3-mediated IAA homeostasis in NaCl-stressed primary root elongation of Arabidopsis seedlings. *Plant Growth Regul.* **2019**, *88*, 151–158. [CrossRef]

84. Ding, Z.J.; Yan, J.Y.; Li, C.X.; Li, G.X.; Wu, Y.R.; Zheng, S.J. Transcription factor WRKY46 modulates the development of Arabidopsis lateral roots in osmotic/salt stress conditions via regulation of ABA signaling and auxin homeostasis. *Plant J.* **2015**, *84*, 56–69. [CrossRef] [PubMed]
85. Xiao, C.; Fang, Y.; Wang, S.; He, K. The alleviation of ammonium toxicity in plants. *J. Integr. Plant Biol.* **2023**, *65*, 1362–1368. [CrossRef] [PubMed]
86. Di, D.W. New molecular mechanisms of plant response to ammonium nutrition. *Appl. Sci.* **2023**, *13*, 11570. [CrossRef]
87. Di, D.W.; Li, G.J.; Sun, L.; Wu, J.J.; Wang, M.; Kronzucker, H.J.; Fang, S.; Chu, J.F.; Shi, W.M. High ammonium inhibits root growth in Arabidopsis thaliana by promoting auxin conjugation rather than inhibiting auxin biosynthesis. *J. Plant Physiol.* **2021**, *261*, 153415. [CrossRef]
88. Di, D.W.; Sun, L.; Zhang, X.N.; Li, G.J.; Kronzucker, H.J.; Shi, W.M. Involvement of auxin in the regulation of ammonium tolerance in rice (*Oryza sativa* L.). *Plant Soil.* **2018**, *432*, 373–387. [CrossRef]
89. Dziewit, K.; Pencik, A.; Dobrzynska, K.; Novak, O.; Szal, B.; Podgorska, A. Spatiotemporal auxin distribution in Arabidopsis tissues is regulated by anabolic and catabolic reactions under long-term ammonium stress. *BMC Plant Biol.* **2021**, *21*, 602. [CrossRef]
90. Di, D.W.; Sun, L.; Wang, M.; Wu, J.; Kronzucker, H.J.; Fang, S.; Chu, J.; Shi, W.; Li, G. WRKY46 promotes ammonium tolerance in Arabidopsis by repressing NUDX9 and indole-3-acetic acid-conjugating genes and by inhibiting ammonium efflux in the root elongation zone. *New Phytol.* **2021**, *232*, 190–207. [CrossRef]
91. Zhang, Z.Q.; Li, Q.; Li, Z.M.; Staswick, P.E.; Wang, M.Y.; Zhu, Y.; He, Z.H. Dual regulation role of GH3.5 in salicylic acid and auxin signaling during Arabidopsis-*Pseudomonas syringae* interaction. *Plant Physiol.* **2007**, *145*, 450–464. [CrossRef]
92. González-Lamothe, R.; El Oirdi, M.; Brisson, N.; Bouarab, K. The conjugated auxin indole-3-acetic acid-aspartic acid promotes plant disease development. *Plant Cell* **2012**, *24*, 762–777. [CrossRef]
93. Fu, J.; Liu, H.B.; Li, Y.; Yu, H.H.; Li, X.H.; Xiao, J.H.; Wang, S.P. Manipulating broad-spectrum disease resistance by suppressing pathogen-induced auxin accumulation in rice. *Plant Physiol.* **2011**, *155*, 589–602. [CrossRef]
94. Ding, X.; Cao, Y.; Huang, L.; Zhao, J.; Xu, C.; Li, X.; Wang, S. Activation of the indole-3-acetic acid-amido synthetase GH3-8 suppresses expansin expression and promotes salicylate- and jasmonate-independent basal immunity in rice. *Plant Cell* **2008**, *20*, 228–240. [CrossRef]
95. Jez, J.M. Connecting primary and specialized metabolism: Amino acid conjugation of phytohormones by GRETCHEN HAGEN 3 (GH3) acyl acid amido synthetases. *Curr. Opin. Plant Biol.* **2022**, *66*, 102194. [CrossRef]
96. Zolman, B.K.; Martinez, N.; Millius, A.; Adham, A.R.; Bartel, B. Identification and characterization of Arabidopsis indole-3-butyric acid response mutants defective in novel peroxisomal enzymes. *Genetics* **2008**, *180*, 237–251. [CrossRef] [PubMed]
97. Ostrowski, M.; Ciarkowska, A. Pea GH3 acyl acid amidosynthetase conjugates IAA to proteins in immature seeds of *Pisum sativum* L.—A new perspective on formation of high-molecular weight conjugates of auxin. *J. Plant Physiol.* **2021**, *256*, 153312. [CrossRef] [PubMed]
98. Seidel, C.; Walz, A.; Park, S.; Cohen, J.D.; Ludwig-Muller, J. Indole-3-acetic acid protein conjugates: Novel players in auxin homeostasis. *Plant Biol.* **2006**, *8*, 340–345. [CrossRef] [PubMed]
99. Aoi, Y.; Tanaka, K.; Cook, S.D.; Hayashi, K.I.; Kasahara, H. GH3 auxin-amido synthetases alter the ratio of indole-3-acetic acid and phenylacetic acid in Arabidopsis. *Plant Cell Physiol.* **2020**, *61*, 596–605. [CrossRef]
100. Mashiguchi, K.; Hisano, H.; Takeda-Kamiya, N.; Takebayashi, Y.; Ariizumi, T.; Gao, Y.; Ezura, H.; Sato, K.; Zhao, Y.; Hayashi, K.I.; et al. *Agrobacterium tumefaciens* enhances biosynthesis of two distinct auxins in the formation of crown galls. *Plant Cell Physiol.* **2019**, *60*, 29–37. [CrossRef] [PubMed]
101. Sugawara, S.; Mashiguchi, K.; Tanaka, K.; Hishiyama, S.; Sakai, T.; Hanada, K.; Kinoshita-Tsujimura, K.; Yu, H.; Dai, X.; Takebayashi, Y.; et al. Distinct characteristics of indole-3-acetic acid and phenylacetic acid, two common auxins in plants. *Plant Cell Physiol.* **2015**, *56*, 1641–1654. [CrossRef]
102. Sherp, A.M.; Lee, S.G.; Schraft, E.; Jez, J.M. Modification of auxinic phenoxyalkanoic acid herbicides by the acyl acid amido synthetase GH3.15 from Arabidopsis. *J. Biol. Chem.* **2018**, *293*, 17731–17738. [CrossRef] [PubMed]
103. Wang, Q.; De Gernier, H.; Duan, X.L.; Xie, Y.M.; Geelen, D.; Hayashi, K.; Xuan, W.; Geisler, M.; ten Tusscher, K.; Beeckman, T.; et al. GH3-mediated auxin inactivation attenuates multiple stages of lateral root development. *New Phytol.* **2023**, *240*, 1900–1912. [CrossRef] [PubMed]
104. Bartel, B.; Fink, G.R. Ilr1, an amidohydrolase that releases active indole-3-acetic-acid from conjugates. *Science* **1995**, *268*, 1745–1748. [CrossRef] [PubMed]
105. Davies, R.T.; Goetz, D.H.; Lasswell, J.; Anderson, M.N.; Bartel, B. IAR3 encodes an auxin conjugate hydrolase from Arabidopsis. *Plant Cell* **1999**, *11*, 365–376. [CrossRef]

plants

MDPI

Article

All-Year High IAA and ABA Contents in Rhizome Buds May Contribute to Natural Four-Season Shooting in Woody Bamboo *Cephalostachyum pingbianense*

Wei Mao [1,†], Changyan Bao [2,†], Qian Cheng [2], Ning Liang [2,3], Lianchun Wang [4,*] and Hanqi Yang [2,3,*]

1 Faculty of Foreign Languages, Southwest Forestry University, Kunming 650233, China; mwei99@sina.com
2 Institute of Highland Forest Science, Chinese Academy of Forestry, Kunming 650233, China; baochangyan@caf.ac.cn (C.B.); cqq08042@163.com (Q.C.); ln417@126.com (N.L.)
3 Key Laboratory of Breeding and Utilization of Resource Insects, National Forestry and Grassland Administration, Kunming 650233, China
4 Forestry College, Southwest Forestry University, Kunming 650233, China
* Correspondence: wanglianchun@swfu.edu.cn (L.W.); yanghanqikm@caf.ac.cn (H.Y.)
† These authors contributed equally to this work.

Abstract: To explore the regulation mechanism of endogenous phytohormones on rhizome bud germination in *Cephalostachyum pingbianense*, the contents of IAA, ABA, GA, and CTK in seven above- and under-ground bamboo structure components were determined using enzyme-linked immunosorbent assays (ELISA). The results showed that a higher content of IAA, GA, and CTK all year was found in above-ground components and dormant rhizome buds. Meanwhile, a higher ABA content in young shoots and a lower ABA content in the culm base and dormant rhizome buds were detected during the peak period of shooting. The amounts of emerging shoots and the grown bamboo culms were positively correlated with the content of IAA and the ratio of IAA/ABA and (IAA + CTK + GA)/ABA, while they were negatively correlated with the ratio of CTK/IAA in dormant rhizome buds. The all-year high contents of IAA (19–31 ng/g) and ABA (114–144 ng/g) in rhizome buds, as well as interactions among four hormones, may be the key physiological mechanisms to maintain rhizome bud germination throughout the year in *C. pingbianense*. As *C. pingbianense* is a special bamboo species of multi-season shoot sprouting, the above results may supplement scientific data for a comprehensive understanding of physiological mechanisms within the bamboo subfamily.

Keywords: *Cephalostachyum pingbianense*; bamboo shoot; endogenous hormones; ABA; IAA; rhizome bud

1. Introduction

Endogenous phytohormones are compounds synthesized by plants in trace amounts that participate in regulating the entire growth process of plants and stress responses to the natural environment. Therefore, endogenous phytohormones play a key role in the life span of plants [1–5]. Tiller from the lateral bud is an important form of plant growth, especially in clonal plants. Various hormones, such as auxin/IAA, cytokinin (CTK), gibberellin (GA), and abscisic acid (ABA), are involved in the process of lateral bud germination [1,5]. For example, ABA can affect the expression of key transcription factors that maintain bud dormancy [6,7], while IAA [8–10], CTK [11,12], and GA [13] cooperatively promote germination and development of lateral buds in multiple plants.

The rhizome bud germination of bamboo, i.e., bamboo shooting, is the main pathway of asexual reproduction in woody bamboos, as well as the foundation of bamboo resource cultivation and bamboo industry [14–16]. According to the type of underground rhizome, woody bamboos can be totally divided into two types: the scattered bamboos and the clump bamboos, which display different shooting phenologies. The scattered bamboo species (such as *Phyllostachys edulis*) are distributed in the subtropical zone and typically experience a shooting peak from March to May. On the other hand, the clump bamboos occur in tropical regions (such as *Dendrocalamus latiflorus*) and have a shoot-sprouting period from June to August. Both types start shooting soon after the beginning of the rainy season [15,16]. Therefore, empirically, the rhizome buds germination of woody bamboos with a single shooting peak is generally considered to be induced by the regulation of endogenous hormone synthesis and transport following an increase in soil temperature and precipitation in the rainy season [8]. The previous studies indicated the mass fractions of various phytohormones fluctuated regularly during the shooting period, implying that antagonism or coordination of endogenous hormones, such as IAA, GA, and ABA, were key signaling factors for bamboo shoot initiation [17–19].

Woody bamboos are typical clonal plants, and endogenous hormones will transport and regulate physiological activity in the whole gene system [20]. In the present study, we adopted the hypothesis that there is extensive in situ synthesis and transport of endogenous hormones among different ramets (or culms) of bamboo genet, and these phytohormones synergistically regulate the germination of rhizome buds and culm growth and development [14]. Although abundant achievements have been made in the mechanism of bamboo rhizome bud germination, there are still some defects in these studies: (1) mostly focused on bamboo species with a single sprouting peak, especially scattered bamboo; (2) Previous studies, as far as we knew, were only based on the shooting season (lasting ca. 3–4 months), without examining the content changes of endogenous hormones throughout the year, e.g., [17–19]; (3) Most studies only focused on rhizome and its lateral buds and disregarded the connection between the above-ground and underground components of bamboo genet, which may be due to the large and complex culm branching and rhizome system of woody bamboos [21]. Thus far, the precise regulatory mechanism of endogenous hormones on rhizome bud germination in bamboo is still largely unclear.

Compared to the infinitely extended rhizome system in the scattered bamboos, the clone ramets of the clump bamboos are clustered together, similar to rice, forming an ideal and closed observation system for in vivo endogenous phytohormones, which is very conducive to studying the biosynthesis and transmission process of endogenous hormones among clone ramets in bamboos. *Cephalostachyum pingbianense* is a small-scaled cluster bamboo and mainly occurs in southern Yunnan, China [22]. *C. pingbianense* is currently the only known bamboo species that can produce bamboo shoots throughout the year under natural conditions. According to existing research, this species has two peak periods for shoot sprouting, namely in autumn and spring [23,24]. Compared with the majority of woody bamboos with a single shooting peak, *C. pingbianense* displays an interesting shooting phenology and provides an ideal experimental material for studying the germination mechanism of bamboos, which is of great significance for understanding the mechanism diversity of rhizome buds germination within woody bamboos.

In this article, we fixed-site observed shooting situations, examined the concentration changes of four key endogenous hormones, i.e., IAA, CTK, GA, and ABA, in the above-ground and underground components of the entire ramet system during four seasons in *C. pingbianense* clumps, and further analyzed the correlation between endogenous hormones and the amounts of the emerging and degraded shoots in each season. Thus, we inferred the possible regulatory mechanism of endogenous hormones on naturally occurring four-season shooting in *C. pingbianense*. This study is helpful to the understanding of the biosynthesis and interactions of endogenous hormones in the ramets of *C. pingbianense*, which provides a reference for analyzing its unique shooting mechanism and improves our comprehensive understanding of the diversity of shooting mechanisms in woody bamboos.

2. Results

2.1. Sprouting Phenology and Quantity of Bamboo Shoots

The phenology observation results of shoot sprouting showed that *C. pingbianense* could produce bamboo shoots throughout the year (Table 1, Figure 1). The least number of shoots appeared in June, with only one shoot out of six clumps. For the convenience of discussion, we assumed that June is the initial stage of the bamboo shooting of *C. pingbianense*. Meanwhile, *C. pingbianense* displayed two peak periods of shooting, with the highest shoot number in October, followed by March (Figure 1A). According to the season division, the highest number of shoots appeared in the autumn, followed by the spring. The season with the highest number of degraded shoots was spring, followed by winter (Figure 1B). The highest survived shoot number was in autumn, making up 38.3% of the total number of survived bamboo shoots, while the lowest was in summer (3.5%). The number of bamboo shoots that survived in the spring was almost equal to that in the winter. Overall, the peak period of degraded bamboo shoots was delayed by about one month compared to the peak period of emergent shoots, and the number of survived bamboo shoots (or young bamboo) was positively correlated with the number of sprouting shoots.

Table 1. Shooting phenology in six sample clumps of *C. pingbianense*.

Season/		**Spring**			**Summer**			**Autumn**			**Winter**		
Month	**2**	**3**	**4**	**5**	**6**	**7**	**8**	**9**	**10**	**11**	**12**	**1**	**Total**
Number of emerging shoots	17	18	7	3	1	4	9	15	20	15	14	12	135
Number of degraded shoots	6	5	6	2	1	2	2	3	6	8	5	3	49
Number of grown bamboo	11	13	1	1	0	2	7	12	14	7	9	9	86
Proportion of emerging shoots (amount/%)		42/31.1%			8/5.9%			44/32.6%			41/30.4%		
Proportion of degraded shoots (amount/%)		17/34.7%			5/10.2%			11/22.4%			16/32.7%		
Proportion of grown bamboo (amount/%)		25/29.1%			3/3.5%			33/38.3%			25/29.1%		

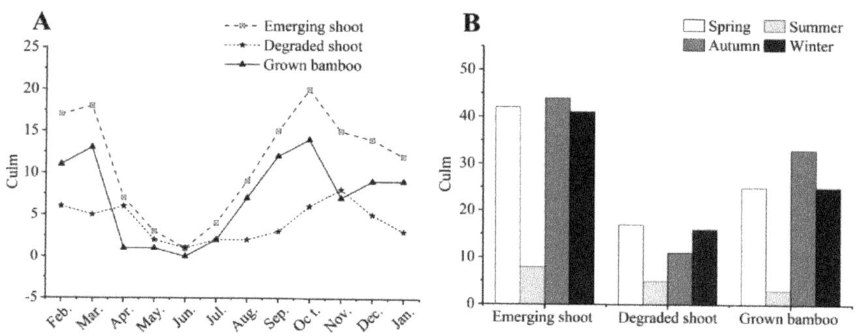

Figure 1. Emergent and degraded bamboo shoot amounts in six sample clumps of *C. pingbianense*. (**A**) Monthly and (**B**) seasonal bamboo shoot amounts.

2.2. Annual Spatial-Temporal Distribution of Endogenous Hormones

The results of content determination with the fresh weight of four endogenous hormones in *C. pingbianense* indicated that the mass fraction of four hormones in different components was significantly different (Figure 2). In each examined component, ABA had the highest content of 89–144 ng/g, with a relative abundance of 74.7–85.7%. The following sequence of content was as follows: IAA 15–31 ng/g (10.6–19.5%), CTK 2.6–6.7 ng/g (1.5–4.5%), and GA 1.4–4.0 ng/g (0.8–2.5%).

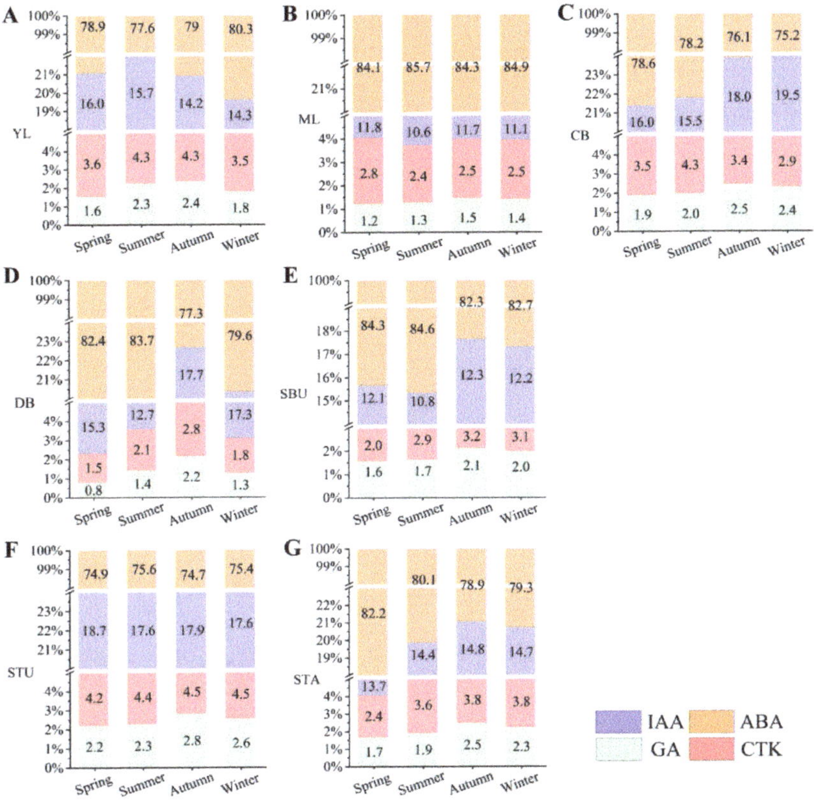

Figure 2. The relative abundance changes of four endogenous hormones in the examined components of *C. pingbianense*. (**A**) Young leaves (YL), (**B**) mature leaves (ML), (**C**) culm base (CB), (**D**) dormant rhizome buds (DB), (**E**) shoot base underground (SBU), (**F**) shoot tips underground (STU), and (**G**) shoot tips above-ground (STA).

The annual content changes of four endogenous hormones in seven components of *C. pingbianense* were as follows (Figures 3 and 4).

IAA: Spatially, the contents of IAA in the underground components (including dormant rhizome buds and shoots) were totally higher than those of the above-ground components (including young leaves, mature leaves, and culm bases), and young tissues such as young leaves and shoot tips had much higher IAA contents than mature tissues such as culm bases and mature leaves. Temporally, IAA contents in the examined components were totally higher in the peak period of shoot sprouting, such as spring, autumn, and winter, than in the initial stage, i.e., from May to July.

ABA: Spatially, the highest average content of ABA was in dormant rhizome buds, and the lowest one was in culm bases. Totally, dormant rhizome buds and mature tissues such as mature leaves and shoot bases underground had a higher average ABA content than shoots, especially in shoot tips. Temporally, among the above-ground components, both young leaves and mature leaves had the highest ABA content in November, while the lowest contents occurred in March in mature leaves and in July in young leaves. Culm bases displayed an opposite trend, namely, the highest value was in March and the lowest value was in November. On the other hand, among the underground components, the shoots had the lowest ABA content in March or May and the highest content in November.

Conversely, dormant rhizome buds had the highest ABA content in March and the lowest value in November.

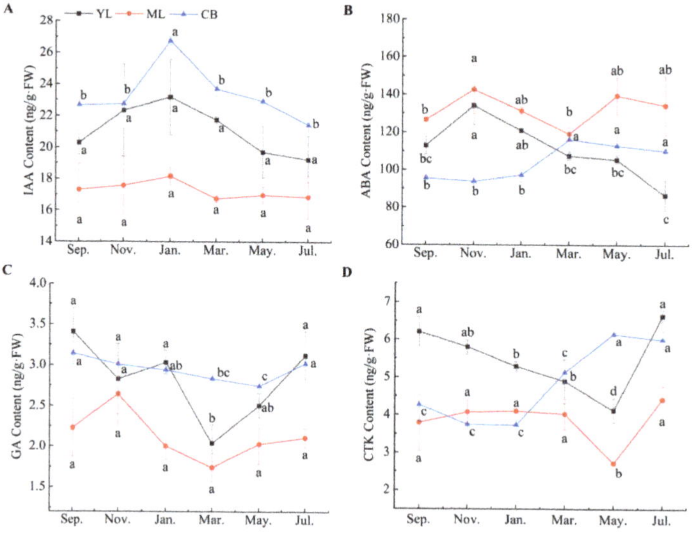

Figure 3. The annual content changes of four endogenous hormones in the above-ground components. Different lowercase letters in each sub-figure indicate significant differences at $p < 0.05$. (**A**) IAA content; (**B**) ABA content; (**C**) GA content; (**D**) CTK content. YL, young leaves; ML, mature leaves; CB, culm base.

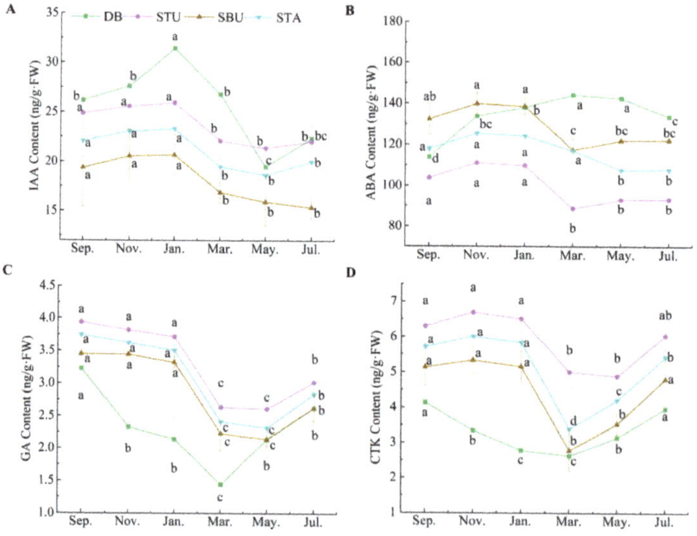

Figure 4. The annual content changes of four endogenous hormones in the underground components. Different lowercase letters in each sub-figure indicate significant differences at $p < 0.05$. (**A**) IAA content; (**B**) ABA content; (**C**) GA content; (**D**) CTK content. DB, dormant rhizome buds; STU, shoot tip underground; SBU, shoot base underground; STA, shoot tip above-ground.

GA: Spatially, dormant rhizome buds had the lowest average GA contents among the examined components, and as a whole, the underground components had a higher GA content than the above-ground components. Among the three above-ground components, mature leaves had the lowest GA contents. Temporally, in the peak period of shoot sprouting around September, the highest GA contents in the examined components were observed, while the lowest contents were found in the initial stage around May. Among them, the lowest GA contents in young leaves, mature leaves, and dormant rhizome buds appeared in March.

CTK: Spatially, taken overall, the CTK contents of underground components were higher than in the above-ground components, and the CTK contents of young tissues such as young leaves and shoot tips were much higher than those in mature tissues. Similar to GA, dormant rhizome buds also had the lowest average CTK contents among the examined components. Temporally, both young leaves and mature leaves had the lowest CTK content in May and the highest value in July, while culm base had the highest CTK content in May and the lowest CTK value in November. Among the underground components, shoots and rhizome buds had the lowest CTK contents around March and the highest contents around November.

2.3. Correlation Analysis of Endogenous Hormones

2.3.1. Endogenous Hormone Content Distribution at the Component Level

To explore possible biosynthesis and transportation of the examined endogenous hormones among different components, we selected the data of endogenous hormone contents at the initial period (May) and peak period (November) of shooting phenology to analyze the distribution correlation of endogenous hormones at the component level. The correlation analysis results were as follows (Figure 5).

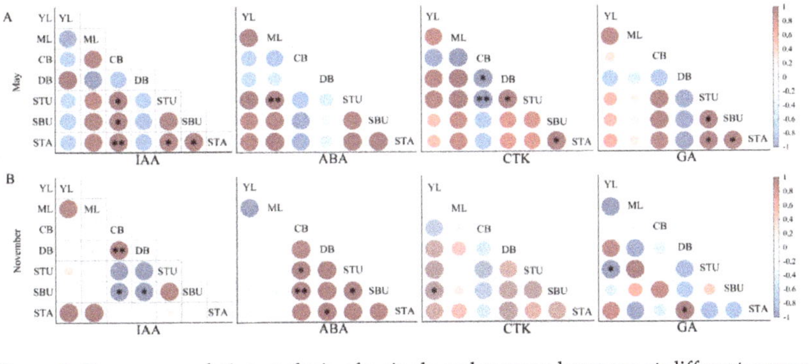

Figure 5. Pearson correlation analysis of a single endogenous hormone at different components. * indicates $p < 0.05$, and ** indicates $p < 0.01$. (**A**) Initial period of shooting in May; (**B**) peak period of shooting in November. YL, young leaves; ML, mature leaves; CB, culm base; DB, dormant rhizome buds; SBU, shoot base underground; STU, shoot tip underground; STA, shoot tip above-ground. The size and shade of the disc were directly proportional to the absolute value of the correlation coefficient.

Initial period (May): IAA contents in culm base (CB) were positively correlated with those in shoot tip underground (STU) and shoot base underground (SBU), respectively, and had a significantly positive correlation with that in shoot tip above-ground (STA) ($p < 0.01$). Meanwhile, the IAA content in STA was positively correlated with that in STU and SBU, respectively. As for ABA content, only one significantly positive correlation was detected between mature leaves (ML) and STU. In terms of CTK content, CB was negatively correlated with dormant rhizome buds (DB) ($p < 0.05$) and STU ($p < 0.01$). Moreover, a positive correlation was found between DB and STU, as well as between SBU and STA, respectively. In terms of GA content, STU was positively correlated with SBU and STA, respectively, and a positive correlation existed between SBU and STA (Figure 5A).

Peak period (November): IAA content in CB was significantly positively correlated with DB, while IAA content in SBU was negatively correlated with CB and DB, respectively. In terms of ABA content, CB was positively correlated with STU ($p < 0.05$) and SBU ($p < 0.01$), respectively. In addition, DB was positively correlated with STA, and STU was also positively correlated with SBU. Regarding CTK content, young leaves (YL) were positively correlated with SBU. As for GA content, YL was negatively correlated with STU, and DB was positively correlated with STA (Figure 5B).

2.3.2. Endogenous Hormone Interactions at the Bamboo Clump Level

To investigate interactions among the examined endogenous hormones, the correlations among four hormones at the initial period (May) and peak period (November) were analyzed at the bamboo clump or genet level. Results showed that correlations between different hormones were significantly different in seven components at two typical periods of shooting phenology (Figure 6).

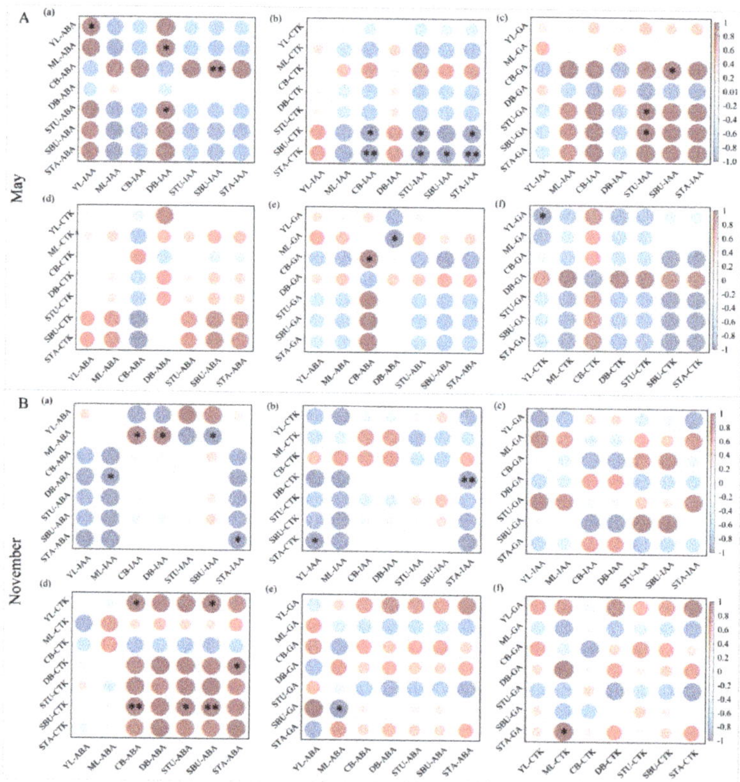

Figure 6. Pearson correlation analysis of four endogenous hormones in seven components. * indicates $p < 0.05$, and ** indicates $p < 0.01$. (**A**) Initial period of shooting in May; (**B**) peak period of shooting in November. In both sub-figures (**A**) and (**B**), (a) correlation between ABA and IAA; (b) correlation between CTK and IAA; (c) correlation between GA and IAA; (d) correlation between CTK and ABA; (e) correlation between GA and ABA; (f) correlation between GA and CTK. YL, young leaves; ML, mature leaves; CB, culm base; DB, dormant rhizome buds; SBU, shoot base underground; STU, shoot tip underground; STA, shoot tip above-ground. The size and shade of the disc were directly proportional to the absolute value of the correlation coefficient.

Initial period (May): IAA had some specific correlations with ABA, CTK, and GA in different components (Figure 6A). As shown in Figure 6A(a), four positive correlations between IAA and ABA were detected, namely both IAA and ABA in YL (young leaves); IAA in DB (dormant rhizome buds) vs. ABA in ML (mature leaves) and STU (shoot tips underground); and IAA in SBU (shoot base underground) vs. ABA in CB (culm base). On the other hand, there were seven negative correlations between IAA and CTK, including IAA in CB vs. CTK in STA (shoot tip above-ground) and SBU; IAA in STU vs. CTK in STA and SBU; IAA in SBU vs. CTK in STA; and IAA in STA vs. CTK in SBU and STA. In addition, three positive correlations between IAA and GA were found, namely IAA in STU vs. GA in STU and SBU; and IAA in SBU vs. GA in CB. No significant correlation was detected between CTK and ABA. Two significant correlations were detected between GA and ABA, i.e., ABA and GA in CB; and ABA in DB vs. GA in ML. Only one significant correlation was found between CTK and GA, i.e., CTK content was negatively correlated with GA in YL (Figure 6A). The above results might imply that IAA had a synergistic effect with ABA and GA in the corresponding components, while an antagonistic effect may exist between IAA and CTK.

Peak period (November): Five significant correlations were detected between IAA and ABA, including two positive (IAA in CB vs. ABA in ML; IAA in DB vs. ABA in ML) and three negatives (IAA in ML vs. ABA in DB; IAA in SBU vs. ABA in ML; IAA in STA vs. ABA in STA). Two negative correlations were found between IAA and CTK, i.e., IAA in YL vs. CTK in STA; and IAA in STA vs. CTK in DB. No significant correlation was detected between IAA and GA. Six positive correlations, including ABA in CB vs. CTK in YL and SBU; ABA in STU vs. CTK in SBU; ABA in SBU vs. CTK in YL and SBU; and ABA in STA vs. CTK in DB, were detected between ABA and CTK, indicating ABA might have a strongly synergistic effect with CTK. Only one negative correlation was found between ABA and GA, i.e., ABA in ML vs. GA in SBU. Meanwhile, only one positive correlation was found between CTK and GA, i.e., CTK in ML vs. GA in STA (Figure 6B).

2.4. Correlation Analysis between Bamboo Shoot Quantity and Endogenous Hormone Contents

The quantity of bamboo shoots is one of the most important goals in bamboo forest cultivation and management. In order to explore the impact of hormone interactions on the phenology of shoot sprouting, we analyzed the correlations between 11 kinds of possible hormone interactions in seven various components: shoot emergency number (SEN), degraded shoot number (DSN), and growing or survived bamboo culm number (GBN) (Figure 7).

The results showed that there were significant differences in the correlation between hormone indicators and SEN, DSN, and GBN at the component level (Figure 7). In the aspect of SEN, six positive correlations (including IAA/ABA in ML; IAA and ABA in YL; IAA and IAA/ABA in CB; (IAA + CTK + GA)/ABA in SBU, with $p < 0.05$ respectively) and seven significantly positive correlations (including (IAA + CTK + GA)/ABA in ML; IAA, IAA/ABA, and (IAA + CTK + GA)/ABA in DB; IAA and IAA/ABA in SBU; and ABA in STA, with $p < 0.01$ respectively) were found. At the same time, one negative correlation (namely GA/IAA in DB) and four significantly negative correlations (including CTK/GA in CB; CTK/IAA in DB; CTK/GA in SBU; and CTK/GA in STA) were detected.

In terms of DSN, one positive correlation (IAA in SBU) and two significantly positive correlations (ABA in YL and ABA in STA) were detected, while four negative correlations (i.e., GA/ABA and (IAA + CTK + GA)/ABA in YL; CTK/IAA and GA/IAA in DB) were found.

As for GBN, one positive correlation ((IAA + CTK + GA)/ABA in SBU) and eight significantly positive correlations (including IAA/ABA and (IAA + CTK + GA)/ABA in ML; IAA, IAA/ABA and (IAA + CTK + GA)/ABA in DB; IAA and IAA/ABA in SBU; ABA in STA) were detected, as well as two negative correlations (ABA in ML and GA/IAA in DB) and four significantly negative correlations (including CTK/GA in CB; CTK/IAA in DB; CTK/GA in SBU; and CTK/GA in STA).

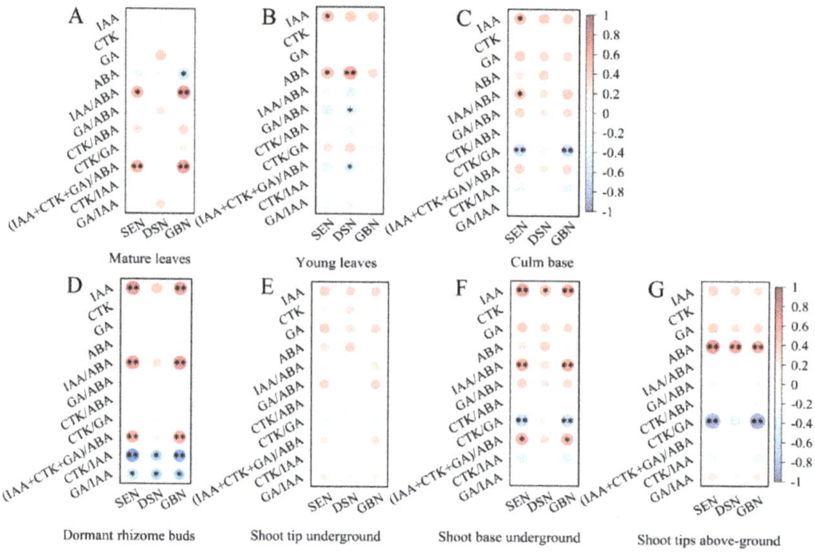

Figure 7. The correlation matrix between the number of bamboo shoots with different growth stages and endogenous hormone indicators from seven components. SEN: shoot emergency number; DSN: degraded shoot number; GBN: growing bamboo number. * indicates $p < 0.05$, and ** indicates $p < 0.01$. (**A**) Mature leaves; (**B**) young leaves; (**C**) base of bamboo culm; (**D**) dormant rhizome buds; (**E**) shoot tip underground; (**F**) shoot base underground; (**G**) shoot tip above-ground. The size and shade of the disc were directly proportional to the absolute value of the correlation coefficient.

According to correlation quantity, dormant rhizome buds (DB, 12 correlations including eight significant correlations) were the most closely related to the shoot sprouting phenology of *C. pingbianense*, followed by shoot base underground (SBU, nine correlations including six significant correlations), shoot tips above-ground (STA, five correlations and all were significant correlations), mature leaves (ML, five correlations including three significant correlations), young leaves (YL, five correlations including one significant correlation), culm base (CB, four correlations including two significant correlations), and shoot tips underground (STU, 0 correlation). Therefore, DB was the closest component related to the shooting phenology of *C. pingbianense*, a situation that may largely reflect the differences in the role of various components in the biological process of bamboo shoot sprouting.

3. Discussion

3.1. Annual Spatio-Temporal Distribution Changes of Endogenous Hormones

The ecological adaptation, growth, and development of plants, including woody bamboo, are influenced by both biological characteristics and the natural environment [25–27]. Among biological characteristics, endogenous hormone regulation is one of the main regulatory mechanisms for clonal reproduction and rapid growth of woody bamboo [28–30]. Due to the vast clonal lineage and the complex above-ground and underground branching systems of woody bamboo, the physiological regulation mechanism of bamboo shooting is greatly hindered. The previous studies have mostly studied content changes in endogenous hormones over a limited time span (usually a shooting period of 3–6 months) and limited components (mainly rhizomes, shoot buds, and above-ground branching), e.g., [16–18]. The peculiar biological characteristic of producing shoots every month in *C. pingbianense* provides valuable experimental materials to explore the physiological mechanism of bamboo shooting [23,24,31]. At the same time, the present study was based on the overall perspective of cloned plants. It measured endogenous hormone changes in the main above-ground and underground components of the entire bamboo genet throughout four seasons, which could more accurately grasp the changing trends of endogenous hormones in various components and provide comprehensive data for hormone biosynthesis.

Empirically, under natural conditions, most woody bamboos can produce shoots during a rhizome bud sprouting period of two to four months, which spans over a period of a season such as spring (mostly popular in the scattered bamboos) or summer (popular in the clump bamboos) [15,21]. Compared with the conventional single-season shooting bamboos, such as the scattered bamboos *Phyllostachys praecox* [17], *Ph. edulis* [19], and the clumps bamboo *Dendrocalamus latiflorus* [30], the contents of the four endogenous hormones in *C. pingbianense* were maintained at a relatively high level throughout the year (Figures 2–4), namely ABA 89–144 ng/g, IAA 15–31 ng/g, GA 1.4–4.0 ng/g, and CTK 2.6–6.7 ng/g. Moreover, all orders of hormone content at seven components in the four seasons were consistent, i.e., ABA > IAA > CTK > GA. In particular, *C. pingbianense* possessed significantly high contents of ABA and IAA throughout the year, which was significantly different from the reported bamboo species, e.g., [17–19]. This may be one of the internal physiological regulatory mechanisms of the four-season shooting in *C. pingbianense* under natural conditions [16].

Moreover, there were significant differences in hormone content among the four components. The contents of IAA, CTK, and GA were higher in the meristem than those in mature tissues, respectively, while ABA content was higher in mature tissues than those in the meristem. It may offer new clues to explore the possible biosynthesis of the examined endogenous hormones among different components; for example, IAA may be synthesized in situ in meristems such as the shoot tips and young leaf tips [20,32,33]. Temporally, four hormone contents changed, apparently at the component level. The examined components had higher IAA, CTK, GA (in dormant buds), and ABA (in meristems) contents at the peak period of shooting than those at the initial stage. However, the ABA contents in the culm base and dormant buds during the initial period of shooting were higher than those during the shooting peak period. It may be related to the physiological activity of ABA in promoting embryonic tissue growth and early seedling development [34,35]. Moreover, the increased amount of ABA in shoot tips above-ground (STA) and shoot tips underground (STU) (Figure 4B) from the initial period (rainy season) to the peak period (dry season) of bamboo shooting may be because of the accumulation of ABA induced by drought in plants [36,37].

3.2. Correlation between Bamboo Shooting Phenology and Endogenous Hormones

The shoot sprouting period in common bamboo often lasts for approximately three months, and bamboo shoot production is often associated with an increase in precipitation and soil temperature [14,16,21]. However, the results of this study indicated that there was not a strong correlation between the shooting phenology of *C. pingbianense* and local natural hydrothermal conditions [24]. For instance, despite the relatively low soil temperature in winter, the shoot quantity reaches its peak. Meanwhile, the summer, with the best hydrothermal conditions, is actually the period of a year with the lowest shoot number. This may reflect the unique biological characteristics of *C. pingbianense* [22]. At the same time, this article verified that shoot sprouting throughout the year was the biological characteristic of *C. pingbianense*. Furthermore, the number of shoots fluctuated from season to season, mainly in relation to spring and summer [23,24]. Therefore, the shoot yield of *C. pingbianense* forest showed a fluctuating trend under natural conditions.

The results of this article further reveal that the three components most closely related to the shot phenology were the dormant rhizome bud, shoot base underground, and shoot tips above-ground. Dormant rhizome buds are the precursor of bamboo shoots, which can develop into shoots underground after germination. During the sprouting process of the rhizome bud, the biosynthesis or transport of endogenous hormones plays an important regulatory role [19,28]. Our result also exhibited that the amounts of shoot and young bamboo culm were positively correlated with IAA content, ratio of IAA/ABA, and (IAA + CTK + GA)/ABA, but were negatively correlated with ratio of CTK/IAA and GA/IAA in dormant rhizome buds. These results indicated that all four endogenous hormones were involved in the physiological regulation processes of rhizome bud germination and young bamboo culm formation. In addition, the above result also confirmed that in the regulation process of breaking shoot dormancy, an appropriate higher mass fraction of IAA may promote shoot sprouting, while higher contents of ABA, CTK, and GA have an inhibitory effect on shoot sprouting [17,19]. Furthermore, it was worth noting that the number of degraded shoots was significantly and positively correlated with ABA contents in young leaves (YL) (Figure 7B) and shoot tip above-ground (STA) (Figure 7G). This correlation was probably related to the function of ABA in promoting carbohydrate transportation and metabolism among components [18,38,39]. In summary, the endogenous hormones in bamboo components often exhibit synergistic or antagonistic interactions, thereby regulating rhizome bud germination and nutrient allocation between components in *C. pingbianense*, leading to phenological phenomena such as shoot sprouting and degraded shoots [16,19,40,41].

4. Materials and Methods

4.1. Plant Material

The sample collection site is located in the central distribution area of *C. pingbianense* in the Daweishan National Nature Reserve in southeastern Yunnan Province (22°54′ N, 103°42′ E). From September 2020 to September 2021, six thriving bamboo clumps (30–40 culms per clump) were randomly selected for the fixed observation and sampling of bamboo clumps. The numbers of sprouting and degraded shoots were recorded once a month, and endogenous hormone samples were collected every two months.

The components for collecting endogenous hormone samples were as follows: (1) young leaves (YL) at the top of the branches; (2) mature leaves (ML) at the lower part of the branches; (3) basal internodes of the adult bamboo culm, namely, culm base (CB); (4) dormant rhizome buds (DB); (5) shoot base underground (SBU); (6) shoot tips underground (STU); (7) shoot tips above-ground (STA), approximately 10–15 cm shoot tips unearthed (Figure 8). Following conventional or empirical views, we regarded each clump as a potential genet and the culms within as ramets of a clone in clump or sympodial bamboos [14,15,21]. The sampling components in this study represented the developmental stages of bamboo culm, including rhizome buds germination, shoot sprouting, and young culm development. Meanwhile, samples also included the genets or mother bamboo system (e.g., leave, culm, rhizome) and ramets (e.g., rhizome buds, new bamboo shoots). The weight of each sample was approximately 1–2 g, and three duplicate samples were collected. The samples were quickly collected and stored in liquid nitrogen. After being transported to the laboratory, all samples were stored in a −80 °C refrigerator for subsequent hormone content determination.

4.2. Hormone Analysis

An enzyme-linked immunosorbent assay (ELISA) kit (China Agricultural University, Beijing, China) [17–19] was used to determine the content of four endogenous hormones with fresh weight: IAA, GA, ABA, and CTK. Each sample was measured three times, repeatedly.

The specific extraction procedures were as follows: (1) weighed 0.2 g of freeze-dried and ground sample, added 5 mL of extraction solution, shook well, and placed the mixture for 4 h at 4 °C according to the instructions of the ELISA kit manufacturer; (2) centrifuged the mixed extraction solution at 3500 rpm/min for 8 min at 4 °C and collected the supernatant with a pipette; then added 1 mL of extraction solution to the remaining precipitate, shook and extracted at 4 °C again, and collected the supernatant again; (3) passed supernatant through a C-18 solid-phase extraction column, dried the sample obtained after passing through the column, and then diluted it to 5 mL with sample diluent from the ELISA kit, then added the standard substance and placed the mixture for 0.5 h at 37 °C; (4) after color development, measured the OD values of the standard substance and each sample at 490 nm with an ELISA spectrophotometer; (5) fitted the standard curve of hormones and calculated the hormone content in each sample using a logit curve based on the measured values.

Figure 8. Habitat and structural components examined in C. pingbianense. (**A**) Habitat; (**B**) young leaves (YL), mature leaves (ML); (**C**) culm base (CB), dormant rhizome buds (DB); (**D**) shoot base underground (SBU), shoot tip underground (STU); (**E**) shoot tip above-ground (STA).

4.3. Seasonal Division

Daweishan National Nature Reserve is located on the southern edge of the subtropical zone. According to local climate characteristics and agricultural habits, the four seasons were divided into: spring (February, March, and April); summer (May, June, and July); autumn (August, September, and October); and winter (November, December, and January).

4.4. Data Processing

Data were pre-processed using Microsoft Excel 2016 (Microsoft Corporation, Redmond, WA, USA) and analyzed using SPSS software (v. 20.0; IBM Corporation, Armonk, NY, USA). One-way analysis of variance and Duncan's test were used to compare the mass fractions and ratios of IAA, ABA, GA, and CTK among developmental periods and bamboo components. Diagrams were visualized using the software Origin 2022 (Origin Lab. Corporation, Northampton, MA, USA).

5. Conclusions

The attribute of producing shoots all year round under natural conditions makes *C. pingbianense* different from woody bamboos. It is also interesting that the emerging bamboo shoot amount is not subject to the increasing precipitation and soil temperature in the habitat of *C. pingbianense*. The above phenological phenomena suggest that endogenous hormones may be the dominant factor in bamboo shoot sprouting in this species. Among four examined endogenous hormones in seven structural components, ABA had the highest mass fraction, followed by IAA, CTK, and GA. The bamboo shoot amount is significantly correlated with IAA content and ratios of IAA/ABA, (IAA + CTK + GA)/ABA, and CTK/IAA in the underground components. In particular, the contents of IAA and ABA in rhizome buds maintained a high level throughout the year, suggesting their crucial roles in bamboo shoot sprouting all year round in *C. pingbianense*.

Author Contributions: Designed experiments, performed data analysis, and wrote the paper, W.M., L.W. and H.Y.; performed field detection, experiments, and data analysis, C.B., Q.C. and N.L.; revised the paper, W.M. and H.Y. All authors have read and agreed to the published version of the manuscript.

Funding: This research was funded by the National Natural Science Foundation of China (No. 31870574) and the Department of Sciences and Technology of the Xizang Autonomous Region (No. XZ201801-GA-11).

Data Availability Statement: Data are contained within the article.

Acknowledgments: We thank the editor and reviewers for their constructive comments on this manuscript and the Pingbian sub-bureau of Dawei Mountain National Nature Reserve for their assistance in experimental sampling.

Conflicts of Interest: The authors declare that they have no conflict of interest. The funders had no role in the design of the study, in the collection, analysis, or interpretation of data, in the writing of the manuscript, or in the decision to publish the results.

References

1. Li, J.; Li, C. Seventy-year major research progress in plant hormones by Chinese scholars. *Sci. Sin. Vitae* **2019**, *49*, 1227–1281. (In Chinese)
2. Mähönen, A.P.; Bishopp, A.; Higuchi, M.; Nieminen, K.M.; Kinoshita, K.; Törmäkangas, K.; Ikeda, Y.; Oka, A.; Kakimoto, T.; Helariutta, Y. Cytokinin signaling and its inhibitor AHP6 regulate cell fate during vascular development. *Science* **2006**, *311*, 94–98. [CrossRef] [PubMed]
3. Guo, Y.; An, L.; Yu, H.; Yang, M. Endogenous hormones and biochemical changes during flower development and florescence in the buds and leaves of *Lycium ruthenicum* Murr. *Forests* **2022**, *13*, 763. [CrossRef]
4. Mir, A.R.; Alam, P.; Hayat, S. Auxin regulates growth, photosynthetic efficiency and mitigates copper induced toxicity via modulation of nutrient status, sugar metabolism and antioxidant potential in *Brassica juncea*. *Plant Physiol. Biochem.* **2022**, *185*, 244–259. [CrossRef]
5. Mishra, B.S.; Sharma, M.; Laxmi, A. Role of sugar and auxin crosstalk in plant growth and development. *Physiol. Plant.* **2022**, *174*, e13546. [CrossRef] [PubMed]
6. Zheng, C.; Acheampong, A.K.; Shi, Z.; Mugzech, A.; Halaly-Basha, T.; Shaya, F.; Sun, Y.; Colova, V.; Mosquna, A.; Ophir, R.; et al. Abscisic acid catabolism enhances dormancy release of grapevine buds. *Plant Cell Environ.* **2018**, *41*, 2490–2503. [CrossRef] [PubMed]
7. Li, X.; Qian, Q.; Fu, Z.; Wang, Y.; Xiong, G.; Zeng, D.; Wang, X.; Liu, X.; Teng, S.; Hiroshi, F.; et al. Control of tillering in rice. *Nature* **2003**, *422*, 618–621. [CrossRef]
8. Wang, H.; Liu, S.; Ma, S.; Wang, Y.; Yang, H.; Liu, J.; Li, M.; Cui, X.; Liang, S.; Cheng, Q.; et al. Characterization of the molecular events underlying the establishment of axillary meristem region in pepper. *Int. J. Mol. Sci.* **2023**, *24*, 12718. [CrossRef]
9. Petra, J. A novel putative auxin carrier family regulates intracellular auxin homeostasis in plants. *Nature* **2012**, *485*, 119–122. [CrossRef]
10. Asim, M.; Zhang, Y.; Liu, W. Editorial: Cooperation of gene regulatory networks and phytohormones in cell development and morphogenesis. *Front. Plant Sci.* **2023**, *14*, 1290538. [CrossRef]
11. Ma, J.; Xie, L.; Zhao, Q.; Sun, Y.; Zhang, D. Cyclanilide induces lateral bud outgrowth by modulating cytokinin biosynthesis and signaling pathways in apple identified via transcriptome analysis. *Int. J. Mol. Sci.* **2022**, *23*, 581. [CrossRef]
12. Müller, D.; Waldie, T.; Miyawaki, K.; To, J.P.C.; Melnyk, C.W.; Kieber, J.J.; Kakimoto, T.; Leyser, O. Cytokinin is required for escape but not release from auxin mediated apical dominance. *Plant J.* **2015**, *82*, 874–886. [CrossRef] [PubMed]
13. Zhang, Q.; Wang, J.; Wang, L.; Wang, J.; Wang, Q.; Yu, P.; Bai, M.; Fan, M. Gibberellin repression of axillary bud formation in Arabidopsis by modulation of DELLA-SPL9 complex activity. *J. Integr. Plant Biol.* **2020**, *62*, 421–432. [CrossRef] [PubMed]
14. Jiang, Z.H. *Bamboo and Rattan in the World;* Liaoning Science and Technology Publishing House: Shenyang, China, 2002; pp. 1–86.
15. Yang, Y.M.; Wang, K.L.; Sun, M.S. *Yunnan Bamboo Flora;* Yunnan Publishing Group Co., Ltd.: Kunming, China, 2019; pp. 1–17, 543–544.
16. Li, L.S.; Xia, T.Z.; Li, B.; Yang, H.Q. Hormone and carbohydrate metabolism associated genes play important roles in rhizome bud full-year germination of *Cephalostachyum pingbianense*. *Physiol. Plantarum* **2022**, *174*, e13674. [CrossRef] [PubMed]
17. Huang, J.Q.; Liu, L.; Zhang, B.S.; Qiu, L.Z. Dynamic changes of endophytohormones in rhizomal buds of *Phyllostachys praecox*. *Sci. Silvae Sin.* **2002**, *38*, 38–41. [CrossRef]
18. Zhang, S.S.; Ding, X.C.; Zhang, Z.Y.; Cai, H.J. Effects of pruning hormonal and single-sugar regulation by hooking shooting on the yield of *Dendrocalamus latiforus*. *Sci. Silvae Sin.* **2018**, *54*, 31–39. [CrossRef]
19. Shen, Z.; Zhang, Y.; Zhang, L.; Li, Y.; Sun, Y.; Li, Z. Changes in the distribution of endogenous hormones in *Phyllostachys Edulis* "Pachyloen" during bamboo shooting. *PLoS ONE* **2020**, *15*, e0241806. [CrossRef] [PubMed]
20. Bai, Y.; Cai, M.; Mu, C.; Cheng, W.; Zheng, H.; Cheng, Z.; Li, J.; Mu, S.; Gao, J. New insights into the local auxin biosynthesis and its effects on the rapid growth of Moso bamboo (*Phyllostachys edulis*). *Front. Plant Sci.* **2022**, *13*, 858686. [CrossRef] [PubMed]

21. Yi, T.P.; Shi, J.Y.; Ma, L.S.; Wang, H.T.; Yang, L. *Iconographia Bambusoidearum Sinicarum*; Science Press: Beijing, China, 2008.
22. Yang, H.Q.; Sun, M.S.; Mao, W.; Yang, Y.M.; Li, D.Z. *Cephalostachyum pingbianense* (Poaceae: Bambusoideae), Comb. Nova. *Ann. Bot. Fenn.* **2008**, *45*, 394–395. [CrossRef]
23. Zheng, X.Q.; Cui, Y.Z.; Chen, L.N.; Yang, H.Q. Study on bamboo shooting and shoot growth of *Cephalostachyum pingbianense*. *For. Res.* **2018**, *31*, 131–136. [CrossRef]
24. Li, L.S.; Chen, L.N.; Li, B.; Xia, T.Z.; Dou, P.T.; Yang, H.Q. Seasonal variations of bamboo shooting and its correlation with meteorological factors on *Cephalostachyum pingbianense*. *J. Northwest Forest. Univ.* **2021**, *36*, 66–72. [CrossRef]
25. Chen, F.; Wang, N.; Zhou, J.; Zhao, Z.; Lv, K.; Huang, Y.; Huang, G.; Qiu, L. Summer dormancy of *Myricaria laxiflora* to escape flooding stress: Changes in phytohormones and enzymes induced by environmental factors. *Plant Physiol. Bioch.* **2022**, *193*, 61–69. [CrossRef] [PubMed]
26. Verma, S.; Negi, N.P.; Pareek, S.; Mudgal, G.; Kumar, D. Auxin response factors in plant adaptation to drought and salinity stress. *Physiol. Plant.* **2022**, *174*, e13714. [CrossRef]
27. Dou, P.; Cheng, Q.; Liang, N.; Bao, C.; Zhang, Z.; Chen, L.; Yang, H. Rhizosphere Microbe Affects Soil Available Nitrogen and Its Implication for the Ecological Adaptability and Rapid Growth of *Dendrocalamus sinicus*, the Strongest Bamboo in the World. *Int. J. Mol. Sci.* **2023**, *24*, 14665. [CrossRef] [PubMed]
28. Bhandawat, A.; Singh, G.; Seth, R.; Singh, P.; Sharma, R.K. Genome-wide transcriptional profiling to elucidate key candidates involved in bud burst and rattling growth in a subtropical bamboo (*Dendrocalamus hamiltonii*). *Front. Plant Sci.* **2017**, *7*, 2038. [CrossRef] [PubMed]
29. Chen, L.N.; Dou, P.T.; Chen, Y.K.; Yang, H.Q. Mutant IAA21 genes from *Dendrocalamus sinicus* Chia et J. L. Sun inhibit stem and root growth in transgenic tobacco by interacting with ARF5. *Plant Physiol. Bioch.* **2023**, *201*, 107827. [CrossRef]
30. Fan, L.; Li, B.; Han, Y.; Chen, L.; He, T.; Zheng, Y.; Rong, J. Lower light intensities increase shoot germination with improved leaf biosynthesis in Ma bamboo (*Dendrocalamus latiflorus Munro*). *Forests* **2022**, *13*, 1723. [CrossRef]
31. Li, L.; Xia, T.; Yang, H. Seasonal patterns of rhizosphere microorganisms suggest carbohydrate-degrading and nitrogen-fixing microbes contribute to the attribute of full-year shooting in woody bamboo *Cephalostachyum pingbianense*. *Front. Microbiol.* **2022**, *13*, 1033293. [CrossRef]
32. Perez, V.C.; Zhao, H.; Lin, M.; Kim, J. Occurrence, function, and biosynthesis of the natural auxin phenylacetic acid (PAA) in plants. *Plants* **2023**, *12*, 266. [CrossRef]
33. Tejeda-Sartorius, O.; Soto-Hernández, R.M.; San Miguel-Chávez, R.; Trejo-Téllez, L.I.; Caamal-Velázquez, H. Endogenous hormone profile and sugars display differential distribution in leaves and pseudobulbs of *Laelia anceps* plants induced and non-induced to flowering by exogenous gibberellic acid. *Plants* **2022**, *11*, 845. [CrossRef]
34. Miao, C.; Xiao, L.; Hua, K.; Zou, C.; Zhao, Y.; Ray, A.; Bressan, R.A.; Zhu, J.K. Mutations in a subfamily of abscisic acid receptor genes promote rice growth and productivity. *Proc. Natl. Acad. Sci. USA* **2018**, *115*, 6058–6063. [CrossRef] [PubMed]
35. Chen, H.; Zhang, J.; Neff, M.M.; Hong, S.; Zhang, H.; Deng, X.; Xiong, L. Integration of light and abscisic acid signaling during seed germination and early seedling development. *Proc. Natl. Acad. Sci. USA* **2008**, *105*, 4495–4500. [CrossRef]
36. Tang, N.; Ma, S.; Zong, W.; Yang, N.; Lv, Y.; Yan, C.; Guo, Z.; Li, J.; Li, X.; Xiang, Y.; et al. MODD mediates deactivation and degradation of OsbZIP46 to negatively regulate ABA signaling and drought resistance in rice. *Plant Cell* **2016**, *28*, 2161–2177. [CrossRef] [PubMed]
37. Gupta, A.; Rico-Medina, A.; Caño-Delgado, A. The physiology of plant responses to drought. *Science* **2020**, *368*, 266–269. [CrossRef] [PubMed]
38. Mega, R.; Abe, F.; Kim, J.S.; Tsuboi, Y.; Tanaka, K.; Kobayashi, H.; Sakata, Y.; Hanada, K.; Tsujimoto, H.; Kikuchi, J.; et al. Tuning water-use efficiency and drought tolerance in wheat using abscisic acid receptors. *Nat. Plants* **2019**, *5*, 153–159. [CrossRef] [PubMed]
39. Lin, Q.; Zhang, Z.; Wu, F.; Feng, M.; Sun, Y.; Chen, W.; Cheng, Z.; Zhang, X.; Ren, Y.; Lei, C.; et al. The APC/CTE E3 ubiquitin ligase complex mediates the antagonistic regulation of root growth and tillering by ABA and GA. *Plant Cell* **2020**, *32*, 1973–1987. [CrossRef]
40. Shou, Y.; Zhu, Y.; Ding, Y. Transcriptome analysis of lateral buds from *Phyllostachys edulis* rhizome during germination and early shoot stages. *BMC Plant Biol.* **2020**, *20*, 229. [CrossRef]
41. Zhao, J.; Gao, P.; Li, C.; Lin, X.; Guo, X.; Liu, S. *PhePEBP* family genes regulated by plant hormones and drought are associated with the activation of lateral buds and seedling growth in *Phyllostachys edulis*. *Tree Physiol.* **2019**, *39*, 1387–1404. [CrossRef]

plants

MDPI

Review

Advances in Plant Auxin Biology: Synthesis, Metabolism, Signaling, Interaction with Other Hormones, and Roles under Abiotic Stress

Jianshuang Gao [1,2], Shunyao Zhuang [1,*] and Weiwei Zhang [1]

[1] State Key Lab of Soil and Sustainable Agriculture, Institute of Soil Science, Chinese Academy of Sciences, Nanjing 210008, China; gaojianshuang@hufe.edu.cn (J.G.); zhangweiwei@issas.ac.cn (W.Z.)
[2] School of Economic Geography, Hunan University of Finance and Economics, Changsha 410205, China
* Correspondence: syzhuang@issas.ac.cn; Tel.: +86-25-86881103; Fax: +86-25-86881000

Abstract: Auxin is a key hormone that regulates plant growth and development, including plant shape and sensitivity to environmental changes. Auxin is biosynthesized and metabolized via many parallel pathways, and it is sensed and transduced by both normal and atypical pathways. The production, catabolism, and signal transduction pathways of auxin primarily govern its role in plant growth and development, and in the response to stress. Recent research has discovered that auxin not only responds to intrinsic developmental signals, but also mediates various environmental signals (e.g., drought, heavy metals, and temperature stresses) and interacts with hormones such as cytokinin, abscisic acid, gibberellin, and ethylene, all of which are involved in the regulation of plant growth and development, as well as the maintenance of homeostatic equilibrium in plant cells. In this review, we discuss the latest research on auxin types, biosynthesis and metabolism, polar transport, signaling pathways, and interactions with other hormones. We also summarize the important role of auxin in plants under abiotic stresses. These discussions provide new perspectives to understand the molecular mechanisms of auxin's functions in plant development.

Keywords: auxin; auxin signal transduction; abiotic stress; hormone

1. Introduction

The hormone auxin, which is composed of various substances with growth-inducing effects, is important in plant physiology [1]. Auxin is an endogenous hormone characterized by the presence of an unsaturated aromatic ring and an acetic acid side chain [2,3]. It was first discovered in coleoptile experiments by Darwin [4]. In addition, auxin refers to a class of small molecule compounds that primarily act through polar transport and signal transduction, and auxin production and metabolism influence how plants grow, develop, and respond to external stimuli to adapt to their changing environment [5]. The establishment and maintenance of polarity, apical dominance, phototropism, gravity, senescence, pathogen response, abiotic stress responses, and fruit formation are only a few of the processes controlled by auxin during plant growth and development [6] (Figure 1). Auxins have been used widely in the field of agricultural production, greatly improving the yield and quality of crops, and creating significant social and economic benefits.

The effects of auxins on plant development have been widely studied. It has been found that auxins play a key role in regulating the formation of adventitious roots, forms, types, and concentrations of auxins also have different impacts on this process [7,8]. In particular, the level of free auxins in plant tissues is crucial for growth and development of plants. *Arabidopsis thaliana* can form adventitious roots from young leaf explant without exogenous auxins, whereas exogenous auxins induce older leaf explants to form adventitious roots [9]. Auxins are often used to induce parthenocarpy, increase fruit set, and inhibit flowering, as well as to improve low fruit maturation rates [10,11]. Moreover, studies have shown that indole-3-acetic acid (IAA) affects plant flowering, but the specific mechanism is still unclear. Most studies suggest that low concentrations of IAA promote flowering, while high concentrations inhibit it [12,13]. The development of neighboring organs is inhibited or retarded by a floral organ that produces high levels of free auxin. During floral bud development, young organs that produce high levels of free IAA inhibit or delay the initiation and development of organ primordia at the shoot tip [12]. Zhao et al. found that higher IAA level resulted in longer hypocotyls and shorter primary roots in *yucca*, an *Arabidopsis* activation-tagged mutant [14]. On the other hand, results of metabolic and transcriptomic analyses indicate that auxin participants in the early stages of fruit development, and a corresponding increase in sugar at the ripening stages [15,16]. Furthermore, the maturity of strawberries was negatively correlated with the IAA content [17,18].

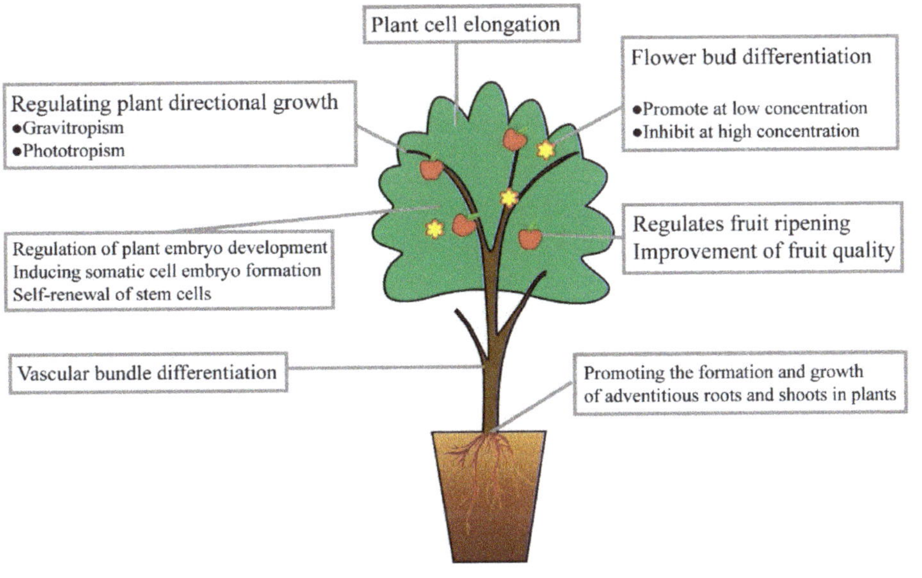

Figure 1. Schematic presentation of auxin function in plants.

Auxins not only respond to intrinsic developmental signals, but also mediate various environmental signals, participating in the regulation of plant growth and development and growth responses, such as gravity and light signals. For example, the asymmetric distribution of auxins is essential for the formation of plant gravitropic responses, which is mainly achieved by auxin polar transport and signal transduction [19–23]. The asymmetric distribution of auxins is also the main cause of plant phototropic growth, which is mainly regulated by various auxin transporters [24–31].

Auxin plays an important role in regulating plant responses to various stressors and has received much research attention in recent years. Abiotic stressors, such as drought, salinity, and low temperature, have imposed increasingly serious constraints on the stable and high yield of grains. Since the beginning of this century, global grain production has been mainly affected by abiotic stresses [32]. Early in the vegetative growth stage, abiotic stress impacts the development and differentiation of plant cells. Abiotic stress during the reproductive growth stage might cause a large reduction in yield or possibly no harvest at all [33]. Auxin is one of the hormones that plants deploy during stress to maintain their homeostasis in vivo and lessen the negative effects of stress on plant growth and development [34,35].

Auxin affects the yield, quality, and resistance of plants by regulating the important signal molecules required for their plastic growth and development, especially of roots.

This paper reviews research on indole-acetic acid (IAA, an auxin) synthesis and metabolism, polar transport, and signal transduction in recent years to provide a reference for the use of IAA in high-quality cultivation and rapid plant propagation.

2. Auxin Types

Currently, there are two main categories of known auxins: endogenous auxins and synthetic auxins (Figure 2). Endogenous auxins mainly include IAA, indole-3-butyric acid (IBA), 4-chloro-indole-3-acetic acid (4-C1-IAA), and phenylacetic acid (PAA). They exist in two forms, bound and in a free state [36–40]. Initially, IBA was discovered in the tuber of the horse bell, but has since been discovered in other plants. According to Campanella et al. (2004), IBA accounts for 25–30% of total Arabidopsis auxin [41]. This auxin is widely used in agricultural production as a rooting agent, and also participates in auxin-mediated leaf formation, cell division, stem bending, and root hair formation [42,43]. Originally, 4-C1-IAA was found in immature pea seeds; however, the model plant *Arabidopsis thaliana* does not produce this form of auxin. The main roles of 4-C1-IAA are the promotion of pea seed coat development and the elongation of corn colloblasts [44]. Phenylacetic acid is the only phenyl-derived endogenous auxin found so far, which is mainly involved in the interaction between roots and soil microorganisms [45]. Auxin used as a plant growth regulator mostly consists of synthetic auxins, such as NAA, 2, 4-D, trichlorophenoxyacetic acid (2, 4, 5-T), and picloram, among others [46]. These synthetic auxins are more stable than IAA [47].

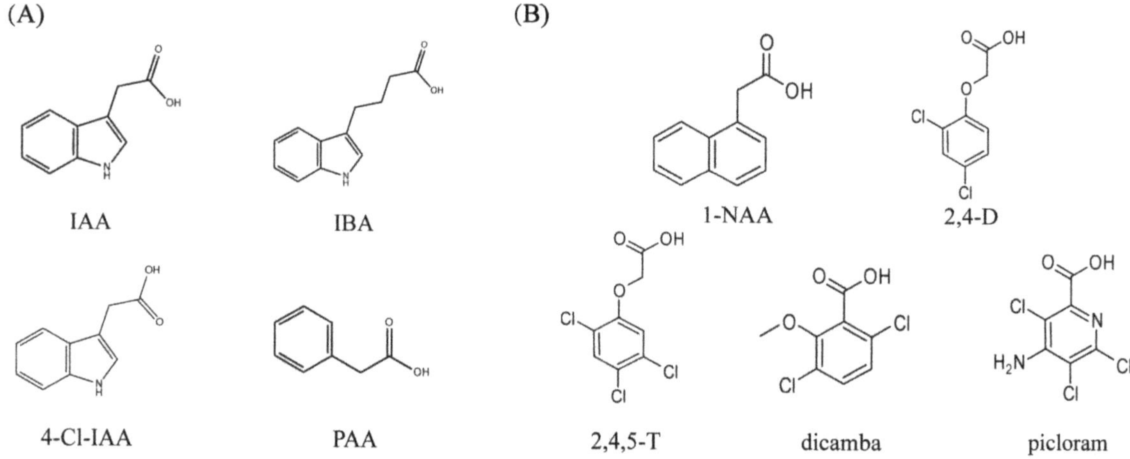

Figure 2. Examples of endogenous auxins (**A**) and some synthetic auxins (**B**) are presented. (**A**) IAA: indole-acetic acid; IBA: indole-3-butyric acid; 4-Cl-IAA: 4-chloroindole-3-acetic acid; and PAA: phenyl-acetic acid. (**B**) 1-NAA: 1-Naphthalene-acetic acid; 2,4-D: 2,4-dichlorophenoxyacetic acid; 2,4,5-T: 2,4,5-trichlorophenoxy-acetic acid; dicamba: 3,6-dichloro-2-methoxybenzoic acid, and picloram: 4-Amino-3,5,6-trichloropicolinic acid [48].

3. Auxin Synthesis

Auxin is synthesized in plants via several pathways [49]. The metabolism of IAA mainly can be divided into two kinds synthesis pathways: tryptophan (Trp)-dependent and tryptophan-independent [50,51] (Figure 3). Wang et al. (2020) suggested that the cytoplasmic enzyme indole synthase (INS) may be a key enzyme in the Trp-independent IAA biosynthetic pathway [52]. However, little is known about the molecular components and physiological functions of the Trp-independent pathway. The tryptophan-dependent auxin synthesis pathway dominates in plants, and IAA synthesis pathways are classified as the indole-3-acetaldoxime (IAOx) pathway, the indole-3-acetamide (IAM) pathway, and the indole-3-pyruvate (IPyA) pathway [53] (Figure 3). The indole synthase gene (*INS*) is the primary gene in the tryptophan-independent auxin synthesis pathway [54]. Indole, also known as indole glycerophospholipid (IGP), is a key node in both the tryptophan-dependent and tryptophan-independent pathways of auxin production [55]. The following focuses on the tryptophan-dependent auxin synthesis pathway. It is mainly mediated by transaminases and decarboxylases [56].

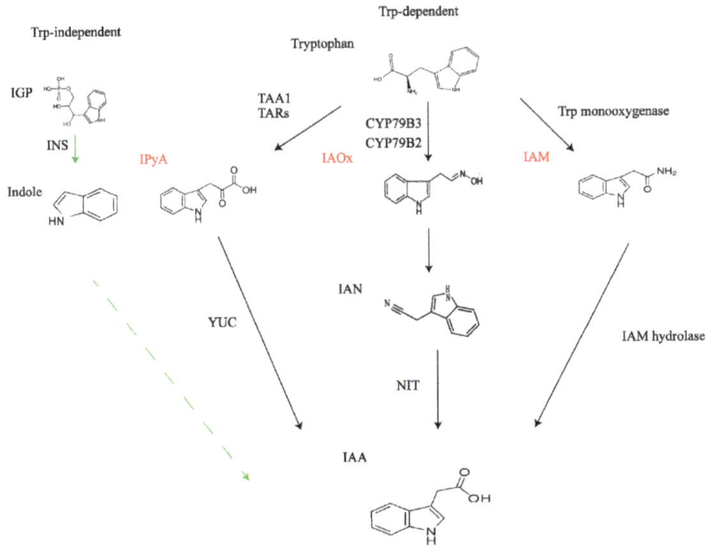

Figure 3. A model of the tryptophan (Trp)-dependent and Trp-independent indole acetic acid (IAA) biosynthetic pathways. IGP, indole-3-glycerol phosphate; INS, indole synthase gene; TAA1, tryptophan aminotransferase; TARs, TAA1-associated proteins; IPyA, indole-3-pyruvate; YUC, YUCCA; IAOx, indole-3-acetaldoxime; IAM, indole-3-acetamide; CYP79B2 and CYP79B3 in the cellular phosphorus P450 (CYP) mono-oxygenase family.

While other redundant processes function in parallel, the IPyA path is a significant and often conserved mechanism for IAA production in plants. It mainly involves two reactions: First, tryptophan is deaminated to IPyA by tryptophan aminotransferase (TAA1) and TAA1-associated proteins (TARs) in Arabidopsis [57,58]. IPyA is then decarboxylated to IAA by an irreversible reaction catalyzed by flavin-containing monooxygenases of the YUCCA (YUC) family. Trp is first converted to IPA by Trp transaminase and subsequently catalyzed by YUCase to produce IAA [2,59]. Studies have shown that ATA1 and YUC co-originals have been found in the genome of plants [60–62]. In *Marchantia polymorpha*, knockout of a single TAA gene results in loss of cell and tissue differentiation leading to severe growth and developmental defects [62]. Therefore, IPyA pathways are the main pathways of IAA biosynthesis in plants.

The IAOx and IAM pathways play only minor roles in IAA homeostasis. IAOx, IAM, and IAN (indole-3-acetonitrile) are intermediates in the biosynthesis of IAA [63]. The conversion of amino acids to IAOx is mediated by the two related enzymes CYP79B2 and CYPs79B3 in the cellular phosphorus P450 (CYP) mono-oxygenase family [64,65]. IAOx is a precursor of indole glucosides (IGs) and camalexin, which act as defense metabolites in plants [66,67]. So far, both IAOx and CYP79B2/3 genes have been found only in Brassica [63], suggesting that this pathway is restricted to Brassicaceae. IAOx is synthesized from Trp catalyzed by the enzyme CYP79B2/B3, converted directly to IAN, and later generates IAA in the presence of nitrilase (NIT) [68]. Tryptophan is converted to IAA by the formation of IAM. The pathway begins with the conversion of tryptophan to IAM and IAM hydrolysis products catalyzed by the Trp monooxygenase, the former in the production of auxin in the presence of IAM hydrolase. The latter is directly mediated by indole precursors [69]. In Arabidopsis, disruption of the major IAM hydrolases IAMH1 and IAMH2 did not result in substantial developmental defects or changes in IAA content. It is suggested that the IAM pathway plays only a secondary role in growth hormone homeostasis [69].

4. Auxin Metabolism

The metabolism of IAA mainly proceeds the following three ways: (1) The formation of auxin conjugates, such as amide conjugates with amino acids and polypeptides, and the formation of ester conjugates with polysaccharides and inositol, which are generally used for the transport and storage of auxin [40,70,71]. (2) Conversion to IBA, which is more stable than IAA and can produce a variety of conjugates [72]. (3) Oxidative decomposition, in which IAA can be decomposed by oxidation of its side chain (decarboxylated) or the indole ring (non-decarboxylated). The decarboxylated oxidation process is more complex, and conjugated IAA is generally decomposed via non-decarboxylated oxidation [73]. This reaction is the oxidation of IAA to 2-oxoindole-3-acetic acid (oxaa), which is then glycosylated to oxaa-glc [74,75].

The predominant metabolic pathway for IAA is oxidative catabolism, as it was shown that oxaa is the most abundant IAA metabolite in *Arabidopsis* [76]. In algae, vascular and non-vascular land plants, oxidative catabolites are present at higher levels than amide-linked catabolic metabolites under normal physiological conditions. It suggests that oxidation is the major pathway for plant IAA catabolism [74,77–79]. IAA oxidase 1 (DIOXYGENASE FOR AUXIN OXIDATION 1, DAO1) is a member of the 2-oxoglutarate and iron (II)-dependent oxygenases superfamily [78]. Both Arabidopsis *AtDAO1* and rice *OsDAO* convert IAA to oxaa in vitro [80,81]. The *dao1-1* mutant exhibited an auxin accumulation phenotype, but plants overexpressing *atdao1* did not exhibit a significant auxin deficiency phenotype [78,81]. Loss of DAO1 function results in only minor developmental defects [80,82]. On the other hand, the GH3 gene encodes an acylamide synthase that catalyzes the coupling reactions of salicylic acid (SA), jasmonic acid (JA), and IAA with amino acids [83]. It was shown that the major natural auxin, IAA, is inactivated mainly through the GH3-ILR1DAO pathway [76]. First, IAA is converted to IAA-amino acid conjugates (IAA-aspartate (IAA-Asp) and IAA-glutamic acid (IAA-Glu) by GH3-type IAA amide synthase. DAO1 dioxygenase irreversibly oxidizes IAA—Asp and IAA—Glu to indole diketone-3 -acetic acid-aspartic acid (oxIAA-Asp) and oxIAA-Glu. oxIAA-Glu is then hydrolyzed by ILR1 to release inactive oxIAA [76,78]. It has been shown that DAO and GH3 enzymes play redundant roles in regulating IAA levels [78,81,84].

5. Auxin Transportation

Auxin can be transferred in higher plants in two ways: long-distance vascular transport and short-distance active transport requiring transport vehicles [85–88]. The latter is important in the asymmetric distribution of auxin, which is also known as auxin polar transport [89]. Three transport proteins are required for polar auxin transport (PAT): auxin-influx carrier AUXIN/LIKE-AUX (AUX1/LAX) family proteins, auxin-efflux carriers PIN-FORMED (PIN) family proteins, ATP-binding cassette B (ABCB) family proteins (Table 1). They are the main family of transporter proteins involved in PAT. Their quantity, polarity, and capacity to transport auxin at the PM influence the pace and directionality of intercellular auxin flow, establishing the pattern of auxin distribution [90,91]. These protein families are frequently functioned in plants to modulate auxin polar transport and distribution [92–95].

The AUX1/LAX family contains four highly homologous genes (*AUX1, LAX1, LAX2,* and *LAX3*) that encode transmembrane proteins in *Arabidopsis* [95,96]. The AUX1/LAX family is involved in a number of developmental processes, including embryogenesis, seed germination, leaf morphogenesis, vascularization, and root and terminal hook development [97,98]. The amount and polarity of the AUX1/LAX protein at the plasma membrane (PM) are strictly controlled. It helps to coordinate the distribution of growth factors essential for normal plant growth and development [99–101]. For example, in roots, asymmetric localization of AUX1 at the apical PM of protodermal cells promotes auxin flow toward the tip (root direction). While AUX1 is positioned at the base of the side roots and epidermal cells, it drives the flux to the base end (in the direction of the stem) [21]. In root columella cells, the increase in cytoplasmic AUX1 content implies a dynamic regulation of PM targeting and AUX1. Rapid subcellular localization and polarity regulation of AUX1 in root tissues can control auxin flow, which in turn regulates root growth in response to gravitational stimuli or other environmental inputs [21].

Two different transporters mediate growth hormone efflux, the ABC and PINs transporters. The ABCB family are nonpolar transporter proteins that are uniformly distributed along the PM [102,103]. Previously, ABCB1, ABCB4, and ABCB19 were considered to be nonpolar. However, it has been suggested that some homologs, including ABCB14 and ABCB15, may have polar membrane localization functions that contribute to the directionality of auxin flow [104,105]. On the other hand, polar-localized transport proteins (PINs) are components of the PAT machinery and have an important influence on the directionality of auxin flow in plant tissues and organs [106,107]. The eight members of the PIN family are transmembrane proteins, PIN1, PIN2, PIN3, PIN4, and PIN7 are localized to the PM, PIN5 and PIN8 are localized to the ER, and PIN6 is localized to the endoplasmic reticulum (ER) and PM [108]. The PIN located in the PM usually contains a long hydrophilic ring that separates multiple transmembrane structural domains, whereas the PIN located in the ER is characterized by a short hydrophilic ring in plants. And it has been shown that PIN transports auxin in unicellular plants [109,110], or in heterologous systems, including mammalian cells or Xenopus oocytes [109,111]. PINs primarily regulate physiology and development, such as embryogenesis, initiation, localization and formation of new organs, and tropic responses [107,112,113]. Interestingly, PINs and ABCB interact and control PAT in plants independently or interdependently [92,94].

Table 1. Types, coding genes, and functions of auxin transport proteins.

Types	Coding Genes	Functions	References
auxin-influx carrier	*AUX1*	*AUX1*: amino acids transporters and auxin permease activity; adventitious root development	[114–123]
	LAX1	*LAX1*: shoot and root pole formation,	
	LAX2	*LAX2*: xylem development, gravitropistic response, auxin distribution	
	LAX3	*LAX3*: auxin distribution, lateral root development, hook formation	
auxin-efflux carrier	*PIN (PIN-FORMED)* gene family,	*PIN1:* downward auxin transport, organ initiation, flower organ formation, leaf vein formation, and stem gravity response; *PIN2:* transport auxin from apex to elongation zone; root gravitropism *PIN3:* mediate auxin flow toward the lower hypocotyl side *PIN4* and *PIN7:* Auxin distribution during plant embryonic development *PIN5:* Transport auxin to the endoplasmic reticulum cavity *PIN6* and *PIN8:* Auxin transport across the plasma membrane	[23,104,106,113,124–130]
	ABCB gene family	*ABCB4, 14, 15, 19* and *21:* Auxin transport functions	[95,104,131]

6. Gravitropism and Phototropism of Auxin

Studies on the effects of auxin on gravitropism and phototropism have been investigated in recent years. A functionally deficient mutant of *AUXIN1* (*AUX1*), an auxin-influx carrier, with reduced gravity response when expressed in lateral root cap and epidermal cells [21]. The AtAUX1 protein acts together with the growth hormone export protein AtPIN2 to regulate the gravitropic response of the root system according to environmental signals and stimulation [20]. In the root columella, PIN3 is rapidly repositioned laterally in response to gravitational stimulation [22]. Plants exhibit gravitropic growth after perceiving gravity signals, and this process can be timely terminated to avoid excessive bending of plant tissues. For example, experiments on the mechanics of hypocotyl bending in *Arabidopsis thaliana* showed that when the hypocotyl stimulated by gravity for a short period of time (2–3 h), it began to show obvious gravitational response, which gradually weakened with the increase in processing time and almost disappeared after 30 h of processing [23]. Rakusova et al. (2016) found that this is due to an essential mechanism for restoring symmetry to PIN3-dependent auxin flow. PIN3 regulates gravity-mediated growth hormone transfer to the lateral hypocotyl and promotes its development. Afterwards PIN3 polarizes to the other side of the cell, accelerating auxin consumption to terminate the bend. However, pharmacological or genetic alterations prevent the PIN3 response from terminating, resulting in hyperbolic hypocotyls [23]. PIN3 and *PIN-FORMED7* (*PIN7*) modulate the directional transport and distribution of auxins on both sides of the root, further affecting the root response angle to gravity, and thus participate in the formation of root gravitropic morphology [19]. The asymmetric distribution of auxins is also the main cause of plant phototropic growth, which is mainly regulated by various auxin transporters. Plant hypocotyl phototropic bending is mainly regulated by auxin polar transport genes *PIN3*, *PIN4*, and *PIN7* [25]. ABCB-mediated (ATP-binding cassette B) auxin polar transport is involved in the regulation of hypocotyl growth by light signals, the regulation of which depends on light signal receptors *cryptochrome1* (*CRY1*), *phototropin1* (*PHOT1*), and *phytochrome1* (*PHYB*) [132,133]. Auxin-mediated light signals not only participate in phototropic responses, but also in plant shade avoidance syndrome (SAS) [26]. Shade-tolerant plants, such as *Arabidopsis thaliana*, need a certain degree of shade to grow normally [28]. SAS is mainly reflected in the morphological changes in plants, such as stem and petiole phototropic bending, delayed leaf development, and downward growth [27]. Changing the ratio of red to far-red light (R:FR) can effectively simulate plant shade responses, and therefore is widely used in shade response research [29,30]. In natural environments (high R:FR), the red light receptor phytochromeB (PHYB) inhibits plant phototropic responses; while under shading conditions, PHYB promotes plant phototropism by regulating the transcriptional activity of three important *basic Helix-Loop-Helix* (*bHLH*) transcription factors phytochrome interacting factors (PIFs) (PIF4, PIF5, and PIF7) [31]. The latest study showed that *SAV4* (*Shade Avoidance 4*) participates in the shade response of plant hypocotyls by regulating ABCB1-mediated auxin polar transport [24].

7. Auxin Signaling Pathway

The auxin signaling pathway consists of auxin/IAA transcriptional inhibitors, auxin response factors (ARFs), and receptor proteins transport inhibitor response 1 (TIR1)/auxin-signaling F-box (AFB) (Figure 4) [134]. When auxin concentrations are low, Aux/IAA inhibitors bind to the ARF transcription factor, thereby inhibiting ARF activity. The binding of auxin to receptor TIR1/AFBs allows TIR1 to easily bind Aux/IAA proteins and induces a ubiquitination reaction [135]. Following degradation by the 26S proteasome, the AUX1/IAAs protein complex releases ARF, thereby initiating the regulation of downstream genes [136]. Research has shown that four Aux/IAA-ARF combinations, known as auxin signaling modules, are related to different stages of lateral root growth [137].

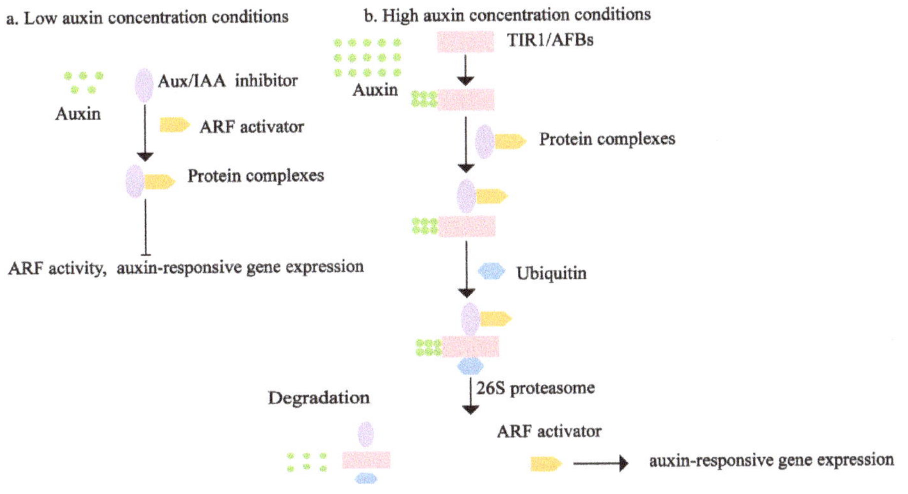

Figure 4. The auxin signaling transduction pathway in plants. Under low auxin concentration conditions, the auxin transduction repressor auxin/indole-acetic acid protein (Aux/IAA) forms a heterodimer with the auxin response factor (ARF), which inhibits the transcriptional activity of ARF, resulting in the suppression of auxin response gene expression. Under high auxin concentration, the auxin receptor transport inhibitor response 1 (TIR1) binds to Aux/IAA, ubiquitinates and degrades AUX/IAA by the action of the 26S proteasome, and ARF is released, activating the expression of auxin-responsive genes.

Signal transduction is an important link in plant auxin research, and there are four main auxin signal transduction pathways: The TIR1/AFB-Aux/IAA-TPL-ARFs pathway, the TMK1-IAA32/34-ARFs pathway, the TMK1/ABP1-ROP2/6-PINs or RICs pathway, and the SKP2-AE2FC/DPB pathway [138]. TIR1/AFB-Aux/IAA-TPL-ARFs is an extensively researched and widely recognized signaling pathway originating from the nucleus [139]. The first two pathways mediate the expression of auxin downstream genes by regulating ARF transcription factors, while the last two pathways directly activate some auxin efflux proteins and mediate the fast non-genomic effects induced by auxin [140]. It has been demonstrated that TIR1/AFB growth hormone signaling possesses a non-transcriptional branch that regulates rapid cellular processes, including cytoplasmic Ca^{2+} spiking and membrane depolarization. These processes have been associated with root growth inhibition [141]. In land plants, the TIR1/AFB receptor has adenylate cyclase (AC) activity, which contributes to root growth regulation by TIR1/AFB signaling and produces cAMP as a second messenger in this process. But this process still requires the involvement of an unknown mechanism [142]. Transmembrane kinase (TMK) has been demonstrated to mediate both transcriptional and non-transcriptional auxin signaling in Arabidopsis. Furthermore, it has been shown to activate Rho GTPase, which in turn controls the cytoskeleton [52,143]. Different TMKs have different roles in auxin signaling. Accumulation of growth hormone on the concave side of the apical hook stimulates TMK1 cleavage, which in turn leads to cytoplasmic and nuclear translocation cations that regulate gene transcription by stabilizing two nonclassical Aux/IAA proteins [144]. TMK4 has been demonstrated to regulate BR-mediated plant development [145] and to be involved in the negative regulation of growth hormone biosynthesis [52]. It has been shown that TMK1 inhibits plant growth by regulating ABI1/2, which mediates ABA signaling enhanced by high concentrations of auxin. Thus, TMKs coordinate growth hormone signaling with other signaling cascades, and TMKs may mediate differential growth hormone responses by phosphorylating different downstream components [146]. S-Phase Kinase-Associated Protein 2A (SKP2A) is a cell cycle-regulated F-box protein that controls the stability of at least two cell division transcriptional factors, E2FC and DPB [147]. Previous study has showed that auxin can regulate cell division through the SKP2A pathway. In the presence of auxin, SKP2A promotes degradation of cell cycle targets; additionally auxin enhances SKP2A protein hydrolysis to impede its excessive functionality [148]. Overexpression of SKP2A results in increased cell division and induces lateral root primordia (LRP) formation, a process known to be dependent on auxin signaling [149]. Although the SKP2A-E2FC/DPB pathway has been proposed, we lack evidence of many of its parts, thus more research is required. For example, SKP2A binding to SCF is able to target degradation of downstream E2FC and DPB; is it regulated by the proteasome in the same way, and what are the effects of degradation on the plant? Why does the mutation SKP2A not have a significant effect on plant growth and development?

8. Interaction of Auxin with Other Hormones

The signaling pathways of various hormones in plants often cross each other, forming a complex regulatory network. To date, the most attention has been paid to the interaction between auxin and cytokinin (CTK), jasmonic acid (JA), and abscisic acid (ABA).

8.1. Interaction between Auxin and CTK

The interaction between auxin and CTK involves both antagonistic and synergistic effects. In Arabidopsis, CTK signaling was shown to regulate the rate of auxin (IAA-indole-3-acetic acid) biosynthesis [150]. CTK regulated the auxin gradient to control the growth of lateral roots [151]. Auxin might modulate the mutual binding of CTK molecules and inactivate CTK. For example, auxin regulates CTK levels in the stem by inducing the expression of *Cytokinin oxidase/dehydrogenase* (CKX), suppressing the expression of *ATP/ADP isopentenyltransferase* (IPTs), and promoting the expression of strigolactone biosynthesis-related genes [152,153]. Auxin might modulate the mutual binding of *CTK* molecules and inactivate CTK [154]. Moreover, Nordstrom found that auxins inhibit CTK biosynthesis mainly through the *isopentenyladenosine-5′-monophosphate (iPMP)-independent* pathway and that this negative regulation is a fast-acting process [155]. During root development, antagonistic effects were observed, with auxin encouraging adventitious root production and exogenously applied physiological CTK inhibiting root formation and reversing IAA's effects [156]. Moreover, the rice auxin response factor OsARF25 can bind to the promoter of the cytokinin oxidase gene *OsCKX4* and activate its expression, thereby enhancing CTK metabolism [157].

8.2. Interaction between Auxin and JA

Numerous investigations have found that JA is involved in PAT and biosynthesis. For example, Li discovered that JA not only regulates auxin production by stimulating *ASA1* expression, but it also influences PAT [158]. Furthermore, JA can stimulate auxin production genes *ASA1* and *YUC2* by increasing the expression of the transcription factor *ERF109*, thus enhancing auxin biosynthesis [159]. Furthermore, both JA and IAA regulate transcription factor *WRKY57*, which can modulate JA and IAA signaling pathways in feedback [160].

8.3. Interaction between Auxin and ABA

It has been shown that there are both antagonistic and synergistic effects between them. As reported previously, PYL8 mediates the synergistic action of ABA and auxin to promote lateral root growth after sprouting [161]. WRKY46 contributes to the forward inhibition of osmotic/salt stress-induced LR inhibition by regulating the ABA pathway and growth hormone homeostasis [162]. During seed dormancy, auxin can stimulate *ABI3* expression by activating ARF10/16, thereby activating the ABA signaling pathway [163]. Additionally, ABA can modulate auxin signaling, and the ABA receptor PYL8 activates MYB77, which increases the production of auxin response genes [161].

8.4. Interaction between Auxin and Ethylene

Ethylene can regulate the synthesis of auxin. Exogenous 1-aminocyclopropane-1-carboxylic acid (ACC) treatment increased the expression of *AUX1*, *PIN3*, and *PIN7* while inhibiting lateral root development [118,164]. Auxin also regulates the production of ethylene. Upon *SlARF2* silencing, tomatoes produced less ethylene and expressed fewer ripening-related genes such as *RIN*, *CNR*, *NOR*, and *TAGL1* [117]. Exogenous auxin treatment of peaches resulted in increased *PpACS1* expression and ethylene production [165,166]. The expression of *MdARF5* was promoted in apples treated with NAA, which combined with the promoters of *MdERF2*, *MdACS3a*, *MdACS1*, and *MdACO1* to induce ethylene biosynthesis [167]. Moreover, *CpARF2* interacts with *CpEIL1* in papaya to promote *CpACS1* and *CpACO1* transcription and regulate fruit ripening [168].

In addition, auxin interacts with other hormones such as gibberellin, brassinosteroids, and salicylic acid. For example, in rice, auxin and gibberellin can regulate the negative gravity response of rice stem by antagonizing the expression of *XET* [169]. During hypocotyl growth, brassinosteroids can activate the auxin signaling pathway by inducing the transcription of *IAA19* and *ARF7* by BRASSINAZOLE-RESISTANT 1 (BZR1) [170]. Salicylic acid can inhibit the auxin biosynthesis induced by H_2O_2 by inhibiting the function of CATALASE2 [171].

9. The Role of Auxin in Stress

9.1. Heat Stress

Auxin plays an important role in heat stress-induced thermal morphogenesis, including stem (hypocotyl) elongation and subleaf glands [172]. Plants respond to high-temperature stress through auxin anabolism, polar transport, and signal transduction. High heat increases the level of free IAA by triggering the dominant two-stage IAA biosynthesis pathway from Trp to 3-IPA via TAA-1, then oxidative decarboxylation of 3-IPA to IAA catalyzed by flavin monooxygenases of the YUC subfamily [57,173,174]. In *Arabidopsis*, the homeostasis, turnover, and distribution of free IAA in the hypocotyl are controlled by the IAA amidosynthetase VAS2-GH3.17 under high temperatures [175]. In the *Arabidopsis* root system, the heat-stimulating protein HSP90 acts as a molecular chaperone for the auxin receptor TIR1 and influences the polar distribution of the auxin transporter protein PIN1 in the plasma membrane, thereby creating a concentration gradient of auxin and regulating the plant root system, growth, and development [176]. In *Sorghum bicolor* L., high temperatures upregulated the expression levels of most *SbARF* genes, and the *SbARF17/24* genes were found to be heavily expressed and accumulated in vascular tissue [177]. The above results indicated that auxin plays an important role in plant resistance to high-temperature stress and thermal stimulus transduction.

9.2. Flood Stress

Flooding prevents oxygen from reaching the roots, inhibits adventitious root (AR) formation, and might lead to moderate to severe root damage [178,179]. The formation of ARs is largely dependent on local auxin biosynthesis and translocation. Qi et al. (2023) found that endogenous auxin levels in hypocotyls increased, while externally applied NAA enhanced AR formation, at 72 h after flooding [180]. In addition, auxin treatment upregulated the expression levels of ethylene biosynthesis genes (*CsACS1*, *CsACS2*, *CsACO5*) and ROS signaling genes (e.g., *CsRBOHB* and *CsRBOHF3*) under flooding stress [181]. Gao et al. (2022, 2023) found that *Aux/IAA* gene expression and the auxin content were downregulated after 8 days of waterlogging, while exogenous spermidine alleviated waterlogging stress in roots and increased the auxin content in *Phyllostachys praecox* [182–184]. These results indicated that auxins play an important role in coping with flooding stress in plants.

9.3. Cold Stress

Cold temperatures limit plant growth mainly by causing cold damage to tissues. Zhu et al. (2015) discovered that low temperatures reduced the expression of *PIN1/3/7* and auxin biosynthesis-related genes and decreased auxin accumulation, which in turn inhibited the division potential of *Arabidopsis* meristematic tissue cells [34]. In the auxin degradation pathway, relative expression levels of *Gretchen Hagen 3* gene (*GH3.3* and *GH3.6*) was upregulated by cold stress in *Cicer arietinum* shoots [185]. Moreover, overexpression of *CsARF5* enhanced cold stress tolerance in cucumber [186]. The cold stress significantly altered transcript levels of *SlSAURs* genes in Solanaceae species [187]. The levels of certain auxin response factors (*ARFs; TaARF8,TaARF9* and *TaARF21*) are reduced at low temperatures [188]. The changes in the expression of these genes suggest that low temperatures altered the expression of genes involved in auxin metabolism, thereby affecting auxin levels and inhibiting plant growth.

9.4. Salt Stress

Excess salt disrupts plant physiological, biochemical, and molecular processes and salt stress is the second most important abiotic factor affecting global agricultural productivity [189]. Auxin improved salt tolerance in cucumber seedlings, and transcriptomic analysis revealed that auxin signaling genes *SAUR*, *Aux/IAA*, and *GH3* were downregulated in salt stress [190]. The expression of 5NG4-like, a key molecular transporter gene induced by auxin, was upregulated in seedlings treated with NaCl with exogenously added silicon [191]. It was hypothesized that auxin signaling genes play a key role in silica-mediated salt tolerance. However, functional studies are required to determine the underlying mechanisms.

9.5. Drought Stress

Drought stress downregulated auxin genes in the auxin sub-pathway, including genes encoding auxin influx proteins, auxin response proteins (AUX/IAA), ARF, and GH3. RNA-Seq-based transcriptome analysis showed that melatonin upregulated 23 genes involved in growth hormone signaling, including *AUX/IAA*, *ARF*, and *SAUR*, in *Davidia involucrate* [192]. In the present study, overexpression of *OsIAA6* and *IbARF5* improved drought tolerance in rice and Arabidopsis, respectively [193,194]. In drought- and CO_2-treated cucumber roots, IAA levels were decreased. By contrast, gibberellin (GA) had a significant inducing effect. Thus, auxin might regulate the response of cucumber to drought stress downstream of GA [181].

9.6. Heavy Metal Stress

Heavy metals damage plant cells by disrupting a variety of physiological processes [195]. Cadmium (Cd) affects auxin biosynthesis and transport, thereby altering the formation of quiescent centers (QC), whereas exogenous auxin restores normal root development [196]. Cd increased the expression of IAA influx carrier *AUX1* and strongly repressed the expression of *PIN5*, and *OsPIN5b* was involved in the regulation of IAA homeostasis, transport, and distribution [197,198]. In contrast, arsenic (As) reduced the expression of the *AUX1* and the efflux carrier *PIN5*. However, both Cd and As affect adventitious root (AR) and lateral root (LR) development through the regulation of auxin carriers in turn [198,199]. A recent study highlighted the role of auxin in the response of cucumbers to cadmium stress. The study showed that exogenous application of selenium significantly inhibited the harmful effects of cadmium. Auxin binding protein (ABP19a-like) levels were higher in Se-treated seedlings than in cadmium-treated seedlings [200]. However, further functional studies are needed to validate auxin's involvement cadmium mitigation or other heavy metal stresses.

10. Conclusions and Perspectives

In recent years, great progress has been made in understanding the mechanisms of auxin sensing and signaling. Moreover, the interaction between auxin and environmental signals in controlling plant growth and development has garnered increased interest. Many studies have shown that environmental signals, particularly abiotic stress, directly regulate some key genes of auxin synthesis and metabolism, polar transport, and signaling. However, the molecular mechanism of auxin regulation under abiotic stress requires further investigation. Based on the current research status, we suggest that future research directions should concentrate on the following areas. Firstly, in the face of the increasingly serious problem of global warming, the possible role of temperature signaling and its molecular mechanisms require further study. For example, in male sterility of plants under high-temperature stress, in addition to changes in the endogenous auxin content, more evidence is required to support which transcription factors are directly involved in the effects of high-temperature stress on plant growth and development. Spatial and temporal regulation of auxin synthesis is critical for plant development. It combines gene editing technology to precisely regulate the expression of key enzymes for auxin synthesis, thus realizing precise control of plant growth. A recent study has shown that ultra-rapid global phosphorylation downstream of auxin sensing on ABP1-TMK cell surfaces allows auxin responses to be completed within seconds. Notably, previously, TIR1/AFB-mediated slow auxin signaling responses tended to be in the 20–30 min. In summary, the auxin fast response may play a pivotal role in plant stress as well as in signaling cascade responses [201]. It deserves to be explored further.

Additionally, current research on Aux/IAA in the auxin signaling pathway has focused on plant growth and development, but little attention has been paid to the role of Aux/IAA in the auxin-mediated response to environmental interactions (e.g., drought, heavy metals, nutrient deficiencies, and other abiotic stresses) and it is critical to understand how auxin interacts with other hormones in this process. Future genetic investigations, together with computational modeling, will enable the identification of novel candidate genes that modulate Aux/IAA, and hence the overall auxin signaling regulation network. Moreover, Wang et al. (2004) [202] found that strigolactones (SLs) reduced the inhibitory effect of *WRKY41* on the expression of CBF/*dehydration response element binding factor 1* (*DREB1*) to promote cold tolerance in plants. It is because the effects of SLs and auxin on *wrky* genes in plants under stress, but the specific mechanism is still unclear. In synergism or antagonism with other hormones and signaling molecules, auxin may affect photosynthesis, plant antioxidants, arbuscular mycorrhizal (AM) symbiosis, etc., thereby mitigating the damage caused by different abiotic stresses on plants. This deserves to be further explored to provide new ideas and approaches for agriculture and biotechnology applications.

Author Contributions: J.G. and S.Z. conceived and designed the article structure. J.G. drafted the manuscript. W.Z. and S.Z. modified the paper. All authors have read and agreed to the published version of the manuscript.

Funding: This work was supported by the National Natural Science Foundation of China [grant number 32471979] and the Central Guidance for Local Science and Technology Development Projects (S2023KJCXD0003-1).

Data Availability Statement: The raw data supporting the conclusions of this article will be made available by the authors on request.

Acknowledgments: The authors would like to express their gratitude to Shunyao Zhuang for supplying us.

Conflicts of Interest: The authors declare no conflicts of interest.

References

1. Chandler, J.W. Auxin Response Factors. *Plant Cell Environ.* **2016**, *39*, 1014–1028. [CrossRef]
2. Mashiguchi, K.; Tanaka, K.; Sakai, T.; Sugawara, S.; Kawaide, H.; Natsume, M.; Hanada, A.; Yaeno, T.; Shirasu, K.; Yao, H.; et al. The Main Auxin Biosynthesis Pathway in *Arabidopsis*. *Proc. Natl. Acad. Sci. USA* **2011**, *108*, 18512–18517. [CrossRef] [PubMed]
3. Ferro, N.; Bredow, T.; Jacobsen, H.-J.; Reinard, T. Route to Novel Auxin: Auxin Chemical Space toward Biological Correlation Carriers. *Chem. Rev.* **2010**, *110*, 4690–4708. [CrossRef]
4. Darwin, C.; Darwin, F. The power of movement in plants. In *The General Considerations on the Movements and Growth of Seedling Plants*; Springer: New York, NY, USA, 2009; pp. 1–9. [CrossRef]

5. Balzan, S.; Johal, G.S.; Carraro, N. The Role of Auxin Transporters in Monocots Development. *Front. Plant Sci.* **2014**, *5*, 393. [CrossRef] [PubMed]
6. Krishnamurthy, A.; Rathinasabapathi, B. Auxin and Its Transport Play a Role in Plant Tolerance to Arsenite-Induced Oxidative Stress in *Arabidopsis thaliana*. *Plant Cell Environ.* **2013**, *36*, 1838–1849. [CrossRef]
7. Fan, Y.-M.; Dang, S.-K.; Wang, W.-J.; Wang, H.-M. Effect of Carbon Source, Auxin and Elicitor on the Growth and Synthesis of Secondary Metabolites of Adventitious Roots of *Cajanus cajan* (Linn.) *Millsp. Bull. Bot. Res.* **2018**, *38*, 391–398. [CrossRef]
8. Ahmad, A.; Andersen, A.; Engvild, K. Rooting, Growth and Ethylene Evolution of Pea Cuttings in Response to Chloroindole Auxins. *Physiol. Plant.* **2006**, *69*, 137–140. [CrossRef]
9. Chen, X.; Qu, Y.; Sheng, L.; Liu, J.; Huang, H.; Xu, L. A Simple Method Suitable to Study de Novo Root Organogenesis. *Front. Plant Sci.* **2014**, *5*, 208. [CrossRef]
10. Sharif, R.; Su, L.; Chen, X.; Qi, X. Hormonal Interactions Underlying Parthenocarpic Fruit Formation in Horticultural Crops. *Hortic. Res.* **2022**, *9*, uhab024. [CrossRef]
11. Maupilé, L.; Chaib, J.; Boualem, A.; Bendahmane, A. Parthenocarpy, a Pollination-Independent Fruit Set Mechanism to Ensure Yield Stability. *Trends Plant Sci.* **2024**, *20*, S1360–S1385. [CrossRef]
12. Aloni, R.; Aloni, E.; Langhans, M.; Ullrich, C.I. Role of Auxin in Regulating Arabidopsis Flower Development. *Planta* **2006**, *223*, 315–328. [CrossRef]
13. Zhang, D.; Ren, L.; Yue, J.; Wang, L.; Zhuo, L.; Shen, X. GA4 and IAA Were Involved in the Morphogenesis and Development of Flowers in Agapanthus Praecox ssp. Orientalis. *J. Plant Physiol.* **2014**, *171*, 966–976. [CrossRef]
14. Zhao, Y.; Christensen, S.K.; Fankhauser, C.; Cashman, J.R.; Cohen, J.D.; Weigel, D.; Chory, J. A Role for Flavin Monooxygenase-like Enzymes in Auxin Biosynthesis. *Science* **2001**, *291*, 306–309. [CrossRef]
15. Sagar, M.; Chervin, C.; Mila, I.; Hao, Y.; Roustan, J.P.; Benichou, M.; Gibon, Y.; Biais, B.; Maury, P.; Latché, A.; et al. SlARF4, an Auxin Response Factor Involved in the Control of Sugar Metabolism during Tomato Fruit Development. *Plant Physiol.* **2013**, *161*, 1362–1374. [CrossRef]
16. Yang, X.M.; Zheng, G.Q.; Xu, X.; Lu, D.; Yang, L. Changes in endogenous hormone contents and cell wall component, degrading Enzyme Activity and their relation in *Lycium barbarum*. *Chin. Bull. Bot.* **2014**, *49*, 30–40. [CrossRef]
17. Mouden, S.; Bac-Molenaar, J.A.; Kappers, I.F.; Beerling, E.A.M.; Leiss, K.A. Elicitor Application in Strawberry Results in Long-Term Increase of Plant Resilience Without Yield Loss. *Front. Plant Sci.* **2021**, *12*, 695908. [CrossRef]
18. Manning, K. Changes in Gene Expression during Strawberry Fruit Ripening and Their Regulation by Auxin. *Planta* **1994**, *194*, 62–68. [CrossRef]
19. Wang, H.Z.; Yang, K.Z.; Zou, J.J.; Zhu, L.L.; Xie, Z.D.; Morita, M.T.; Tasaka, M.; Friml, J.; Grotewold, E.; Beeckman, T.; et al. Transcriptional Regulation of *PIN* Genes by *FOUR LIPS* and *MYB88* during *Arabidopsis* Root Gravitropism. *Nat. Commun.* **2015**, *6*, 8822. [CrossRef]
20. Sun, H.; Xu, F.; Guo, X.; Wu, D.; Zhang, X.; Lou, M.; Luo, F.; Zhao, Q.; Xu, G.; Zhang, Y. A Strigolactone Signal Inhibits Secondary Lateral Root Development in Rice. *Front. Plant Sci.* **2019**, *10*, 1527. [CrossRef]
21. Swarup, R.; Friml, J.; Marchant, A.; Ljung, K.; Sandberg, G.; Palme, K.; Bennett, M. Localization of the Auxin Permease *AUX1* Suggests Two Functionally Distinct Hormone Transport Pathways Operate in the *Arabidopsis* Root Apex. *Genes Dev.* **2001**, *15*, 2648–2653. [CrossRef]
22. Friml, J.; Wiśniewska, J.; Benková, E.; Mendgen, K.; Palme, K. Lateral Relocation of Auxin Efflux Regulator PIN3 Mediates Tropism in *Arabidopsis*. *Nature* **2002**, *415*, 806–809. [CrossRef]
23. Rakusová, H.; Abbas, M.; Han, H.; Song, S.; Robert, H.S.; Friml, J. Termination of Shoot Gravitropic Responses by Auxin Feedback on PIN3 Polarity. *Curr. Biol.* **2016**, *26*, 3026–3032. [CrossRef]
24. Ge, Y.; Yan, F.; Zourelidou, M.; Wang, M.; Ljung, K.; Fastner, A.; Hammes, U.Z.; Di Donato, M.; Geisler, M.; Schwechheimer, C.; et al. *SHADE AVOIDANCE 4* Is Required for Proper Auxin Distribution in the Hypocotyl. *Plant Physiol.* **2017**, *173*, 788–800. [CrossRef]
25. Willige, B.C.; Ahlers, S.; Zourelidou, M.; Barbosa, I.C.R.; Demarsy, E.; Trevisan, M.; Davis, P.A.; Roelfsema, M.R.G.; Hangarter, R.; Fankhauser, C.; et al. D6PK AGCVIII Kinases Are Required for Auxin Transport and Phototropic Hypocotyl Bending in *Arabidopsis*. *Plant Cell* **2013**, *25*, 1674–1688. [CrossRef]
26. Hornitschek, P.; Kohnen, M.V.; Lorrain, S.; Rougemont, J.; Ljung, K.; López-Vidriero, I.; Franco-Zorrilla, J.M.; Solano, R.; Trevisan, M.; Pradervand, S.; et al. Phytochrome Interacting Factors 4 and 5 Control Seedling Growth in Changing Light Conditions by Directly Controlling Auxin Signaling. *Plant J.* **2012**, *71*, 699–711. [CrossRef]
27. Hersch, M.; Lorrain, S.; de Wit, M.; Trevisan, M.; Ljung, K.; Bergmann, S.; Fankhauser, C. Light Intensity Modulates the Regulatory Network of the Shade Avoidance Response in *Arabidopsis*. *Proc. Natl. Acad. Sci. USA* **2014**, *111*, 6515–6520. [CrossRef]
28. Casal, J.J. Photoreceptor Signaling Networks in Plant Responses to Shade. *Annu. Rev. Plant Biol.* **2013**, *64*, 403–427. [CrossRef]
29. Tao, Y.; Ferrer, J.-L.; Ljung, K.; Pojer, F.; Hong, F.; Long, J.A.; Li, L.; Moreno, J.E.; Bowman, M.E.; Ivans, L.J.; et al. Rapid Synthesis of Auxin via a New Tryptophan-Dependent Pathway Is Required for Shade Avoidance in Plants. *Cell* **2008**, *133*, 164–176. [CrossRef]
30. Krouk, G.; Lacombe, B.; Bielach, A.; Perrine-Walker, F.; Malinska, K.; Mounier, E.; Hoyerova, K.; Tillard, P.; Leon, S.; Ljung, K.; et al. Nitrate-Regulated Auxin Transport by NRT1.1 Defines a Mechanism for Nutrient Sensing in Plants. *Dev. Cell* **2010**, *18*, 927–937. [CrossRef]

31. Goyal, A.; Karayekov, E.; Galvão, V.C.; Ren, H.; Casal, J.J.; Fankhauser, C. Shade Promotes Phototropism through Phytochrome B-Controlled Auxin Production. *Curr. Biol.* **2016**, *26*, 3280–3287. [CrossRef]
32. Da Costa, V.A.; Cothren, J.T.; Bynum, J.B. Abiotic Stress Effects on Plant Growth and Yield Components of 1-MCP Treated Cotton Plants. *Agron. J.* **2011**, *103*, 1591–1596. [CrossRef]
33. Weber, V.S.; Melchinger, A.E.; Magorokosho, C.; Makumbi, D.; B?Nziger, M.; Atlin, G.N. Efficiency of Managed-Stress Screening of Elite Maize Hybrids under Drought and Low Nitrogen for Yield under Rainfed Conditions in Southern Africa. *Crop Sci.* **2012**, *52*, 1011–1020. [CrossRef]
34. Zhu, J.; Zhang, K.-X.; Wang, W.-S.; Gong, W.; Liu, W.-C.; Chen, H.-G.; Xu, H.-H.; Lu, Y.-T. Low Temperature Inhibits Root Growth by Reducing Auxin Accumulation via ARR1/12. *Plant Cell Physiol.* **2015**, *56*, 727–736. [CrossRef]
35. Zhang, C.; Li, G.; Chen, T.; Feng, B.; Fu, W.; Yan, J.; Islam, M.R.; Jin, Q.; Tao, L.; Fu, G. Heat Stress Induces Spikelet Sterility in Rice at Anthesis through Inhibition of Pollen Tube Elongation Interfering with Auxin Homeostasis in Pollinated Pistils. *Rice* **2018**, *11*, 14. [CrossRef]
36. Simon, S.; Petrášek, J. Why Plants Need More than One Type of Auxin. *Plant Sci.* **2011**, *180*, 454–460. [CrossRef]
37. Korasick, D.A.; Enders, T.A.; Strader, L.C. Auxin Biosynthesis and Storage Forms. *J. Exp. Bot.* **2013**, *64*, 2541–2555. [CrossRef]
38. Strader, L.C.; Nemhauser, J.L. Auxin 2012: A Rich Mea Ho'oulu. *Development* **2013**, *140*, 1153–1157. [CrossRef] [PubMed]
39. Tognetti, V.B.; Muhlenbock, P.; Breusegems, F.V. Stress Homeostasis-the Redox and Auxin Perspective. *Plant Cell Environ.* **2011**, *35*, 321–333. [CrossRef]
40. Ludwig-Müller, J.; Jülke, S.; Bierfreund, N.M.; Decker, E.L.; Reski, R. Moss (*Physcomitrella patens*) GH3 Proteins Act in Auxin Homeostasis. *New Phytol.* **2009**, *181*, 323–338. [CrossRef]
41. Campanella, J.J.; Olajide, A.F.; Magnus, V.; Ludwig-Müller, J. A Novel Auxin Conjugate Hydrolase from Wheat with Substrate Specificity for Longer Side-Chain Auxin Amide Conjugates. *Plant Physiol.* **2004**, *135*, 2230–2240. [CrossRef]
42. Laubscher, C.P.; Ndakidemi, P.A. Rooting Success Using IBA Auxin on Endangered *Leucadendron laxum* (PROTEACEAE) in Different Rooting Mediums. *Afr. J. Biotechnol.* **2008**, *7*, 3437–3442. [CrossRef]
43. Rigas, S.; Ditengou, F.A.; Ljung, K.; Daras, G.; Tietz, O.; Palme, K.; Hatzopoulos, P. Root Gravitropism and Root Hair Development Constitute Coupled Developmental Responses Regulated by Auxin Homeostasis in the *Arabidopsis* Root Apex. *New Phytol.* **2013**, *197*, 1130–1141. [CrossRef] [PubMed]
44. Kageyama, E.; Katayama, M.; Masui, Y.; Kageyama, K.; Kawabata, Y. Synthesis and Plant Growth-Regulating Activities of L-Lactic Acid Derivatives of 4-Chloroindole-3-Acetic Acid. *J. Pestici Sci.* **2006**, *31*, 130–138. [CrossRef]
45. Somers, E.; Ptacek, D.; Gysegom, P.; Srinivasan, M.; Vanderleyden, J. *Azospirillum brasilense* Produces the Auxin-Like Phenylacetic Acid by Using the Key Enzyme for Indole-3-Acetic Acid Biosynthesis. *Appl. Environ. Microbiol.* **2005**, *71*, 1803–1810. [CrossRef]
46. Wójcik, A.M.; Wójcik; Wójcikowska, B.; Gaj, M.D. Current Perspectives on the Auxin-Mediated Genetic Network That Controls the Induction of Somatic Embryogenesis in Plants. *Int. J. Mol. Sci.* **2020**, *21*, 1333. [CrossRef] [PubMed]
47. Mataa, M.; Tominaga, S.; Kozaki, I. Effects of Exogenous Growth Regulator Applications on Source-Leaf Carbohydrate Accumulation Patterns in Ponkan (Citrus Reticulata Blanco). *J. Jpn. Soc. Hortic. Sci.* **2008**, *66*, 245–251. [CrossRef]
48. Sauer, M.; Robert, S.; Kleine-Vehn, J. Auxin: Simply Complicated. *J. Exp. Bot.* **2013**, *64*, 2565–2577. [CrossRef]
49. Zhao, Y.D. Auxin Biosynthesis and Its Role in Plant Development. *Annu. Rev. Plant Biol.* **2010**, *61*, 49–64. [CrossRef]
50. Woodward, A.W.; Bartel, B. Auxin: Regulation, Action, and Interaction. *Ann. Bot.* **2005**, *95*, 707–735. [CrossRef]
51. Normanly, J. Approaching Cellular and Molecular Resolution of Auxin Biosynthesis and Metabolism. *Cold Spring Harb. Perspect. Biol.* **2010**, *2*, a001594. [CrossRef]
52. Wang, Q.; Qin, G.; Cao, M.; Chen, R.; He, Y.; Yang, L.; Zeng, Z.; Yu, Y.; Gu, Y.; Xing, W.; et al. A Phosphorylation-Based Switch Controls TAA1-Mediated Auxin Biosynthesis in Plants. *Nat. Commun.* **2020**, *11*, 679. [CrossRef] [PubMed]
53. Dimkpa, C.O.; Zeng, J.; McLean, J.E.; Britt, D.W.; Zhan, J.X.; Anderson, A.J. Anderson Production of Indole-3-Acetic Acid via the Indole-3-Acetamide Pathway in the Plant-Beneficial Bacterium *Pseudomonas chlororaphis* O6 Is Inhibited by ZnO Nanoparticles but Enhanced by CuO Nanoparticles. *Appl. Environ. Microbiol.* **2012**, *78*, 1404–1410. [CrossRef]
54. Wang, B.; Chu, J.; Yu, T.; Xu, Q.; Li, J. Tryptophan-Independent Auxin Biosynthesis Contributes to Early Embryogenesis in *Arabidopsis*. *Proc. Natl. Acad. Sci. USA* **2015**, *112*, 4821–4826. [CrossRef]
55. Bagautdinov, B.; Yutani, K. Structure of Indole-3-glycerol Phosphate Synthase from Thermus Thermophilus HB8: Implications for Thermal Stability. *Acta Crystallogr. D Biol. Crystallogr.* **2011**, *67*, 1054–1064. [CrossRef] [PubMed]
56. Gomes, G.L.B.; Scortecci, K.C. Auxin and Its Role in Plant Development: Structure, Signalling, Regulation and Response Mechanisms. *Plant Biol. J.* **2021**, *23*, 894–904. [CrossRef] [PubMed]
57. Stepanova, A.N.; Robertson-Hoyt, J.; Yun, J.; Benavente, L.M.; Xie, D.-Y.; Dolezal, K.; Schlereth, A.; Jürgens, G.; Alonso, J.M. TAA1-Mediated Auxin Biosynthesis Is Essential for Hormone Crosstalk and Plant Development. *Cell* **2008**, *133*, 177–191. [CrossRef]
58. Yamada, M.; Greenham, K.; Prigge, M.J.; Jensen, P.J.; Estelle, M. *The TRANSPORT INHIBITOR RESPONSE2* Gene Is Required for Auxin Synthesis and Diverse Aspects of Plant Development. *Plant Physiol.* **2009**, *151*, 168–179. [CrossRef]
59. Stepanova, A.N.; Yun, J.; Robles, L.M.; Novak, O.; He, W.; Guo, H.; Ljung, K.; Alonso, J.M. The *Arabidopsis YUCCA1* Flavin Monooxygenase Functions in the Indole-3-Pyruvic Acid Branch of Auxin Biosynthesis. *Plant Cell* **2011**, *23*, 3961–3973. [CrossRef]
60. Matthes, M.S.; Best, N.B.; Robil, J.M.; Malcomber, S.; Gallavotti, A.; McSteen, P. Auxin EvoDevo: Conservation and Diversification of Genes Regulating Auxin Biosynthesis, Transport, and Signaling. *Mol. Plant* **2019**, *12*, 298–320. [CrossRef]

61. Poulet, A.; Kriechbaumer, V. Bioinformatics Analysis of Phylogeny and Transcription of *TAA/YUC* Auxin Biosynthetic Genes. *Int. J. Mol. Sci.* **2017**, *18*, 1791. [CrossRef]
62. Eklund, D.M.; Ishizaki, K.; Flores-Sandoval, E.; Kikuchi, S.; Takebayashi, Y.; Tsukamoto, S.; Hirakawa, Y.; Nonomura, M.; Kato, H.; Kouno, M.; et al. Auxin Produced by the Indole-3-Pyruvic Acid Pathway Regulates Development and Gemmae Dormancy in the Liverwort Marchantia Polymorpha. *Plant Cell* **2015**, *27*, 1650–1669. [CrossRef]
63. Sugawara, S.; Hishiyama, S.; Jikumaru, Y.; Hanada, A.; Nishimura, T.; Koshiba, T.; Zhao, Y.; Kamiya, Y.; Kasahara, H. Biochemical Analyses of Indole-3-Acetaldoxime-Dependent Auxin Biosynthesis in *Arabidopsis*. *Proc. Natl. Acad. Sci. USA* **2009**, *106*, 5430–5435. [CrossRef]
64. Mikkelsen, M.D.; Hansen, C.H.; Wittstock, U.; Halkier, B.A. Cytochrome P450 *CYP79B2* from *Arabidopsis* Catalyzes the Conversion of Tryptophan to Indole-3-Acetaldoxime, a Precursor of Indole Glucosinolates and Indole-3-Acetic Acid. *J. Biol. Chem.* **2000**, *275*, 33712–33717. [CrossRef]
65. Zhao, Y.; Hull, A.K.; Gupta, N.R.; Goss, K.A.; Alonso, J.; Ecker, J.R.; Normanly, J.; Chory, J.; Celenza, J.L. Trp-Dependent Auxin Biosynthesis in *Arabidopsis*: Involvement of Cytochrome P450s CYP79B2 and CYP79B3. *Genes Dev.* **2002**, *16*, 3100–3112. [CrossRef]
66. Hull, A.K.; Vij, R.; Celenza, J.L. Arabidopsis Cytochrome P450s That Catalyze the First Step of Tryptophan-Dependent Indole-3-Acetic Acid Biosynthesis. *Proc. Natl. Acad. Sci. USA* **2000**, *97*, 2379–2384. [CrossRef] [PubMed]
67. Nafisi, M.; Goregaoker, S.; Botanga, C.J.; Glawischnig, E.; Olsen, C.E.; Halkier, B.A.; Glazebrook, J. *Arabidopsis* Cytochrome P450 Monooxygenase 71A13 Catalyzes the Conversion of Indole-3-Acetaldoxime in *Camalexin Synthesis*. *Plant Cell* **2007**, *19*, 2039–2052. [CrossRef]
68. Malka, S.K.; Cheng, Y. Possible Interactions between the Biosynthetic Pathways of Indole Glucosinolate and Auxin. *Front. Plant Sci.* **2017**, *8*, 2131. [CrossRef]
69. Gao, Y.; Dai, X.; Aoi, Y.; Takebayashi, Y.; Yang, L.; Guo, X.; Zeng, Q.; Yu, H.; Kasahara, H.; Zhao, Y. Two Homologous INDOLE-3-ACETAMIDE (IAM) HYDROLASE Genes Are Required for the Auxin Effects of IAM in *Arabidopsis*. *J. Genet. Genom.* **2020**, *47*, 157–165. [CrossRef]
70. Pencík, A.; Casanova-Sáez, R.; Pilarová, V.; Žukauskaite, A.; Pinto, R.; Micol, J.L.; Ljung, K.; Novák, O. Ultra-Rapid Auxin Metabolite Profiling for High-Throughput Mutant Screening in Arabidopsis. *J. Exp. Bot.* **2018**, *69*, 2569–2579. [CrossRef]
71. Brunoni, F.; Collani, S.; Casanova-Sáez, R.; Šimura, J.; Karady, M.; Schmid, M.; Ljung, K.; Bellini, C. Conifers Exhibit a Characteristic Inactivation of Auxin to Maintain Tissue Homeostasis. *New Phytol.* **2020**, *226*, 1753–1765. [CrossRef]
72. Fattorini, L.; Veloccia, A.; Della Rovere, F.; D'Angeli, S.; Falasca, G.; Altamura, M.M. Indole-3-Butyric Acid Promotes Adventitious Rooting in *Arabidopsis thaliana* Thin Cell Layers by Conversion into Indole-3-Acetic Acid and Stimulation of Anthranilate Synthase Activity. *BMC Plant Biol.* **2017**, *17*, 121. [CrossRef]
73. Revelou, P.K.; Kokotou, M.G.; Constantinou-Kokotou, V. Identification of Auxin Metabolites in *Brassicaceae* by Ultra-Performance Liquid Chromatography Coupled with High-Resolution Mass Spectrometry. *Molecules* **2019**, *24*, 2615. [CrossRef]
74. Pencík, A.; Simonovik, B.; Petersson, S.V.; Henyková, E.; Simon, S.; Greenham, K.; Zhang, Y.; Kowalczyk, M.; Estelle, M.; Zazímalová, E.; et al. Regulation of Auxin Homeostasis and Gradients in *Arabidopsis* Roots through the Formation of the Indole-3-Acetic Acid Catabolite 2-Oxindole-3-Acetic Acid. *Plant Cell* **2013**, *25*, 3858–3870. [CrossRef] [PubMed]
75. Kubeš, M.; Yang, H.; Richter, G.L.; Cheng, Y.; Młodzińska, E.; Wang, X.; Blakeslee, J.J.; Carraro, N.; Petrášek, J.; Zažímalová, E.; et al. The *Arabidopsis* Concentration-Dependent Influx/Efflux Transporter *ABCB4* Regulates Cellular Auxin Levels in the Root Epidermis. *Plant J.* **2012**, *69*, 640–654. [CrossRef]
76. Hayashi, K.; Arai, K.; Aoi, Y.; Tanaka, Y.; Hira, H.; Guo, R.; Hu, Y.; Ge, C.; Zhao, Y.; Kasahara, H.; et al. The Main Oxidative Inactivation Pathway of the Plant Hormone Auxin. *Nat. Commun.* **2021**, *12*, 6752. [CrossRef] [PubMed]
77. Novák, O.; Hényková, E.; Sairanen, I.; Kowalczyk, M.; Pospíšil, T.; Ljung, K. Tissue-Specific Profiling of the *Arabidopsis thaliana* Auxin Metabolome. *Plant J.* **2012**, *72*, 523–536. [CrossRef] [PubMed]
78. Porco, S.; Pěnčík, A.; Rashed, A.; Voß, U.; Casanova-Sáez, R.; Bishopp, A.; Golebiowska, A.; Bhosale, R.; Swarup, R.; Swarup, K.; et al. Dioxygenase-Encoding *AtDAO1* Gene Controls IAA Oxidation and Homeostasis in *Arabidopsis*. *Proc. Natl. Acad. Sci. USA* **2016**, *113*, 11016–11021. [CrossRef] [PubMed]
79. Žižková, E.; Kubeš, M.; Dobrev, P.I.; Přibyl, P.; Šimura, J.; Zahajská, L.; Záveská Drábková, L.; Novák, O.; Motyka, V. Control of Cytokinin and Auxin Homeostasis in Cyanobacteria and Algae. *Ann. Bot.* **2017**, *119*, 151–166. [CrossRef]
80. Zhao, Z.; Zhang, Y.; Liu, X.; Zhang, X.; Liu, S.; Yu, X.; Ren, Y.; Zheng, X.; Zhou, K.; Jiang, L.; et al. A Role for a Dioxygenase in Auxin Metabolism and Reproductive Development in Rice. *Dev. Cell* **2013**, *27*, 113–122. [CrossRef]
81. Zhang, J.; Lin, J.E.; Harris, C.; Campos Mastrotti Pereira, F.; Wu, F.; Blakeslee, J.J.; Peer, W.A. DAO1 Catalyzes Temporal and Tissue-Specific Oxidative Inactivation of Auxin in *Arabidopsis thaliana*. *Proc. Natl. Acad. Sci. USA* **2016**, *113*, 11010–11015. [CrossRef]
82. Zhang, J.; Peer, W.A. Auxin Homeostasis: The DAO of Catabolism. *J. Exp. Bot.* **2017**, *68*, 3145–3154. [CrossRef] [PubMed]
83. Wojtaczka, P.; Ciarkowska, A.; Starzynska, E.; Ostrowski, M. The GH3 Amidosynthetases Family and Their Role in Metabolic Crosstalk Modulation of Plant Signaling Compounds. *Phytochemistry* **2022**, *194*, 113039. [CrossRef]
84. Mellor, N.; Band, L.R.; Pěnčík, A.; Novák, O.; Rashed, A.; Holman, T.; Wilson, M.H.; Voß, U.; Bishopp, A.; King, J.R.; et al. Dynamic Regulation of Auxin Oxidase and Conjugating Enzymes *AtDAO1* and *GH3* Modulates Auxin Homeostasis. *Proc. Natl. Acad. Sci. USA* **2016**, *113*, 11022–11027. [CrossRef] [PubMed]
85. Vernoux, T.; Besnard, F.; Traas, J. Auxin at the Shoot Apical Meristem. *Cold Spring Harb. Perspect. Biol.* **2010**, *2*, a001487. [CrossRef]

86. Friml, J. Auxin Transport-Shaping the Plant. *Curr. Opin. Plant Biol.* **2003**, *6*, 7–12. [CrossRef] [PubMed]
87. Overvoorde, P.; Fukaki, H.; Beeckman, T. Auxin Control of Root Development. *Cold Spring Harb. Perspect. Biol.* **2010**, *2*, a001537. [CrossRef]
88. Adamowski, M.; Narasimhan, M.; Kania, U.; Glanc, M.; De Jaeger, G.; Friml, J. A Functional Study of *AUXILIN-LIKE1* and 2, Two Putative Clathrin Uncoating Factors in *Arabidopsis*. *Plant Cell* **2018**, *30*, 700–716. [CrossRef]
89. Scanlon, M.J. The Polar Auxin Transport Inhibitor N-1-Naphthylphthalamic Acid Disrupts Leaf Initiation, KNOX Protein Regulation, and Formation of Leaf Margins in Maize. *Plant Physiol.* **2003**, *133*, 597–605. [CrossRef]
90. Adamowski, M.; Friml, J. PIN-Dependent Auxin Transport: Action, Regulation, and Evolution. *Plant Cell* **2015**, *27*, 20–32. [CrossRef]
91. Semeradova, H.; Montesinos, J.C.; Benkova, E. All Roads Lead to Auxin: Post-Translational Regulation of Auxin Transport by Multiple Hormonal Pathways. *Plant Comm.* **2020**, *1*, 100048. [CrossRef]
92. Bandyopadhyay, A.; Blakeslee, J.J.; Lee, O.R.; Mravec, J.; Sauer, M.; Titapiwatanakun, B.; Makam, S.N.; Bouchard, R.; Geisler, M.; Martinoia, E.; et al. Interactions of PIN and PGP Auxin Transport Mechanisms. *Biochem. Soc. Trans.* **2007**, *35*, 137–141. [CrossRef] [PubMed]
93. Blakeslee, J.J.; Bandyopadhyay, A.; Lee, O.R.; Mravec, J.; Titapiwatanakun, B.; Sauer, M.; Makam, S.N.; Cheng, Y.; Bouchard, R.; Adamec, J.; et al. Interactions among PIN-FORMED and P-Glycoprotein Auxin Transporters in Arabidopsis. *Plant Cell* **2007**, *19*, 131–147. [CrossRef] [PubMed]
94. Titapiwatanakun, B.; Blakeslee, J.J.; Bandyopadhyay, A.; Yang, H.; Mravec, J.; Sauer, M.; Cheng, Y.; Adamec, J.; Nagashima, A.; Geisler, M.; et al. *ABCB19/PGP19* Stabilises *PIN1* in Membrane Microdomains in *Arabidopsis*. *Plant J.* **2009**, *57*, 27–44. [CrossRef] [PubMed]
95. Yang, H.; Murphy, A.S. Functional Expression and Characterization of Arabidopsis *ABCB*, *AUX 1* and *PIN* Auxin Transporters in *Schizosaccharomyces pombe*. *Plant J.* **2009**, *59*, 179–191. [CrossRef] [PubMed]
96. Carrier, D.J.; Bakar, N.T.A.; Swarup, R.; Callaghan, R.; Napier, R.M.; Bennett, M.J.; Kerr, I.D. The Binding of Auxin to the *Arabidopsis* Auxin Influx Transporter *AUX1*. *Plant Physiol.* **2008**, *148*, 529–535. [CrossRef]
97. Mazzoni-Putman, S.M.; Brumos, J.; Zhao, C.; Alonso, J.M.; Stepanova, A.N. Auxin Interactions with Other Hormones in Plant Development. *Cold Spring Harb. Perspect. Biol.* **2021**, *13*, a039990. [CrossRef]
98. Petrášek, J.; Malínská, K.; Zažímalová, E. Auxin Transporters Controlling Plant Development. In *Transporters and Pumps in Plant Signaling*; Geisler, M., Venema, K., Eds.; Springer: Berlin/Heidelberg, Germany, 2011; pp. 255–290, ISBN 978-3-642-14369-4.
99. Swarup, R.; Kargul, J.; Marchant, A.; Zadik, D.; Rahman, A.; Mills, R.; Yemm, A.; May, S.; Williams, L.; Millner, P.; et al. Structure-Function Analysis of the Presumptive *Arabidopsis* Auxin Permease *AUX1*. *Plant Cell* **2004**, *16*, 3069–3083. [CrossRef]
100. Jonsson, K.; Boutté, Y.; Singh, R.K.; Gendre, D.; Bhalerao, R.P. Ethylene Regulates Differential Growth via BIG ARF-GEF-Dependent Post-Golgi Secretory Trafficking in *Arabidopsis*. *Plant Cell* **2017**, *29*, 1039–1052. [CrossRef]
101. Péret, B.; Swarup, K.; Ferguson, A.; Seth, M.; Yang, Y.; Dhondt, S.; James, N.; Casimiro, I.; Perry, P.; Syed, A.; et al. *AUX/LAX* Genes Encode a Family of Auxin Influx Transporters That Perform Distinct Functions during *Arabidopsis* Development. *Plant Cell* **2012**, *24*, 2874–2885. [CrossRef]
102. Fukui, K.; Hayashi, K. Manipulation and Sensing of Auxin Metabolism, Transport and Signaling. *Plant Cell Physiol.* **2018**, *59*, 1500–1510. [CrossRef]
103. Kaneda, M.; Schuetz, M.; Lin, B.S.; Chanis, C.; Hamberger, B.; Western, T.L.; Ehlting, J.; Samuels, A.L. ABC transporters coordinately expressed during lignification of Arabidopsis stems include a set of ABCBs associated with auxin transport. *J. Exp. Bot.* **2011**, *62*, 2063–2077. [CrossRef]
104. Cho, M.; Cho, H.-T. The Function of ABCB Transporters in Auxin Transport. *Plant Signal. Behav.* **2013**, *8*, e22990. [CrossRef]
105. Geisler, M.; Aryal, B.; di Donato, M.; Hao, P. A Critical View on ABC Transporters and Their Interacting Partners in Auxin Transport. *Plant Cell Physiol.* **2017**, *58*, 1601–1614. [CrossRef] [PubMed]
106. Friml, J.; Benková, E.; Blilou, I.; Wisniewska, J.; Hamann, T.; Ljung, K.; Woody, S.; Sandberg, G.; Scheres, B.; Jürgens, G.; et al. *AtPIN4* Mediates Sink-Driven Auxin Gradients and Root Patterning in *Arabidopsis*. *Cell* **2002**, *108*, 661–673. [CrossRef] [PubMed]
107. Benková, E.; Michniewicz, M.; Sauer, M.; Teichmann, T.; Seifertová, D.; Jürgens, G.; Friml, J. Local, Efflux-Dependent Auxin Gradients as a Common Module for Plant Organ Formation. *Cell* **2003**, *115*, 591–602. [CrossRef] [PubMed]
108. Zhou, J.-J.; Luo, J. The *PIN-FORMED* Auxin Efflux Carriers in Plants. *Int. J. Mol. Sci.* **2018**, *19*, 2759. [CrossRef]
109. Petrášek, J.; Mravec, J.; Bouchard, R.; Blakeslee, J.J.; Abas, M.; Seifertová, D.; Wiśniewska, J.; Tadele, Z.; Kubeš, M.; Čovanová, M.; et al. PIN Proteins Perform a Rate-Limiting Function in Cellular Auxin Efflux. *Science* **2006**, *312*, 914–918. [CrossRef]
110. Barbez, E.; Laňková, M.; Pařezová, M.; Maizel, A.; Zažímalová, E.; Petrášek, J.; Friml, J.; Kleine-Vehn, J. Single-Cell-Based System to Monitor Carrier Driven Cellular Auxin Homeostasis. *BMC Plant Biol.* **2013**, *13*, 20. [CrossRef]
111. Zourelidou, M.; Absmanner, B.; Weller, B.; Barbosa, I.C.; Willige, B.C.; Fastner, A.; Streit, V.; Port, S.A.; Colcombet, J.; de la Fuente van Bentem, S.; et al. Auxin Efflux by PIN-FORMED Proteins Is Activated by Two Different Protein Kinases, D6 PROTEIN KINASE and PINOID. *eLife* **2014**, *3*, e02860. [CrossRef]
112. Blilou, I.; Xu, J.; Wildwater, M.; Willemsen, V.; Paponov, I.; Friml, J.; Heidstra, R.; Aida, M.; Palme, K.; Scheres, B. The PIN Auxin Efflux Facilitator Network Controls Growth and Patterning in Arabidopsis Roots. *Nature* **2005**, *433*, 39–44. [CrossRef]
113. Zhang, Y.; Xiao, G.; Wang, X.; Zhang, X.; Friml, J. Evolution of Fast Root Gravitropism in Seed Plants. *Nat. Commun.* **2019**, *10*, 3480. [CrossRef]

114. Liu, E.; Zhu, S.; Du, M.; Lyu, H.; Zeng, S.; Liu, Q.; Wu, G.; Jiang, J.; Dang, X.; Dong, Z.; et al. *LAX1*, Functioning with MADS-Box Genes, Determines Normal Palea Development in Rice. *Gene* **2023**, *883*, 147635. [CrossRef] [PubMed]
115. Robert, H.S.; Grunewald, W.; Sauer, M.; Cannoot, B.; Soriano, M.; Swarup, R.; Weijers, D.; Bennett, M.; Boutilier, K.; Friml, J. Plant Embryogenesis Requires AUX/LAX-Mediated Auxin Influx. *Development* **2015**, *142*, 702–711. [CrossRef]
116. da Costa, C.T.; Offringa, R.; Fett-Neto, A.G. The Role of Auxin Transporters and Receptors in Adventitious Rooting of *Arabidopsis thaliana* Pre-Etiolated Flooded Seedlings. *Plant Sci.* **2020**, *290*, 110294. [CrossRef] [PubMed]
117. Hao, Y.; Hu, G.; Breitel, D.; Liu, M.; Zouine, M. Auxin Response Factor SlARF2 Is an Essential Component of the Regulatory Mechanism Controlling Fruit Ripening in Tomato. *PLoS Genet.* **2015**, *11*, e1005649. [CrossRef] [PubMed]
118. Lewis, D.R.; Negi, S.; Sukumar, P.; Muday, G.K. Ethylene Inhibits Lateral Root Development, Increases IAA Transport and Expression of *PIN3* and *PIN7* Auxin Efflux Carriers. *Development* **2011**, *138*, 3485–3495. [CrossRef] [PubMed]
119. Vandenbussche, F.; Petrásek, J.; Zádníková, P.; Hoyerová, K.; Pesek, B.; Raz, V.; Swarup, R.; Bennett, M.; Zazímalová, E.; Benková, E.; et al. The Auxin Influx Carriers AUX1 and LAX3 Are Involved in Auxin-Ethylene Interactions during Apical Hook Development in *Arabidopsis thaliana* Seedlings. *Development* **2010**, *137*, 597–606. [CrossRef]
120. Cabello, J.V.; Chan, R.L. *Arabidopsis* and Sunflower Plants with Increased Xylem Area Show Enhanced Seed Yield. *Plant J.* **2019**, *99*, 717–732. [CrossRef]
121. Moreno-Piovano, G.S.; Moreno, J.E.; Cabello, J.V.; Arce, A.L.; Otegui, M.E.; Chan, R.L. A Role for *LAX2* in Regulating Xylem Development and Lateral-Vein Symmetry in the Leaf. *Ann. Bot.* **2017**, *120*, 577–590. [CrossRef]
122. Tidy, A.; Abu Bakar, N.; Carrier, D.; Kerr, I.D.; Hodgman, C.; Bennett, M.J.; Swarup, R. Mechanistic Insight into the Role of *AUXIN RESISTANCE4* in Trafficking of *AUXIN1* and *LIKE AUX1-2*. *Plant Physiol.* **2024**, *194*, 422–433. [CrossRef]
123. Mora, C.C.; Perotti, M.F.; González-Grandío, E.; Ribone, P.A.; Cubas, P.; Chan, R.L. *AtHB40* Modulates Primary Root Length and Gravitropism Involving CYCLINB and Auxin Transporters. *Plant Sci.* **2022**, *324*, 111421. [CrossRef] [PubMed]
124. Okada, K.; Ueda, J.; Komaki, M.K.; Bell, C.J.; Shimura, Y. Requirement of the Auxin Polar Transport System in Early Stages of *Arabidopsis* Floral Bud Formation. *Plant Cell* **1991**, *3*, 677–684. [CrossRef] [PubMed]
125. Shi, Z.; Jiang, Y.; Han, X.; Liu, X.; Cao, R.; Qi, M.; Xu, T.; Li, T. *SlPIN1* Regulates Auxin Efflux to Affect Flower Abscission Process. *Sci. Rep.* **2017**, *7*, 14919. [CrossRef]
126. Friml, J.; Vieten, A.; Sauer, M.; Weijers, D.; Schwarz, H.; Hamann, T.; Offringa, R.; Jürgens, G. Efflux-Dependent Auxin Gradients Establish the Apical–Basal Axis of *Arabidopsis*. *Nature* **2003**, *426*, 147–153. [CrossRef]
127. Mravec, J.; Skůpa, P.; Bailly, A.; Hoyerová, K.; Krecek, P.; Bielach, A.; Petrásek, J.; Zhang, J.; Gaykova, V.; Stierhof, Y.-D.; et al. Subcellular Homeostasis of Phytohormone Auxin Is Mediated by the ER-Localized PIN5 Transporter. *Nature* **2009**, *459*, 1136–1140. [CrossRef] [PubMed]
128. Simon, S.; Skůpa, P.; Viaene, T.; Zwiewka, M.; Tejos, R.; Klíma, P.; Čarná, M.; Rolčík, J.; De Rycke, R.; Moreno, I.; et al. PIN6 Auxin Transporter at Endoplasmic Reticulum and Plasma Membrane Mediates Auxin Homeostasis and Organogenesis in *Arabidopsis*. *New Phytol.* **2016**, *211*, 65–74. [CrossRef] [PubMed]
129. Lv, S.; Wang, L.; Zhang, X.; Li, X.; Fan, L.; Xu, Y.; Zhao, Y.; Xie, H.; Sawchuk, M.G.; Scarpella, E.; et al. *Arabidopsis NHX5* and *NHX6* Regulate *PIN6*-Mediated Auxin Homeostasis and Growth. *J. Plant Physiol.* **2020**, *255*, 153305. [CrossRef]
130. Seifu, Y.W.; Pukyšová, V.; Rýdza, N.; Bilanovičová, V.; Zwiewka, M.; Sedláček, M.; Nodzyński, T. Mapping the Membrane Orientation of Auxin Homeostasis Regulators *PIN5* and *PIN8* in *Arabidopsis thaliana* Root Cells Reveals Their Divergent Topology. *Plant Methods* **2024**, *20*, 84. [CrossRef]
131. Yoshihisa, K.; Kazuyoshi, T.; Masafumi, H.; Kojiro, T.; Shoju, F.; Nobukazu, S.; Akifumi, S.; Hideyuki, S.; Daisuke, S.; Bangjun, W. *Arabidopsis ABCB21* Is a Facultative Auxin Importer/Exporter Regulated by Cytoplasmic Auxin Concentration. *Plant Cell Physiol.* **2012**, *53*, 2090–2100. [CrossRef]
132. Wu, G.; Cameron, J.N.; Ljung, K.; Spalding, E.P. A Role for ABCB19-Mediated Polar Auxin Transport in Seedling Photomorphogenesis Mediated by Cryptochrome 1 and Phytochrome B. *Plant J.* **2010**, *62*, 179–191. [CrossRef]
133. Yu, X.; Liu, H.; Klejnot, J.; Lin, C. The Cryptochrome Blue Light Receptors. *Arab. Book* **2010**, *8*, e0135. [CrossRef]
134. Rademacher, E.H.; Möller, B.; Lokerse, A.S.; Llavata-Peris, C.I.; van den Berg, W.; Weijers, D. A cellular expression map of the Arabidopsis *AUXIN RESPONSE FACTOR* gene family. *Plant J.* **2011**, *68*, 597–606. [CrossRef] [PubMed]
135. Dharmasiri, N.; Estelle, M. Auxin Signaling and Regulated Protein Degradation. *Trends Plant Sci.* **2004**, *9*, 302–308. [CrossRef]
136. Sato, A.; Yamamoto, K.T. What's the Physiological Role of Domain II-Less Aux/IAA Proteins? *Plant Signal. Behav.* **2008**, *3*, 496–497. [CrossRef] [PubMed]
137. Goh, T.; Kasahara, H.; Mimura, T.; Kamiya, Y.; Fukaki, H. Multiple *AUX/IAA–ARF* Modules Regulate Lateral Root Formation: The Role of *Arabidopsis* SHY2/IAA3-Mediated Auxin Signalling. *Philos. Trans. R. Soc. Lond. B Biol. Sci.* **2012**, *367*, 1461–1468. [CrossRef] [PubMed]
138. Tromas, A.; Paque, S.; Stierlé, V.; Quettier, A.-L.; Muller, P.; Lechner, E.; Genschik, P.; Perrot-Rechenmann, C. Auxin-Binding Protein 1 Is a Negative Regulator of the $SCF^{(TIR1/AFB)}$ Pathway. *Nat. Commun.* **2013**, *4*, 2496. [CrossRef]
139. Calderón Villalobos, L.I.A.; Lee, S.; De Oliveira, C.; Ivetac, A.; Brandt, W.; Armitage, L.; Sheard, L.B.; Tan, X.; Parry, G.; Mao, H. A Combinatorial TIR1/AFB-Aux/IAA Co-Receptor System for Differential Sensing of Auxin. *Nat. Chem. Biol.* **2016**, *8*, 477–485. [CrossRef]
140. Wang, D.; Pei, K.; Fu, Y.; Sun, Z.; Li, S.; Liu, H.; Tang, K.; Han, B.; Tao, Y. Genome-Wide Analysis of the *Auxin Response Factors (ARF)* Gene Family in Rice (*Oryza Sativa*). *Gene* **2007**, *394*, 13–24. [CrossRef]

141. Serre, N.B.C.; Kralík, D.; Yun, P.; Slouka, Z.; Shabala, S.; Fendrych, M. AFB1 Controls Rapid Auxin Signalling through Membrane Depolarization in *Arabidopsis thaliana* Root. *Nat. Plants* **2021**, *7*, 1229–1238. [CrossRef]
142. Friml, J. Fourteen Stations of Auxin. *CSH Perspect. Biol.* **2022**, *14*, a039859. [CrossRef]
143. Xu, T.; Dai, N.; Chen, J.; Nagawa, S.; Cao, M.; Li, H.; Zhou, Z.; Chen, X.; De Rycke, R.; Rakusová, H.; et al. Cell Surface ABP1-TMK Auxin-Sensing Complex Activates ROP GTPase Signaling. *Science* **2014**, *343*, 1025–1028. [CrossRef]
144. Cao, M.; Chen, R.; Li, P.; Yu, Y.; Zheng, R.; Ge, D.; Zheng, W.; Wang, X.; Gu, Y.; Gelova, Z. TMK1-Mediated Auxin Signalling Regulates Differential Growth of the Apical Hook. *Nature* **2019**, *568*, 240–243. [CrossRef]
145. Kim, M.H.; Kim, Y.; Kim, J.W.; Lee, H.-S.; Lee, W.S.; Kim, S.-K.; Wang, Z.-Y.; Kim, S.-H. Identification of *Arabidopsis BAK1-Associating Receptor-like Kinase 1 (BARK1)* and Characterization of Its Gene Expression and Brassinosteroid-Regulated Root Phenotypes. *Plant Cell Physiol.* **2013**, *54*, 1620–1634. [CrossRef] [PubMed]
146. Yang, J.; He, H.; He, Y.; Zheng, Q.; Li, Q.; Feng, X.; Wang, P.; Qin, G.; Gu, Y.; Wu, P.; et al. TMK1-Based Auxin Signaling Regulates Abscisic Acid Responses via Phosphorylating ABI1/2 in Arabidopsis. *Proc. Natl. Acad. Sci. USA* **2021**, *118*, e2102544118. [CrossRef] [PubMed]
147. Jurado, S.; Díaz-Triviño, S.; Abraham, Z.; Manzano, C.; Gutierrez, C.; del Pozo, C. SKP2A, an F-Box Protein That Regulates Cell Division, Is Degraded via the Ubiquitin Pathway. *Plant J.* **2008**, *53*, 828–841. [CrossRef]
148. Jurado, S.; Abraham, Z.; Manzano, C.; López-Torrejón, G.; Pacios, L.F.; Del Pozo, J.C. The Arabidopsis Cell Cycle F-Box Protein SKP2A Binds to Auxin. *Plant Cell* **2010**, *22*, 3891–3904. [CrossRef]
149. del Pozo, J.C.; Boniotti, M.B.; Gutierrez, C. Arabidopsis E2Fc Functions in Cell Division and Is Degraded by the Ubiquitin-SCF(*AtSKP2*) Pathway in Response to Light. *Plant Cell* **2002**, *14*, 3057–3071. [CrossRef]
150. Jones, B.; Gunnerås, S.A.; Petersson, S.V.; Tarkowski, P.; Graham, N.; May, S.; Dolezal, K.; Sandberg, G.; Ljung, K. Cytokinin Regulation of Auxin Synthesis in *Arabidopsis* Involves a Homeostatic Feedback Loop Regulated via Auxin and Cytokinin Signal Transduction. *Plant Cell* **2010**, *22*, 2956–2969. [CrossRef] [PubMed]
151. Marhavý, P.; Bielach, A.; Abas, L.; Abuzeineh, A.; Duclercq, J.; Tanaka, H.; Pařezová, M.; Petrášek, J.; Friml, J.; Kleine-Vehn, J.; et al. Cytokinin Modulates Endocytic Trafficking of *PIN1* Auxin Efflux Carrier to Control Plant Organogenesis. *Dev. Cell* **2011**, *21*, 796–804. [CrossRef]
152. Hayward, A.; Stirnberg, P.; Beveridge, C.; Leyser, O. Interactions between Auxin and Strigolactone in Shoot Branching Control. *Plant Physiol.* **2009**, *151*, 400–412. [CrossRef]
153. Xie, L.; Chen, F.; Du, H.; Zhang, X.; Wang, X.; Yao, G.; Xu, B. Graphene Oxide and Indole-3-Acetic Acid Cotreatment Regulates the Root Growth of *Brassica napus* L. via Multiple Phytohormone Pathways. *BMC Plant Biol.* **2020**, *20*, 101. [CrossRef] [PubMed]
154. Cerny, M.; Jedelsky, P.L.; Novak, J.; Schlosser, A. Cytokinin Modulates Proteomic, Transcriptomic and Growth Responses to Temperature Shocks in *Arabidopsis*. *Plant Cell Environ.* **2014**, *37*, 1641–1655. [CrossRef]
155. Nordstrom, A. Cytokinins in *Arabidopsis*, Tools, Pathways and Interaction with Auxin. Ph.D. Thesis, Acta Universitatis Agriculturae Sueciae Silvestria, Uppsala, Sverige, 2004.
156. Lloret, P.G.; Casero, P.J. Lateral root initiation. In *Plant Roots-The Hidden Half*; Waisel, Y., Eshel, A., Kafkafi, U., Eds.; Marcel Dekker: New York, NY, USA, 2002; pp. 127–155.
157. Gao, S.; Fang, J.; Xu, F.; Wang, W.; Sun, X.H. *CYTOKININ OXIDASE/DEHYDROGENASE4* Integrates Cytokinin and Auxin Signaling to Control Rice Crown Root Formation1. *Plant Physiol.* **2014**, 1035–1046. [CrossRef]
158. Sun, J.; Xu, Y.; Ye, S. Arabidopsis ASA1 Is Important for Jasmonate-Mediated Regulation of Auxin Biosynthesis and Transport during Lateral Root Formation. *Plant Cell* **2009**, *21*, 1495–1511. [CrossRef]
159. Cai, X.T.; Xu, P.; Zhao, P.X.; Liu, R.; Yu, L.H.; Xiang, C.B. *Arabidopsis ERF109* Mediates Cross-Talk between Jasmonic Acid and Auxin Biosynthesis during Lateral Root Formation. *Nat. Commun.* **2014**, *5*, 5833. [CrossRef] [PubMed]
160. Jiang, Y.; Liang, G.; Yang, S.; Yu, D. *Arabidopsis* WRKY57 functions as a node of convergence for jasmonic acid- and auxin-mediated signaling in jasmonic acid-induced leaf senescence. *Plant Cell* **2014**, *26*, 230–245. [CrossRef] [PubMed]
161. Zhao, Y.; Xing, L.; Wang, X.; Hou, Y.J.; Gao, J.; Wang, P.; Duan, C.G.; Zhu, X.; Zhu, J.K. The ABA Receptor PYL8 Promotes Lateral Root Growth by Enhancing MYB77-Dependent Transcription of Auxin-Responsive Genes. *Sci. Signal.* **2014**, *7*, ra53. [CrossRef] [PubMed]
162. Ding, Z.J.; Yan, J.Y.; Li, C.X.; Li, G.X.; Wu, Y.R.; Zheng, S.J. Transcription Factor *WRKY46* Modulates the Development of Arabidopsis Lateral Roots in Osmotic/Salt Stress Conditions via Regulation of ABA Signaling and Auxin Homeostasis. *Plant J.* **2015**, *84*, 56–69. [CrossRef]
163. Liu, X.; Zhang, H.; Zhao, Y.; Feng, Z.; Li, Q.; Yang, H.Q.; Luan, S.; Li, J.; He, Z.H. Auxin Controls Seed Dormancy through Stimulation of Abscisic Acid Signaling by Inducing ARF-Mediated *ABI3* Activation in *Arabidopsis*. *Proc. Natl. Acad. Sci. USA* **2013**, *110*, 15485–15490. [CrossRef]
164. Negi, S.; Ivanchenko, M.G.; Muday, G.K. Ethylene Regulates Lateral Root Formation and Auxin Transport in *Arabidopsis thaliana*. *Plant* **2008**, *55*, 175–187. [CrossRef]
165. Casadoro, T.G. The Involvement of Auxin in the Ripening of Climacteric Fruits Comes of Age: The Hormone Plays a Role of Its Own and Has an Intense Interplay with Ethylene in Ripening Peaches. *J. Exp. Bot.* **2007**, *58*, 3299–3308. [CrossRef]
166. Miho, T.; Naoko, N.; Hiroshi, F.; Takehiko, S.; Michiharu, N.; Ken-Ichiro, H.; Hiroko, H.; Hirohito, Y.; Yuri, N. Increased Levels of IAA Are Required for System 2 Ethylene Synthesis Causing Fruit Softening in Peach (*Prunus persica* L. Batsch). *J. Exp. Bot.* **2013**, *64*, 1049–1059. [CrossRef]

167. Yue, P.T.; Lu, Q.; Lv, T.X.; Li, X.Y. Auxin-activated *MdARF5* Induces the Expression of Ethylene Biosynthetic Genes to Initiate Apple Fruit Ripening. *New Phytol.* **2020**, *226*, 1782–1795. [CrossRef]
168. Zhang, T.; Li, W.; Xie, R.; Xu, L.; Zhou, Y.; Li, H.; Yuan, C.; Zheng, X.; Xiao, L.; Liu, K. *CpARF2* and *CpEIL1* Interact to Mediate Auxin–Ethylene Interaction and Regulate Fruit Ripening in Papaya. *Plant J.* **2020**, *103*, 1318–1337. [CrossRef]
169. Cui, D.; Neill, S.J.; Tang, Z.; Cai, W. Gibberellin-Regulated XET Is Differentially Induced by Auxin in Rice Leaf Sheath Bases during Gravitropic Bending. *J. Exp. Bot.* **2005**, *56*, 1327–1334. [CrossRef]
170. Zhou, X.-Y.; Song, L.; Xue, H.-W. Brassinosteroids Regulate the Differential Growth of Arabidopsis Hypocotyls through Auxin Signaling Components *IAA19* and *ARF7*. *Mol. Plant* **2013**, *6*, 887–904. [CrossRef] [PubMed]
171. Yuan, H.-M.; Liu, W.-C.; Lu, Y.-T. CATALASE2 Coordinates SA-Mediated Repression of Both Auxin Accumulation and JA Biosynthesis in Plant Defenses. *Cell Host Microbe* **2017**, *21*, 143–155. [CrossRef]
172. Küpers, J.J.; Oskam, L.; Pierik, R. Photoreceptors Regulate Plant Developmental Plasticity through Auxin. *Plants* **2020**, *9*, 940. [CrossRef]
173. de Wit, M.; Lorrain, S.; Fankhauser, C. Auxin-Mediated Plant Architectural Changes in Response to Shade and High Temperature. *Physiol. Plant* **2014**, *151*, 13–24. [CrossRef]
174. Franklin, K.A.; Lee, S.H.; Patel, D.; Kumar, S.V.; Spartz, A.K.; Gu, C.; Ye, S.; Yu, P.; Breen, G.; Cohen, J.D.; et al. *PHYTOCHROME-INTERACTING FACTOR 4 (PIF4)* Regulates Auxin Biosynthesis at High Temperature. *Proc. Natl. Acad. Sci. USA* **2011**, *108*, 20231–20235. [CrossRef]
175. Zheng, Z.; Guo, Y.; Novák, O.; Chen, W.; Ljung, K.; Noel, J.P.; Chory, J. Local Auxin Metabolism Regulates Environment-Induced Hypocotyl Elongation. *Nat. Plants* **2016**, *2*, 16025. [CrossRef] [PubMed]
176. Samakovli, D.; Roka, L.; Dimopoulou, A.; Plitsi, P.K.; Žukauskait, A.; Georgopoulou, P.; Novák, O.; Milioni, D.; Hatzopoulos, P. *HSP90* affects root growth in *Arabidopsis* by regulating the polar distribution of *PIN1*. *New Phytol.* **2021**, *231*, 1814–1831. [CrossRef]
177. Chen, D.; Wang, W.; Wu, Y.; Xie, H.; Zhan, Y. Expression and Distribution of the Auxin Response Factors in Sorghum Bicolor During Development and Temperature Stress. *Int. J. Mol. Sci.* **2019**, *20*, 4816. [CrossRef] [PubMed]
178. Ayi, Q.; Zeng, B.; Liu, J.; Li, S.; van Bodegom, P.M.; Cornelissen, J.H.C. Oxygen Absorption by Adventitious Roots Promotes the Survival of Completely Submerged Terrestrial Plants. *Ann. Bot.* **2016**, *118*, 675–683. [CrossRef] [PubMed]
179. Sauter, M. Root Responses to Flooding. *Curr. Opin. Plant Biol.* **2013**, *16*, 282–286. [CrossRef]
180. Qi, X.; Hu, Z.; Chen, X.; Zhang, M.; Nakazono, M. Involvement of Phytohormones in Flooding Stress Tolerance in Plants. In *Plant Hormones and Climate Change*; Ahammed, G.J., Yu, J., Eds.; Springer: Singapore, 2023. [CrossRef]
181. Sharif, R.; Su, L.; Chen, X.; Qi, X. Involvement of auxin in growth and stress response of cucumber. *Veg. Res.* **2022**, *2*, 13. [CrossRef]
182. Gao, J.; Qian, Z.; Zhang, Y.; Zhuang, S. Exogenous Spermidine Regulates the Anaerobic Enzyme System through Hormone Concentrations and Related-Gene Expression in *Phyllostachys praecox* Roots under Flooding Stress. *Plant Physiol. Biochem.* **2022**, *186*, 182–196. [CrossRef]
183. Gao, J.; Zhuang, S.; Zhang, Y.; Qian, Z. Exogenously Applied Spermidine Alleviates Hypoxia Stress in *Phyllostachys praecox* Seedlings via Changes in Endogenous Hormones and Gene Expression. *BMC Plant Biol.* **2022**, *22*, 200. [CrossRef]
184. Gao, J.; Zhuang, S.; Gui, R. Subsurface Aeration Mitigates Organic Material Mulching-Induced Anaerobic Stress via Regulating Hormone Signaling in *Phyllostachys praecox* Roots. *Front. Plant Sci.* **2023**, *14*, 1121604. [CrossRef]
185. Hu, W.; Yan, H.; Luo, S.; Pan, F.; Wang, Y.; Xiang, Y. Genome-Wide Analysis of Poplar SAUR Gene Family and Expression Profiles under Cold, Polyethylene Glycol and Indole-3-Acetic Acid Treatments. *Plant Physiol. Biochem.* **2018**, *128*, 50–65. [CrossRef]
186. Zhang, X.; Fu, X.; Liu, F.; Wang, Y.; Bi, H.; Ai, X. Hydrogen Sulfide Improves the Cold Stress Resistance through the *CsARF5-CsDREB3* Module in Cucumber. *Int. J. Mol. Sci.* **2021**, *22*, 13229. [CrossRef] [PubMed]
187. Wu, J.; Liu, S.; He, Y.; Guan, X.; Zhu, X.; Cheng, L.; Wang, J.; Lu, G. Genome-Wide Analysis of *SAUR* Gene Family in *Solanaceae* species. *Gene* **2012**, *509*, 38–50. [CrossRef] [PubMed]
188. Xu, L.; Wang, D.; Liu, S.; Fang, Z.; Su, S.; Guo, C.; Zhao, C.; Tang, Y. Comprehensive Atlas of Wheat (*Triticum aestivum* L.) *AUXIN RESPONSE FACTOR* Expression During Male Reproductive Development and Abiotic Stress. *Front. Plant Sci.* **2020**, *11*, 586144. [CrossRef] [PubMed]
189. Mwando, E.K.; Zhou, G.; Angessa, T.T.; Li, C.; Han, Y. Quantitative Trait Loci Mapping for Vigour and Survival Traits of Barley Seedlings after Germinating under Salinity Stress. *Agronomy* **2021**, *11*, 103. [CrossRef]
190. Du, C.; Li, H.; Liu, C.; Fan, H. Understanding of the Postgerminative Development Response to Salinity and Drought Stresses in Cucumber Seeds by Integrated Proteomics and Transcriptomics Analysis. *J. Proteom.* **2021**, *232*, 104062. [CrossRef]
191. Zhu, Y.; Yin, J.; Liang, Y.; Liu, J.; Jia, J.; Huo, H.; Wu, Z.; Yang, R.; Gong, H. Transcriptomic Dynamics Provide an Insight into the Mechanism for Silicon-Mediated Alleviation of Salt Stress in Cucumber Plants. *Ecotoxicol. Environ. Safe* **2019**, *174*, 245–254. [CrossRef]
192. Liu, Q.; Feng, Z.; Xu, W.; Vetukuri, R.R.; Xu, X. Exogenous Melatonin-Stimulated Transcriptomic Alterations of Davidia Involucrata Seedlings under Drought Stress. *Trees* **2021**, *35*, 1025–1038. [CrossRef]
193. Kang, C.; He, S.; Zhai, H.; Li, R.; Zhao, N.; Liu, Q. A Sweetpotato *Auxin Response Factor Gene* (*IbARF5*) Is Involved in Carotenoid Biosynthesis and Salt and Drought Tolerance in Transgenic *Arabidopsis*. *Front. Plant Sci.* **2018**, *9*, 1307. [CrossRef]
194. Jung, H.; Lee, D.-K.; Choi, Y.D.; Kim, J.-K. *OsIAA6*, a Member of the Rice Aux/IAA Gene Family, Is Involved in Drought Tolerance and Tiller Outgrowth. *Plant Sci.* **2015**, *236*, 304–312. [CrossRef]

195. Zheng, L.; Zhang, M.; Zhuo, Z.; Wang, Y.; Gao, X.; Li, Y.; Liu, W.; Zhang, W. Transcriptome Profiling Analysis Reveals Distinct Resistance Response of Cucumber Leaves Infected with Powdery Mildew. *Plant Biol.* **2021**, *23*, 327–340. [CrossRef]
196. Fattorini, L.; Ronzan, M.; Piacentini, D.; Della Rovere, F.; De Virgilio, C.; Sofo, A.; Altamura, M.M.; Falasca, G. Cadmium and Arsenic Affect Quiescent Centre Formation and Maintenance in *Arabidopsis thaliana* Post-Embryonic Roots Disrupting Auxin Biosynthesis and Transport. *Environ. Exp. Bot.* **2017**, *144*, 37–48. [CrossRef]
197. Lu, G.; Coneva, V.; Casaretto, J.A.; Ying, S.; Mahmood, K.; Liu, F.; Nambara, E.; Bi, Y.-M.; Rothstein, S.J. OsPIN5b Modulates Rice (*Oryza sativa*) Plant Architecture and Yield by Changing Auxin Homeostasis, Transport and Distribution. *Plant J.* **2015**, *83*, 913–925. [CrossRef] [PubMed]
198. Ronzan, M.; Piacentini, D.; Fattorini, L.; Rovere, F.D.; Eiche, E.; Riemann, M.; Altamura, M.M.; Falasca, G. Cadmium and Arsenic Affect Root Development in Oryza Sativa L. Negatively Interacting with Auxin. *Environ. Exp. Bot.* **2018**, *151*, 64–75. [CrossRef]
199. Bruno, L.; Pacenza, M.; Forgione, I.; Lamerton, L.R.; Greco, M.; Chiappetta, A.; Bitonti, M.B. In *Arabidopsis thaliana* Cadmium Impact on the Growth of Primary Root by Altering *SCR* Expression and Auxin-Cytokinin Cross-Talk. *Front. Plant Sci.* **2017**, *8*, 1323. [CrossRef] [PubMed]
200. Sun, H.; Dai, H.; Wang, X.; Wang, G. Physiological and Proteomic Analysis of Selenium-Mediated Tolerance to Cd Stress in Cucumber (*Cucumis sativus* L.). *Ecotoxicol. Environ. Safe* **2016**, *133*, 114–126. [CrossRef]
201. Fiedler, L.; Friml, J. Rapid Auxin Signaling: Unknowns Old and New. *Curr. Opin. Plant Biol.* **2023**, *75*, 102443. [CrossRef]
202. Qi, J.; Mao, Y.; Cui, J.; Lu, X.; Xu, J.; Liu, Y.; Zhong, H.; Yu, W.; Li, C. The Role of Strigolactones in Resistance to Environmental Stress in Plants. *Physiol. Plant.* **2024**, *176*, e14419. [CrossRef]

plants

MDPI

Article

The Effects of Auxin Transport Inhibition on the Formation of Various Leaf and Vein Patterns

Carol L. Wenzel [1,*], David M. Holloway [2] and Jim Mattsson [3]

1 Biotechnology Department, British Columbia Institute of Technology, 3700 Willingdon Avenue, Burnaby, BC V5G 3H2, Canada
2 Mathematics Department, British Columbia Institute of Technology, 3700 Willingdon Avenue, Burnaby, BC V5G 3H2, Canada; david_holloway@bcit.ca
3 Biology Department, Simon Fraser University, 8888 University Drive, Burnaby, BC V5A 1G3, Canada; jim_mattsson@sfu.ca
* Correspondence: cwenzel1@bcit.ca

Abstract: Polar auxin transport (PAT) is a known component controlling leaf complexity and venation patterns in some model plant species. Evidence indicates that PAT generates auxin converge points (CPs) that in turn lead to local leaf formation and internally into major vein formation. However, the role of PAT in more diverse leaf arrangements and vein patterns is largely unknown. We used the pharmacological inhibition of PAT in developing pinnate tomato, trifoliate clover, palmate lupin, and bipinnate carrot leaves and observed dosage-dependent reduction to simple leaves in these eudicots. Leaf venation patterns changed from craspedodromous (clover, carrot), semi-craspedodromous (tomato), and brochidodromous (lupin) to more parallel patterning with PAT inhibition. The visualization of auxin responses in transgenic tomato plants showed that discrete and separate CPs in control plants were replaced by diffuse convergence areas near the margin. These effects indicate that PAT plays a universal role in the formation of different leaf and vein patterns in eudicot species via a mechanism that depends on the generation as well as the separation of auxin CPs. Computer simulations indicate that variations in PAT can alter the number of CPs, corresponding leaf lobe formation, and the position of major leaf veins along the leaf margin in support of experimental results.

Keywords: convergent points; leaf complexity; polar auxin transport (PAT); venation patterning

1. Introduction

Plants have a spectacular diversity in leaf shape and complexity, both of which can vary between plant species and even within individual plants (e.g., [1,2]). Much of the diversity in leaf shape is dependent upon the extent of the outgrowth of the margin, in particular the size of the marginal blastozone and the duration of its morphogenetic potential [3–8] as well as the extent and overall distribution of growth in the leaf lamina (e.g., [9,10]).

Growth axes in leaves frequently co-align with the primary and secondary veins, suggesting a link between growth and venation and that developing procambial cells may provide a driving force behind localized leaf expansion [11,12]. Computer simulations in which growth is catalyzed at vein initiation sites in the margin indeed produce a wide range of observed leaf shapes [13]. However, there are many unanswered questions regarding how vein patterning and growth may be coordinated, particularly at the molecular level.

Molecular mechanisms are likely to include the phytohormone auxin (indole acetic acid, IAA), which is involved in both the vein patterning and growth aspects of leaf development. Primary/secondary veins and localized outgrowths of leaf margins are separated by many-cell distances of ~20–50 μm to mms (depending on stage). Understanding auxin's role in catalyzing structures at such length scales requires understanding how auxin is spatially distributed and localized to future vascular or high-growth cells. Auxin has unique transport properties that influence this. In particular, feedback between auxin and the PINFORMED (PIN) efflux proteins ([14,15]; primarily PIN1 in *Arabidopsis*) creates an intercellular polar auxin transport (PAT) that contributes to long-range patterning.

In vein development, PAT is involved in the formation of discrete auxin concentration maxima (convergence points, CPs) in the leaf margin, from which provascular tracks extend into the leaf. Co-localized PIN1-auxin distributions have been characterized for the CP of the primary vein, as the leaf primordium emerges from the shoot apical meristem [16–20], as well as for the subsequent CPs associated with (i) secondary veins, teeth, and lobes in the margins of simple leaves [14,15,21] and (ii) the midveins of individual leaflets in compound leaves, for example, in *Cardamine hirsuta* [22] and tomato [23]. Following CP formation, PAT is also involved in inward vein extension and canalization (in which the early broad distributions of auxin and PIN1 narrow to produce the provascular track, e.g., [14,15,24]).

Pharmacological PAT inhibition (PATi)—for example, by treatment with NPA, N-(1-Naphthyl) phthalamic acid [25,26], which competes with auxin binding to PIN1 [27]—or *PIN* mutations [28] have been shown to affect both of these aspects of vein development in *Arabidopsis* simple leaves: affecting CP formation [25,26] and delaying vein extension [25,26] and canalization [14,15]. In compound leaves, PATi has been shown to affect CPs in *Cardamine* (auxin; [22]) and in tomato (PIN1; [23]).

PATi also influences leaf shape, rounding the simple leaves of *Arabidopsis* [25,26,28] and inducing more marked effects in compound leaves, for example, reducing leaf complexity in tomato [23,29,30], *Cardamine* [22], and pea [31], producing simpler leaves with fewer or no leaflets.

These results point to a shared mechanism for the auxin localization involved in vein patterning and morphogenesis in plants with simple and compound leaves, but questions remain. These include: does vein patterning in compound leaves have a similar response to PATi as in simple leaves; what is the correspondence between morphological changes and vein pattern changes; and can PATi responses in morphologically complex compound leaves provide more information on this relation than in simple leaves?

To begin to address these questions, we conducted PATi experiments, recording vein pattern and morphological changes, across four dicot species with morphologically distinct types of compound leaves: pinnate tomato; bipinnate carrot; palmate lupin; and trifoliate clover. Results over the four morphologies corroborated the general effect of PATi on leaf complexity reported in tomato [23,29,30], *Cardamine* [22], and pea [31]. We also found a dosage dependence, with an overall stronger reduction in complexity at higher concentrations of PAT inhibitors. Previous work in tomato [23] and *Cardamine* [22] showed an effect of PATi on initial CP formation. We extend this to show how PATi affects vein development and vein network patterns in leaves and leaflets, across the diverse morphologies studied. The results are interpreted with the aid of a quantitative computer model of PAT and growth in the margin and lamina. The generation of some of the key experimental observations supports this as a conceptual framework for how auxin transport jointly mediates leaf complexity and venation in diverse types of compound leaves.

2. Results

2.1. Polar Auxin Transport Is Involved in Compound Leaflet Morphogenesis in Different Leaf Complexity Types

Plants from the four different leaf complexity types were exposed to PAT inhibitors. While there was variability within species and treatment levels, all showed an overall reduction in leaf complexity, measured as leaflet count (Table 1, Figure 1). In general, leaf complexity decreased with increasing PAT inhibitor dosage. Different species and leaf numbers (L1, L2, L3) showed varying susceptibilities to different PAT inhibitors.

Table 1. Polar auxin transport inhibitors reduced leaflet count. Leaflet count for the first three mature leaves (L1, L2, L3) at circa 4 weeks in control conditions or exposed to different concentrations (μM) of NPA (N), HFCA (H), or TIBA (T). Cells show: Row 1 = mean ± S.D. and (n); Row 2 = *p*-value for mean treated value being equal to control, * significantly different from control ($p < 5 \times 10^{-2}$) or ** highly significantly different from control ($p < 1 \times 10^{-2}$); Row 3 = % below control median, note increase in percentage with treatment. Colors show: Green = treatment sample mean below control; White = no difference; Orange = treatment sample mean above control. ND = not done. Undefined = *p*-value undefined due to zero S.D.

Plant	Leaf	Control	0.1 H	1 H	0.1 N	1 N	10 N	0.1 T	1 T	10 T
tomato	L1	2.8 ± 0.6 (21)	2.2 ± 0.8 (18)	1.1 ± 0.3 (9)	1.9 ± 0.7 (17)	1.3 ± 0.6 (22)	1.0 ± 0 (17)	2.7 ± 0.8 (15)	3.0 ± 0.6 (13)	2.0 ± 0.9 (6)
	p-value		8.0×10^{-3} **	3.0×10^{-10} **	1.4×10^{-4} **	1.1×10^{-9} **	1.1×10^{-11} **	7.6×10^{-1}	3.7×10^{-1}	7.9×10^{-2}
	% with < median	19	61	100	82	91	100	20	15	67
	L2	3.5 ± 0.8 (22)	2.5 ± 1.1 (17)	1.1 ± 0.3 (11)	2.7 ± 0.6 (17)	1.8 ± 0.9 (22)	1.0 ± 0 (17)	3.6 ± 0.9 (14)	3.3 ± 0.6 (14)	2.3 ± 1.0 (6)
	p-value		4.9×10^{-3} **	4.3×10^{-13} **	1.8×10^{-3} **	4.4×10^{-8} **	2.4×10^{-12} **	7.0×10^{-1}	4.8×10^{-1}	4.4×10^{-2} *
	% with < median	5	47	100	24	73	100	7	0	33
	L3	4.5 ± 0.9 (22)	2.9 ± 0.9 (18)	1.5 ± 0.5 (6)	2.9 ± 0.6 (16)	2.1 ± 1.0 (22)	1.0 ± 0 (14)	4.5 ± 1.0 (14)	4.0 ± 1.0 (13)	3.2 ± 0.4 (6)
	p-value		2.5×10^{-6} **	1.8×10^{-7} **	1.4×10^{-7} **	1.1×10^{-10} **	1.2×10^{-14} **	7.1×10^{-1}	1.4×10^{-1}	7.5×10^{-3} *
	% with < median	32	94	100	100	100	100	21	62	100
clover	L2	2.8 ± 0.6 (13)	2.9 ± 0.3 (9)	1.6 ± 0.5 (7)	ND	5 ± 0 (5)	2.0 ± 0.8 (4)	2.7 ± 0.6 (3)	1.8 ± 0.5 (4)	ND
	p-value		8.2×10^{-1}	2.4×10^{-4} **		8.5×10^{-9} **	1.3×10^{-1}	6.6×10^{-1}	1.1×10^{-2} *	
	% with < median	8	11	100		0	75	33	100	
	L3	3.0 ± 0 (13)	2.4 ± 0.7 (9)	1.9 ± 1.0 (8)	ND	3.0 ± 0 (5)	2.5 ± 0.8 (6)	3.0 ± 0 (3)	2.3 ± 1.0 (4)	ND
	p-value		5.1×10^{-2}	1.5×10^{-2} *		undefined	2.0×10^{-1}	undefined	2.1×10^{-1}	
	% with < median	0	44	63		0	33	0	50	
lupin	L2	6.6 ± 0.8 (47)	ND	ND	ND	6.3 ± 0.6 (3)	4.0 ± 2.3 (57)	ND	ND	ND
	p-value					4.9×10^{-1}	1.4×10^{-11} **			
	% with < median	45				67	86			
	L3	7.0 ± 0.7 (48)	ND	ND	ND	6.3 ± 0.6 (3)	5.3 ± 2.4 (55)	ND	ND	ND
	p-value					1.7×10^{-1}	5.2×10^{-6} **			
	% with < median	25				67	69			
carrot	L1	5.6 ± 1.2 (17)	5.2 ± 0.6 (13)	4.2 ± 2.1 (8)	5.4 ± 0.9 (5)	1.7 ± 1.6 (6)	ND	5.4 ± 1.3 (9)	5.0 ± 0 (9)	ND
	p-value		1.9×10^{-1}	1.3×10^{-1}	7.1×10^{-1}	1.0×10^{-3} **		7.9×10^{-1}	5.6×10^{-2}	
	% with < median	6	0	25	0	83		0	0	
	L2	6.4 ± 1.4 (14)	7.9 ± 1.0 (11)	6.4 ± 2.5 (7)	6.6 ± 1.7 (5)	5.8 ± 1.8 (5)	ND	7.5 ± 1.4 (8)	7.0 ± 1.2 (7)	ND
	p-value		4.9×10^{-3} **	9.5×10^{-1}	7.8×10^{-1}	5.5×10^{-1}		9.1×10^{-2}	2.9×10^{-1}	
	% with < median	50	0	29	40	40		13	14	
	L3	6.9 ± 1.1 (12)	9.0 ± 1.0 (9)	7.4 ± 2.6 (5)	ND	ND	ND	8.0 ± 1.1 (6)	8.6 ± 0.9 (5)	ND
	p-value		2.4×10^{-4} **	7.1×10^{-1}				7.5×10^{-2}	8.8×10^{-3} **	
	% with < median	17	0	20				0	0	

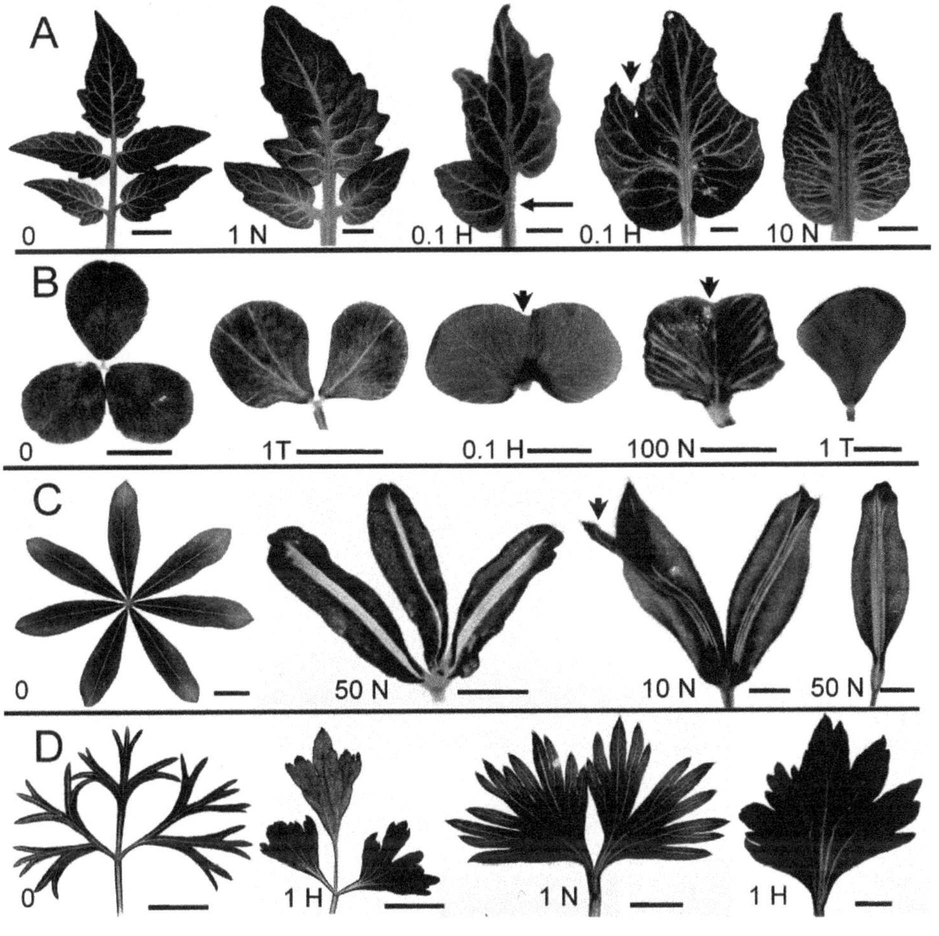

Figure 1. PAT inhibition reduced leaf complexity in ca 4 week old tomato plants ((**A**); leaf 3) and ca 2 weeks old clover ((**B**); leaf 2 and 3), lupin ((**C**); leaf 2 and 3), and carrot ((**D**); leaf 1 and 2). Images show control leaves (0) or leaves exposed to various concentrations (μM) of NPA (N), HFCA (H), or TIBA (T). The apparent absence of a lateral leaflet (arrow in (**A**)) and fusion between adjacent leaflets (arrowheads in (**A–C**)) are indicated. Scale bars are 5 mm ((**A**); control and NPA-treated), 2 mm ((**A**); HFCA-treated), or 0.5 mm ((**B–D**); except that (i) fourth image from left in (**B**) and (ii) third and fourth images from left in (**D**) are 0.2 mm).

Tomato showed a significant mean change from the control leaflet count at NPA and HFCA concentrations as low as 0.1 μM. For all leaf numbers (L1–L3), an increased dosage of NPA or HFCA steadily reduced leaf complexity, from pinnately compound control leaves to simple leaves with a single blade (Figure 1A, L3 shown; Table 1); the simple leaf results were in greater contrast at L3 (which have higher leaflet count in control) than in L1 (lower count in control). For TIBA, only the highest tested concentration (10 μM) changed mean leaf complexity, and only in L2 and L3 (Table 1).

Clover and lupin likewise showed reduced leaf complexity with PATi. In clover, trifoliate control leaves showed reduction to two leaflets, one fused leaf with two laminar blades or a simple leaf depending on inhibitor concentration (Figure 1B). Significant reductions in mean leaflet number from the control were seen at the highest concentrations tested of HFCA and TIBA; while $p > 0.1$ for NPA, leaflet reduction may be indicated by the median leaflet count for L2 (Table 1). Lupin showed a significant change in leaflet count at the highest levels of NPA (Table 1; HFCA and TIBA not tested). The leaflet number decreased from palmately compound control leaves through simpler compound leaves to simple leaves (Figure 1C). In tomato, clover, and lupin, a decreased leaflet count appeared to occur either from inhibited leaflet formation (e.g., arrow in Figure 1A) or from the fusion of adjacent leaflets to form a single leaflet (e.g., arrowheads in Figure 1A–C).

Carrot showed a reduction from bipinnately compound to pinnately compound or even simple leaves with PAT inhibition (Figure 1D). While non-smooth margins complicated the quantification of the leaflet count and may have contributed to apparent increases in the mean leaflet number in some treatments, a significant reduction in mean complexity was seen with high NPA in L1 (Table 1).

2.2. *Polar Auxin Transport Inhibitors Induced More Parallel Leaf/Leaflet Venation Patterning*

Here, we use the terminology in [1] to describe venation patterning based on the positioning of the secondary veins relative to the single centrally localized primary vein in the leaf/leaflet lamina. Secondary vein types include secondaries that join with the primary and (i) run directly from the margin (craspedodromous), (ii) branch just within the margin, with one branch extending from the margin and the other joining with an adjacent secondary (semicraspedodromous), and (iii) joining with adjacent secondaries to form prominent arches (brochidodromous).

All four species tested showed strong effects of PATi on vein patterning (Figure 2). These results indicate similarities in primary and secondary vein patterning in the leaflets of compound leaves to the PATi effects previously noted in simple leaves (e.g., [26]). PATi increased the number of secondary-like veins, with a distalization of the initiation points of these veins, such that the overall vein pattern tended to be more parallel to the proximodistal axis than in control. PATi interfered with the normal connection of veins, such as the secondary to primary connections in control, resulting in the supernumerary veins running in parallel bundles (see especially the central mid-leaflet bundles in PATi tomato and lupin, Figure 2).

Tomato leaflets have a semi-craspedodromous venation pattern, with secondary veins either joining adjacent secondaries or extending from the margin (e.g., Figures 1A and 2). Control tomato leaflets had a hierarchical venation patterning with a central midvein, secondary veins forming loops alongside the midvein or extending from the marginal lobes towards the midvein, and higher-order marginal and areolar veins (Figures 1A and 2). Weak PAT inhibition resulted in vascular overgrowth in the central lamina, appearing to involve the bundling of supernumerary secondary-like veins. In slightly stronger phenotypes, the distal region of the leaflet or leaf had a more diffuse region of multiple secondary-like veins that ran mostly parallel to the longitudinal axis, coalescing in more basal regions of the leaf/leaflet (Figure 2).

PAT inhibition had a similar effect on lupin leaf venation. Mature control lupin leaflets had hierarchical brochidodromous venation patterning with a primary midvein, secondary vein arches, and higher-order venation, whereas PAT inhibition resulted in more parallel venation in the center of the leaf/leaflet (Figure 2). Weaker PATi lupin phenotypes had a broad distal vein initiation region with multiple strands running parallel along the proximodistal axis, whereas stronger phenotypes had more secondary-like initiation sites at the apex that coalesced and bundled in the basal regions (Figure 2). Very strong PATi lupin phenotypes predominantly had parallel-type venation throughout the narrow lamina (Figure 2 inset).

Figure 2. PAT inhibition induced more parallel-like leaf vein patterning. Dark-field images of tomato (Tiny Tom), clover, lupin, or carrot venation patterning on individual leaflets (control) or entire PAT-inhibited leaves (weak and strong phenotypes). Inset for lupin (100 N) shows very strong phenotype with parallel venation throughout lamina. Images show control leaves (0) or leaves exposed to various concentrations (µM) of NPA (N), HFCA (H) or TIBA (T). Scale bars are 1 mm.

Mature clover and carrot control leaves had a craspedodromous venation pattern—a primary midvein with secondary veins extending towards the marginal teeth (clover) or lobes (carrot) (Figure 2). Compared to control leaf patterns, PAT-inhibited clover and carrot leaves produced more parallel secondary veins, but with less distal coalescence and central bundling than observed in tomato and lupin, although carrot veins did coalesce near the petioles (Figure 2). All four species, with their different control venation patterns, showed a tendency towards more parallel secondary venation patterning with PAT inhibition.

We examined the temporal development of leaflet shape and venation patterns in tomato. In control plants, the expansion of leaflets was evident by 6 DAG (days after germination), whereas NPA treatment inhibited leaflet outgrowth, especially at 10 µM NPA

(Figure 3A). In control leaves, differentiated veins first became apparent along the leaf main axis (petiole/rachis), and then the semi-craspedodromous secondary venation of the terminal leaflet formed (Figure 3A). Eventually, differentiated midveins also extended along lateral leaflets to join the original leaf midvein (Figure 3A); these leaflets later also formed a semi-craspedodromous secondary venation pattern. 1 μM NPA treatment increased vascular development in the distal regions of the terminal leaflet and lateral leaflets as they emerged (Figure 3A). The supernumerary veins in PATi leaflets were poorly organized compared to control, filling the leaf lamina and showing decreased extension towards the proximal regions of the leaf. At 10 μM NPA, treated leaves had reduced complexity (lacked lateral leaflets) and showed increased distal initiation and a poorer extension of veins towards the leaf base compared to lower doses of NPA (Figure 3A). At maturity, 10 μM NPA-treated leaves showed simpler morphologies and supernumerary distally initiated, centrally-bundled veins, which could eventually connect to the rachis and/or petiole (Figures 2 and 3A). Overall, these observations in veins corresponded to presumed auxin distributions as indicated by auxin response markers (Figures 3B and 4A–C). In control plants, emerging leaf primordia and later also the lateral leaflets had discrete auxin maxima at their distal apex in the CPs (arrows in Figure 4A–C), as well as in procambial strands (Figures 3B and 4B,C). In contrast, NPA treatment resulted in a large, diffuse auxin response zone at the emerging leaf apex and, if present, in lateral leaflet apices (Figure 4A).

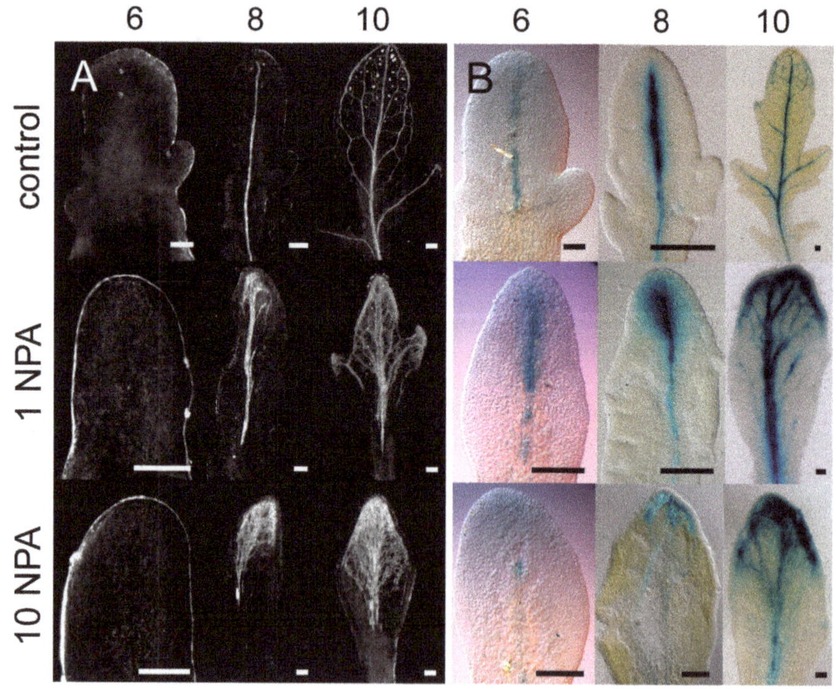

Figure 3. Leaf primordia venation and *pIAA2*::GUS expression in tomato (variety Alicia Craig). Tomato leaf primordia of plants grown for 6, 8, or 10 days in media with 0, 1, or 10 μM NPA. Primordia are viewed with darkfield optics to see differentiated venation patterns (**A**) or with brightfield optics to view *pIAA2*::GUS expression (**B**). NPA decreased leaflet outgrowth and induced vascular overgrowth and IAA2 accumulation near the leaf apex. (**A**) all 8 DAG primordia had a portion of the lamina cut off (left side of image) to assist visualization. Scale bars are 50 μm (6 DAG) or 100 μm (8 and 10 DAG).

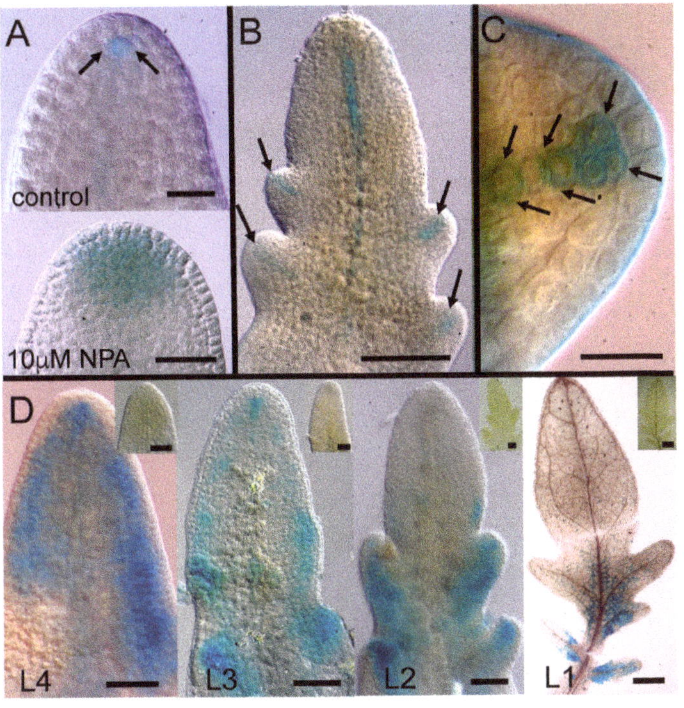

Figure 4. *pDR5*::GUS expression in tomato (variety Alicia Craig) leaf primordia. NPA treatment induced broad *pDR5*::GUS expression in the apex of 6 DAG L2 leaf primordia ((**A**), lower image) compared to the discrete auxin response maximum in control plants ((**A**), arrows in upper image). 7 DAG control L2 leaf primordia had discrete auxin response maxima in the emerging leaflets (arrows in (**B**)). Each developing leaflet had *pDR5*::GUS expression in a discrete auxin maxima below the epidermis as well as in the developing primary provascular cells of the midvein (arrows in (**C**)). Exposure of 8 DAG seedlings to 10 μM 2,4-D induced broad *pDR5*::GUS expression along most of the margin of L4 leaf primordia, with expression becoming restricted to more basal regions in older L3, L2, and L1 primordia ((**D**); insets show controls). Scale bars are 20 μm (**A**,**C**), 100 μm (**B**), or 50 μm ((**D**), except that L1 is 500 μm).

2.3. A PAT-Growth Model Generates the Observed Vein and Morphological Responses to PATi

To aid in understanding these results, auxin transport and growth mechanisms were combined into a computer model. Simulations generating the observed reduction in leaf complexity and alterations in venation provide insight into how PATi induced both effects. Simulations correspond to the initial developmental sequence of the terminal and first lateral lobes and their midveins (such as in Figure 3A).

Modeling PAT inhibitor treatment as a decrease in auxin transmission through PIN1 (decreased T parameter) resulted in a smoothing of the leaf margin and loss of distinct sinuses (Figures 5 and S1), corresponding to the experimentally observed reduction in leaf complexity (Figure 1). This was due to the effect of PATi on the spatial distribution of the auxin CPs, which catalyze cell growth at the lobe tips. In normal conditions, high auxin levels that were localized in the CPs induced an extension of distinct lobes. Reduced transport through PIN1 (corresponding to treatment with PAT inhibitors) impaired the normal auxin flow involved in maintaining a high auxin concentration in CPs and the spacing between CPs. With reduced flow, CPs drew auxin from fewer neighbouring cells, creating more CPs (black asterisks, Figure 5) that were more closely spaced than in normal conditions. The normal auxin distribution along the terminal and lateral margin in Figure 5A can be approximated by a sine wave with three peaks and a long wavelength separating fast growing lobes from slow growing sinuses; with reduced PAT (e.g., Figure 5D) the wavelength was shortened, with less space separating lobes and sinuses. This resulted in a more uniform auxin distribution than normal (i.e., local averages of auxin in PATi conditions tended to have a mix of high and low concentration cells, rather than either all high cells or all low cells as seen in control conditions). As auxin catalyzed growth, PATi produced more uniform leaf outgrowth, without the strong distinction between high and low growth rates between lobes and sinuses characteristic of complex leaves.

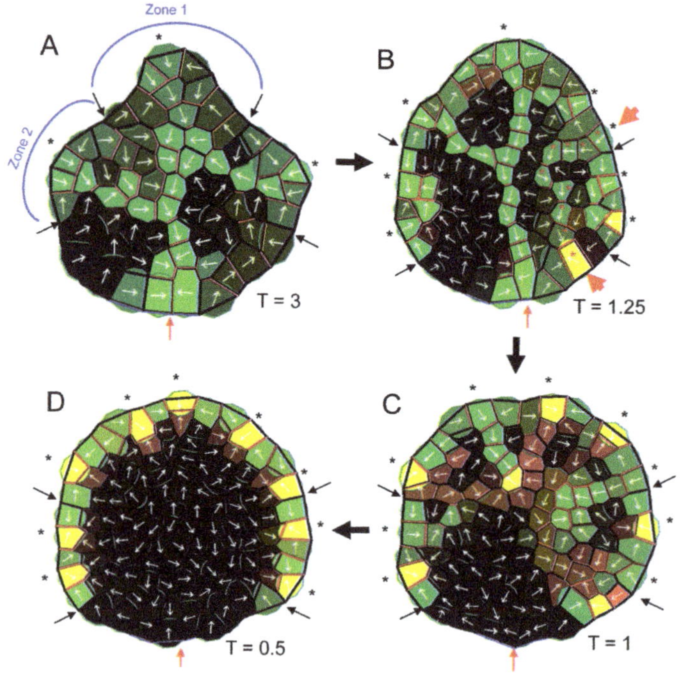

Figure 5. Computer simulations of the effect of PAT inhibitor treatment on leaf margin complexity. Images show the terminal lobe (Zone 1) and first two lateral lobes (Zone 2; only left side is marked in (**A**)), with black arrows delimiting the zone boundaries. Auxin-synthesizing cells in the margin grew in proportion to auxin concentration (green intensity). PIN1 concentration (red) was either cytosolic or in the membrane (shown at the walls). Yellow indicates high PIN1 and auxin concentrations within a cell; black indicates a lack of auxin and PIN1. White arrows indicate the net auxin flux for each cell. All results are shown at the same time (5 h 30 min computational units). PIN1-auxin dynamics in normal conditions (**A**) formed CPs (black asterisks) from which the midvein extended to the base of the leaf and the existing plant vasculature (represented by the blue-walled cells, red arrow) and from which secondary veins extended to the midvein. CPs are local auxin maxima in the margin (with marginal PIN1 polarization into the CPs), causing lobed outgrowth. As transmission through PIN1 (T parameter) was decreased (**B–D**), corresponding to increasing PAT inhibitor concentration, the leaf margin changed from lobed (**A**) to smooth (**B–D**), extra CPs formed (black asterisks), and secondary veins could run parallel to the midvein (red asterisks and red arrowheads in (**B**)). At low T (**C**), the mid-vein was lost, and the extra CPs extended short 'strands'. At very low T (**D**), vein extension was lost. The smoother margins reflect the more uniform marginal auxin distributions (less localization to a few 'discrete' CPs) as PIN1 transmission was lost.

Altering PAT affected vein patterning in the lamina in addition to affecting CP spacing. Mild PAT inhibition (a reduction in the *T* transmissivity parameter) could produce secondary veins that run more parallel to the primary vein (red asterisks in Figures 5B and S1F) rather than connecting to the primary vein (cf. Figure 5A). A further reduction in PAT (decrease in T) reduced vein extension (Figures 5C,D and S1G–I), as seen experimentally (Figure 3). The reduced lobe extension in the lamina, loss of connection to the proximal sink (representing existing vasculature in the plant; red arrows, Figure 5), and impaired drainage of auxin leading to accumulation in the margin all correspond to experimental observations for auxin (Figure 3B) and vasculature (Figures 2 and 3A). The model indicates that PAT dynamics and auxin-catalyzed growth can account for the simultaneous increase in vein number, altered lamina vein patterning, reduction in vein extension, marginal vein overgrowth, and loss of morphological complexity observed in PATi experiments.

2.4. Exogenous Auxin Application Altered Tomato Leaf Complexity but Not Venation Patterning

Prior work reported cases of ectopic blade growth or extra leaflets with the local application of exogenous auxin application to tomato leaves [23]. We confirm this trend statistically: significant reductions in leaflet number with IAA application were found across three separate experiments (Table 2). In control plants, tomato leaflets always attached to the rachis via the short, cylindrical petiolule (Figure 6A). With IAA treatment, leaflets often had ectopic laminar growth on the petiole or rachis, which could extend along the rachis between leaflets (arrows in Figure 6B,C). Adjacent leaflets could be fused together, forming a single leaflet with broad attachments to the petiole/rachis (Figure 6D–F). Consequently, exogenous IAA treatment decreased the leaflet count (Figure 6D–F; Table 2). In some cases, the decreased leaflet count may have been the result of inhibited leaflet formation rather than a fusion of adjacent leaflets (Figure 6G). In general, 1% *w*/*w* IAA treatment resulted in less severe perturbations in growth, with leaflets often forming broad bases of attachment to the petiole or rachis rather than fusions of adjacent leaflets. Thus, plants treated with 1% *w*/*w* IAA generally had more leaflets than plants treated with 10% *w*/*w* IAA (Table 2).

Table 2. Exogenous IAA application reduced leaflet count in tomato. Leaflet number in control plants or after the addition of IAA. Cells show: Row 1 = mean ± S.D. and (n); Row 2 = p-value for mean treated value being equal to control, * significantly different from control ($p < 5 \times 10^{-2}$) or ** highly significantly different from control ($p < 1 \times 10^{-2}$); Row 3 = % below control median, note increase in percentage with treatment; Row 4 = % of plants with fused leaflets. ND = not done. Tiny Tom 1, 2 = experiment 1 or 2 with Tiny Tom. G = Glamour variety.

Leaf	TT 1; Control	TT 1; 10% IAA	TT 2; Control	TT 2; 1% IAA	TT 2; 10% IAA	G; Control	G; 10% IAA
Leaf 2	3.5 ± 0.8 (48)	3.1 ± 1.1 (59)					
p-value		1.5×10^{-2} *	ND	ND	ND	ND	ND
% < median	6	22					
% fused	0	20					
Leaf 3	4.5 ± 0.9 (48)	3.4 ± 1.3 (59)					
p-value		5.0×10^{-7} **	ND	ND	ND	ND	ND
% < median	33	75					
% fused	0	36					
Leaf 4	5.1 ± 0.6 (48)	3.7 ± 1.5 (56)	5.1 ± 0.4 (24)	5.0 ± 0.2 (26)	4.5 ± 0.9 (24)	10.0 ± 1.2 (5)	5.0 ± 1.9 (5)
p-value		3.1×10^{-9} **		9.7×10^{-2}	2.6×10^{-3} **		8.2×10^{-4} **
% < median	6	54	0	4	29	20	100
% fused	0	32	0	0	8	0	0
Leaf 5	5.1 ± 0.5 (48)	4.6 ± 0.9 (47)	5.3 ± 0.5 (24)	5.0 ± 0.2 (26)	4.1 ± 1.2 (24)	15.4 ± 1.9 (5)	5.0 ± 2.5 (5)
p-value		1.0×10^{-3} **		3.8×10^{-2} *	1.0×10^{-4} *		4.8×10^{-5} **
% < median	0	23	0	0	46	20	100
% fused	0	4	0	12	8	0	20
Leaf 6			5.3 ± 0.7 (23)	4.8 ± 1.2 (26)	4.0 ± 1.4 (22)	15.2 ± 1.6 (5)	6.4 ± 3.5 (5)
p-value	ND	ND		4.7×10^{-2} *	2.9×10^{-4} **		1.3×10^{-3} **
% < median			4	12	45	40	100
% fused			0	27	23	0	20
Leaf 7			5.6 ± 0.8 (24)	4.4 ± 1.9 (25)	3.3 ± 2.3 (15)	16.0 ± 0 (5)	10.4 ± 5.1 (5)
p-value	ND	ND		4.2×10^{-3} **	8.9×10^{-4} **		3.5×10^{-2} *
% < median			0	32	67	0	80
% fused			0	56	73	0	60

Figure 6. Exogenous IAA affected tomato (variety Tiny Tom) leaflet formation. Images show tomato leaves from ca 10-week-old control plants (**A**) or plants treated with 1% (**B**,**F**,**G**) or 10% (**C**–**E**) *w*/*w* IAA. IAA application resulted in leaves having broad bases fused along the petiole or rachis (arrows in (**B**,**C**) and/or fused with other leaflets (arrows in (**C**–**F**)). Scale bars are 10 mm.

Nevertheless, both 1% and 10% *w*/*w* IAA treatments resulted in substantial percentages of leaves with leaflets ectopically fused along the petiole or rachis and/or to each other (Table 2). While fusions were never observed in control leaves, fusions were observed frequently in treatments with exogenous IAA (Table 2). In general, younger leaf primordia (higher leaf number) showed a stronger effect from IAA treatment (Table 2; particularly Tiny Tom 2), with the formation of callus-like structures or a complete inhibition of growth in very young leaf primordia exposed to exogenous IAA. A similar tendency towards the fusion of the leaflets to the petiole/rachis and/or to adjacent leaflets with IAA treatment was observed in the tomato variety Glamour (Table 2; Figure S2). These observations indicate that exogenous IAA application disrupted the boundaries delimiting lobes in developing tomato leaflets.

The exposure of 8 DAG seedling leaves to the synthetic auxin analog 2,4-D induced broad ectopic auxin response maxima (Figure 4D), corresponding to the lobe broadening observed for exogenous IAA (Figure 6). In young L4 leaf primordia exposed to 2,4-D, *pDR5*:GUS expression was observed particularly at the margins of the lamina, whereas expression was progressively restricted to more basal regions in older L3, L2, and L1 leaves (Figure 4D). This corresponds to the smaller effect on complexity in older leaves with exogenous IAA, particularly for the percentage of leaves with fused leaflets in IAA Experiment 2 (Tiny Tom 2; Table 2). Little or no *pDR5*:GUS expression was observed in control leaves grown in liquid media at this developmental stage (insets, Figure 4D).

Because the application of exogenous IAA reduced developing complex tomato leaves to more simple leafed structures (Figure 6; Table 2), similar to the morphological effect with PAT inhibition, we wanted to determine whether the semi-craspedodromous venation patterning was also simplified to more parallel-like venation. This was not the case. Exogenous IAA application to developing tomato leaves retained normal venation patterning in regions of ectopic laminar growth and/or laminar fusion between leaflets (Figure 7). Secondary veins extended from the ectopic laminar region into the midvein, joining with other secondary veins to form loops. Additional veins branched from the marginal lobes towards these secondary loops. Each marginal lobe had a semi-craspedodromous venation patterning, with a primary-like midvein extending up to or near the lobe tip, secondary-like vein loops on either side, and higher order venation (Figure 7). Hence, unlike PAT inhibition, a reduction in tomato leaf complexity with exogenous IAA was not correlated with the induction of parallel-like venation patterning.

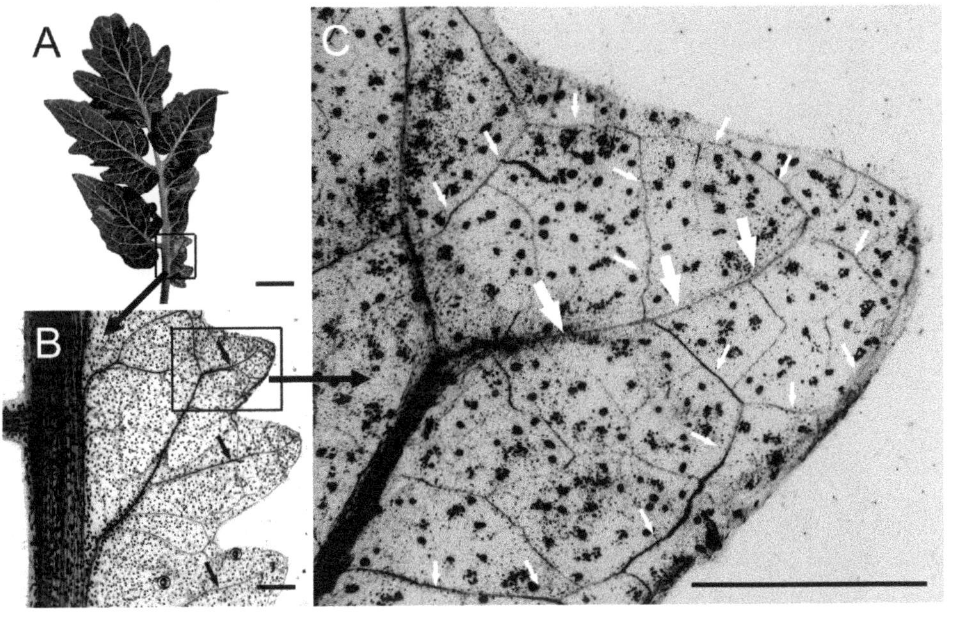

Figure 7. IAA treatment did not alter overall tomato leaf venation patterning. Magnified images (**B**,**C**) of a leaf (**A**) treated with 10% *w*/*w* IAA. Darkfield images were inverted to show leaf venation patterning, and numerous starch granules are also present. IAA treatment induced primary-like (arrows in (**B**); large arrows in (**C**)) veins extending towards the lobe apex and secondary-like (small arrows in (**C**)) vein loops. Scale bars are 10 mm (**A**) or 1 mm (**B**,**C**).

2.5. Modeling Suggests That Vein Developmental Stage Affects the Different Morphological and Venation Responses to Exogenous IAA Exposure

To understand this different response in terms of the PAT mechanism, we simulated exogenous IAA application by a boost of the auxin precursor level, A_{prec}, parameter in the computer model. This increased the marginal auxin supply in specific zones, which catalyzed the additional growth of the lobes. As auxin was higher throughout the lobes and not as focused in the CPs, this produced broader lobes and less distinct sinuses than normal (Figure 8), corresponding to a reduction in complexity. Consistent with the concentration effects in Table 2, higher A_{prec} (corresponding to a higher added IAA concentration) produced broader lobes (Figure 8A–C). Different effects of exogenous IAA by leaf age (L-stage, Table 2) may reflect the maturity of the vasculature: in Figure 8D, PIN1 is boosted in the terminal margin and midvein (to 150 from normal values of up to 120) to represent more mature vasculature (localized PIN1 is observed to increase in veins over the course of development, e.g., [14,15,24]. In this case, the same added auxin is more efficiently drained away and does not affect morphology as much as in the terminal lobe of Figure 8C (normal PIN1 levels, representing the less developed vasculature of a younger leaf). Hence, exposure to increased IAA in leaves with less mature vasculature resulted in a broader terminal lobe with less distinct sinuses (Figure 8C) compared to a leaf with more mature vascular cells capable of transporting auxin away from the margin (Figure 8D).

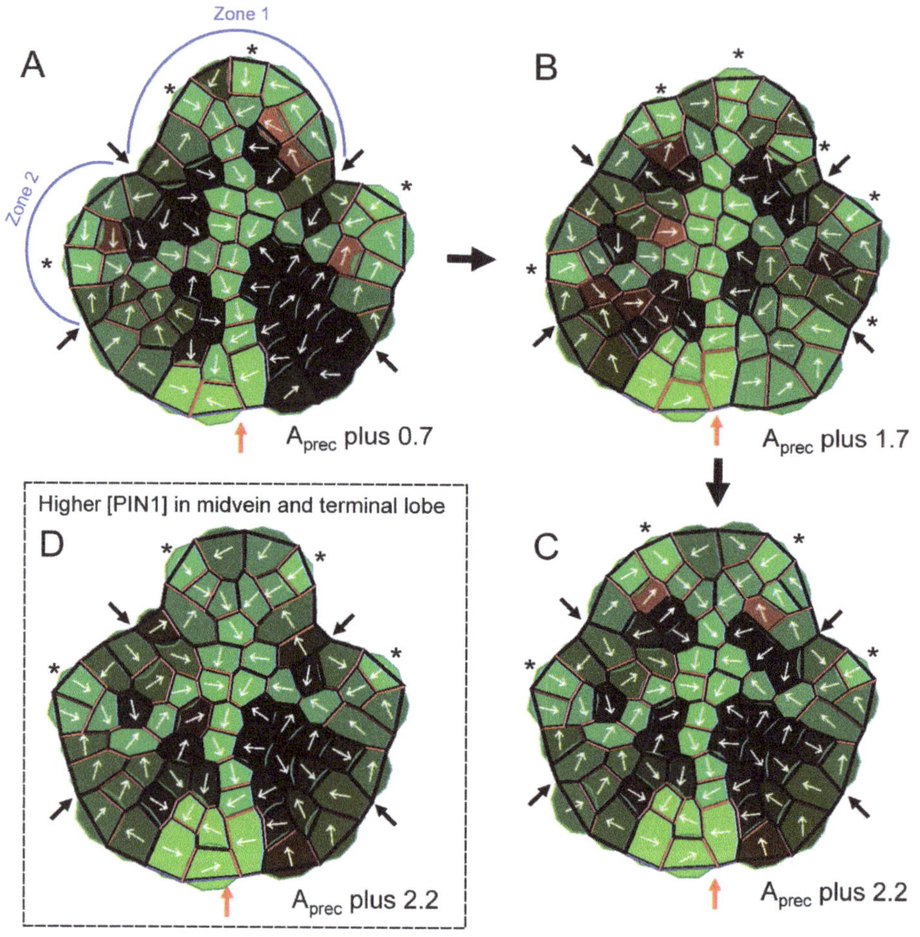

Figure 8. Computer simulations of the effect of exogenous IAA on leaf margin complexity. Model parameters and visualization are as in Figure 5. (**A–C**) Increasing auxin production, corresponding to exogenous IAA application, produced smoother margins (auxin precursor concentration, A_{prec}, was boosted at the time the lateral lobes became auxin-synthesizing). Larger boosts created broader lobes and shallower clefts: (**A**) 0.7 added to A_{prec}; (**B**) 1.7 added to A_{prec}; (**C**) 2.2 added to A_{prec}. (**A–C**) Transport from the margin was not sufficient to reduce auxin levels to normal, resulting in the extra growth of the lobes. (**D**) The leaflets had the same A_{prec} boost as (**C**), but PIN1 was also boosted in the terminal lobe and midvein to simulate established vasculature in more mature leaflets. In this case, the high doses of exogenous auxin could be sufficiently drained, resulting in a narrower, more normal lobe shape (compare region between black arrows delimiting zone 1 in (**C**,**D**)). Asterisks indicate CPs.

While the above results suggest a role of vascular maturation (and hence auxin transport capabilities) on lobing, altering leaf morphology with exogenous IAA had no effect on vascular patterning. Models with normal levels of transmission through PIN1 showed that added auxin may spur marginal growth and lobe broadening, but PAT flows remained sufficient to maintain normal CPs and a midvein in each lobe (Figure 8A–C). Modelling suggests that the difference in results between PATi and exogenous IAA application stems from different aspects of the PAT-growth mechanism. The relative independence of morphogenesis and vein patterns with IAA application reflect that while added auxin may catalyze growth, it was also efficiently transported by normal PIN1 function and did not alter vein patterns (Figure 8). In contrast, PATi-impeded auxin flow shortened CP–CP distances, multiplying the number of CPs and veins, in addition to altering marginal growth and therefore lobing (Figure 5).

3. Discussion

3.1. PATi Reduced Complexity in Diverse Compound Leaf Types

We chose species with a variety of leaf complexity types. PATi could reduce compound leaves as diverse as tomato, with deep serrations and leaflets, to the repeated bifurcations of carrot, to the radially arranged and completely separated leaflets of lupin and clover to simple leaves. Both leaf shapes (Figure 1) and vein patterns (Figure 2) were affected, indicating a role for auxin transport in these processes. Different responses to PATi and to exogenous IAA indicate features of the auxin patterning mechanism involved in morphogenesis and venation. The common general response across these species indicates the role of auxin in compound leaves is at least some 120 million years old, when the Asterids (tomato and carrot) and Rosids (clover and lupin) diverged [32]. Aspects of the PIN–auxin module are likely even more ancient, appearing in mosses [33].

Differences in the species' sensitivities to PATi agents and dosage (Table 1) indicate between-species differences in auxin transport that have arisen more recently. These differences may involve a differing inhibitor affinity for binding sites between species. The crystal structure for the *A. thaliana* PIN1 protein has been resolved, and it is known that NPA competes with and binds to a similar position as the natural auxin IAA [27]. Different responses to different NPA dosages in other species may therefore indicate the affinity of IAA binding to PIN1 relative to *Arabidopsis*. Response strengths to other PAT inhibitors may also indicate varying binding strengths relative to the *Arabidopsis* IAA/NPA sites.

3.2. PATi Produced Supernumary Veins with More Parallel Patterning

The four species studied also represent different vein pattern types: craspedodromous (clover and carrot); semi-craspedodromous (tomato); and brochidodromous (lupin). One of the main distinguishing features of leaf venation is the positioning of the primary and secondary strands with respect to the main leaf axis and to each other. Craspedodromous and brochidodromous secondary veins join with the central primary vein throughout the leaf lamina. PATi affected both the number of secondaries and their angle of approach towards the central lamina.

In the four species tested, PATi increased the number of secondary veins, reflecting the increased number of marginal CP sites (Figure 2). These responses are similar to those observed in *Arabidopsis* simple leaves grown in mild to moderate PATi conditions [25,26]. In *Arabidopsis*, very strong PAT inhibition often results in more vascular tissue around the margins where auxin pools, forming thicker (or bundled) veins with retarded vein extension from the margin towards the leaf base [25,26]. PATi also retarded vein extension in tomato leaf primordia (Figure 3A), although they did eventually extend into the basal lamina by leaf maturation (Figure 2). PATi caused less marginal vascularization than in *Arabidopsis*, possibly because under our experimental conditions these species may have retained enough vascular function to effectively drain auxin from the margins, thus preventing excess marginal vascularization. All four species did show an increased thickness (or bundling) of the secondary veins with PATi (Figure 2), similar to *Arabidopsis*.

PATi in all four species usually eliminated the single dominant midvein (primary vein) of the leaf/leaflet, producing more parallel venation patterning, with secondary-like veins running more parallel to the proximodistal leaf/leaflet axis (Figure 2). There appear to be two aspects to this parallelization: (a) the marginal sources of the secondaries (CPs) appear to become more distal with PATi compared to control; and (b) PATi appears to enhance proximal-wards bending (i.e., a decreased angle of approach) in secondaries as they approach the central lamina, producing parallel bundles as compared to the secondary to primary vein connections observed in controls. The marginal auxin sources of secondaries appeared more distally focused in tomato and lupin, producing bundling along the length of the leaf/leaflet center. In contrast, auxin source distalization was broader and less centralized in clover and carrot, with the leaf/leaflet vascular bending and bundling towards the center occurring more proximally, near the petiole. The angle of approach of secondaries to the midvein region may suggest that PATi interferes with the normal connection process of veins (bundling is also observed in more marginal positions, in the thickening or bundling of secondaries with PATi). Potentially, as the attraction or ability to connect to established veins (such as secondary to primary) is decreased, the increased relative attraction of the leaf base (older venation within the stem) induced the proximal-wards turn in the secondaries.

The diminished primary vein strength with *Arabidopsis* simple leaves similarly shows some of these two parallelization aspects under PATi [25,26]. Compared to the compound leaf types studied here, *Arabidopsis* simple leaves tend to exhibit broader auxin sources (less CP distalization) and more vein bending. The variety of source distalization and vein bending of secondaries observed in Figure 2 may indicate that different constraints on PAT may exist in different compound leaf geometries or developmental programs.

3.3. The Effects of Exogenous IAA Varies with Developmental Stage and Can Differ from PATi

The application of exogenous IAA to developing tomato leaf primordia disrupted leaflet boundary formation, resulting in ectopic laminar growth along the rachis/petiole and fusion between adjacent leaflets that also reduced overall leaf complexity (Figure 6, Table 2). The competency to respond to exogenous IAA depended on the developmental stage. In tomato, auxin response domains were observed throughout the margin of very young leaf primordia and only in more basal (younger) regions of older leaf primordia (Figure 4D). Modeling suggests that exogenous IAA does not impact the morphology of older lobes with more mature veins that can efficiently transport the excess IAA away from the margin (Figure 8). This would explain why fused leaflets were mostly observed between more basal (younger) leaflets if auxin was applied to older leaf primordia (e.g., Figure 6B–E), and the extent of fusion was greater in more newly formed leaves (particularly Experiment 2, Table 2). These younger regions had immature veins that did not remove the excess IAA from the margin during leaf shape formation, resulting in diffuse marginal auxin and leaflet fusion.

While exogenous IAA application can alter the normal pattern of discrete auxin maxima in the margin, particularly in new leaf primordia (Figure 4D), vein patterning appeared largely unaffected in mature leaves even in fused regions (Figure 7). Normal vein patterning suggests that, even in the presence of exogenous IAA, functional veins were eventually formed that could effectively transport the excess IAA from the margin, such that vein patterning was not altered. With PATi, auxin pooled at the margin because it disrupted vein formation and their auxin transport efficiency, creating more marginal CPs and hence supernumerary veins. This suggests that the PATi induction of supernumerary more parallel venation is the result of reduced PAT and not just the altered auxin levels in the margin.

3.4. A PAT-Growth Model for the Morphological and Venation Effects of PATi and Exogenous IAA

The mathematical model of PAT and auxin-dependent growth dynamics captures key features of the experimental results. These include: the proliferation of CPs and generation of supernumerary veins with PATi, with the corresponding reduction in morphological complexity (Figure 5); and the morphological effects of exogenous IAA, which can occur without alterations of the venation patterns (Figure 8).

Modeling indicates that the different venation responses of PATi and exogenous IAA could be due to differences in vein development. Both cases had higher than normal auxin levels in the margin, leading to the formation of more CPs with fewer cells feeding auxin into them. PATi-induced PIN1 disruption affected early plastic stages of vascular development, as the marginal auxin levels remained high, thereby creating more CPs that drive the formation of new secondary veins. With exogenous IAA, veins can still develop normally (as PIN1 localization was not affected) and were able to mitigate the excess auxin in the margin, such that venation patterns were not significantly altered. Vein maturation status at the time of exogenous IAA exposure may also play a role—older leaf primordia / lobes with more mature vascular cells were able to drain auxin more efficiently and therefore had less growth defects compared to younger leaves/lobes with less developed vasculature (e.g., Figure 8D compared to Figure 8C).

The similar effects of lobe broadening for PATi or IAA boosting may occur for different reasons. PATi reduced auxin drainage, resulting in higher auxin in the margin. This created supernumerary CPs that caused more spatially averaged auxin distribution and more uniform marginal growth because the auxin was not being drained away. Exogenous IAA boosted overall marginal auxin in the lobe, lessening the relative high concentration/localization of the CP at the lobe tip, leading to more uniform marginal growth and additional CPs with exogenous IAA.

Previous models included the CP formation and growth of the margin [9,13,34] but did not compute PAT in the lamina of the leaf. By modeling PAT for the margin and the interior of the leaf, our model can address the more complete picture of how PATi interferes with the normal 'reverse fountain' patterning in which PAT plays a role in both CP formation and flow at the margin, as well as in the inward extension and canalization of provascular auxin traces. In particular, this allows the model to capture the effects of auxin drainage from the margin sources, key to interpreting the results of the exogenous IAA experiments. For PATi, the model generates the observed increase in CPs corresponding to supernumerary veins.

We note that our model only has uniform and auxin-dependent growth. The inclusion of additional factors, such as CUC pattern restraining growth in sinuses [9,34], would be expected to enhance lobe–sinus differences and support clearer boundaries in the resulting morphologies. Questions remain regarding the causes of the venation effects of (a) more distalized marginal sources and (b) secondary vein turning rather than joining prior veins (see discussion above) under PATi. Modeling could further explore (a) the PAT influence of marginal auxin synthesis zones and (b) mechanisms for turning the secondaries in the mid-leaf. The model does not presently include (a), but a slight turning (b) was achieved (Figure 5B). As discussed in [35], vein turning could be due to PATi effects on the up-the-gradient (UTG) allocation of PIN1 needed to connect an extending (secondary) vein to an established (primary) vein with high auxin. In this case, the different degrees of vein turning observed between the present four species could represent different degrees of the role of UTG in vein connection across these species. It is likely, though, that other genetic or transport factors (such as plasmodesmata, [36,37]) also influence vein connections and would be needed for a more complete description of PATi effects on vein bending.

4. Materials and Methods

4.1. Plant Material and Growth Conditions

Four plant species were used: (1) tomato, *Lycopersicon esculentum* L. (varieties Glamour and Tiny Tom, Stokes Seeds Ltd., St. Catharines, ON, Canada; PATi experiments on Glamour, IAA experiments on both varieties); (2) white clover, *Trifolium repens* L. (Dawson Seed Co. Ltd., Surrey, BC, Canada); (3) carrot, *Daucus carota* (variety Crona; Svalöv/Weibull AB, Svalöv, Sweden); and (4) lupin, *Lupinus polyphyllus* (wild seed collected from Burnaby Mountain Park, Burnaby, BC, Canada).

PAT inhibitors were used on all four species. Seeds were rinsed in water for about 30 min to remove commercial fungicides (where applicable), sterilized for 1 min in 70% v/v ethanol and then 10 min in 50% v/v commercial bleach, followed by three rinses in sterile water. Lupin seeds were scarified prior to sterilization. Seeds were plated on solid ATS medium [38] supplemented with either N-(1-Naphthyl) phthalamic acid (NPA; TCI, Tokyo, Japan), 9-hydroxyfluorene-9-carboxylic acid (HFCA; Sigma, St. Louis, MO, USA), or 2,3,5-triiodobenzoic acid (TIBA; Sigma, St. Louis, MO, USA) dissolved in DMSO to a final concentration of 0.1, 1, 10, 50 or 100 µM, or an equivalent volume of DMSO for control material. Seeds were plated in 9 cm diameter Petri dishes (ca 20 seeds/plate), stratified at 4 °C overnight, and then grown at about 20 °C, 16 h illumination (ca 150 to 200 µmol m^{-2} s^{-1}) for 2 weeks. At this time, seedlings were transferred to larger containers with sterile media and grown for another 2 to 3 weeks. For analyses of the early development of tomato leaf primordia, sterilized seeds were plated (ca 10 seeds/plate) in 9 cm Petri dishes containing 10 mL of liquid ATS (control, or supplemented with NPA to final volumes of 1 or 10 µM), stratified at 4 °C overnight, and then gently agitated at ca 20 °C with constant illumination for up to 10 d growth.

The effect of exogenous indole acetic acid (IAA) was examined in tomato. Seeds were immersed in water and stratified at 4 °C overnight and then planted in soil (Pro-Mix general purpose, Premier Horticulture Inc., Quakertown, PA, USA) and grown in a greenhouse. Three experiments were performed where IAA (Sigma, St. Louis, MO, USA) was mixed with lanolin at 0, 1, or 10% *w*/*w* concentrations and a small drop placed at the base of the following leaf primordia: (i) leaf 2 (L2) and L3 in 2-week-old Tiny Tom plants (Experiment 1); (ii) L5 and L6 in 6-week-old Tiny Tom plants (Experiment 2); (iii) L4 and L5 in 4-week-old Glamour plants (Experiment 3). Leaf complexity and venation patterning were assessed in mature leaves: L2 to L5 in 6-week-old plants (Experiment 1); L4 to L7 leaves in 10-week-old plants (Experiments 2, 3).

Two auxin responsive markers, *pDR5*::GUS (obtained from G. Hagen, University of Missouri, Columbia, MO, USA; [39]) and *pAtIAA2*::GUS (gift from J. Normanly, University of Massachusetts, Amherst, MA, USA; [40]), were introduced into tomato (variety Ailsa Craig) by *Agrobacterium*-mediated transformation [41]. To test the effect of auxin transport inhibition on auxin response maxima, transformed tomato plants were grown as above in solid ATS or $\frac{1}{2}$ Murashige Skoog media supplemented with 0, 1, or 10 µM NPA, and leaves 1 and 2 excised at 6, 8, and 10 d growth. Leaves were stained for GUS expression as in [25] or using a modified GUS staining solution (100 mM phosphate buffer, 5 mM $K_3(Fe(CN_6))$, 5 mM $K_4(Fe(CN_6))$, 10 mM EDTA, 10% *v*/*v* methanol, 0.01% *v*/*v* Triton X 100, 5 mM X-GLUC (Gold Biotechnology Inc., St. Louis, MO, USA)). To determine the competency of cellular response to exogenous auxin, *pDR5*::GUS seedlings were grown for 8 days in liquid ATS medium, and then excised leaves were transferred to new media containing 0 or 10 µM 2,4-D (Sigma, St. Louis, MO, USA), agitated gently for 16 h, and stained for GUS expression as in [25].

4.2. Analyses of Leaf Complexity and Venation Patterning

We use the terminology from [1] to describe leaf complexity types and leaf venation patterns. To examine the effect of exogenous auxin on tomato leaf complexity, the number of leaflets was counted, and the extent of fusion of leaflets along the rachis and/or to each other was scored 6 weeks after the application of exogenous IAA to leaf primordia. To examine the impact of auxin transport inhibition on leaf complexity, the number of leaflets (for tomato, white clover, carrot, and lupin) was counted for the first three mature leaves (L1, L2, L3) of 4- to 5-week-old seedlings grown on solid media with or without auxin transport inhibitors. Selected mature leaves of all species grown with or without inhibitors were also cleared and analyzed for vascular patterning as in [26]. For tomato seedlings grown in liquid media, developing L1 and L2 leaf primordia in 6–10-day-old seedlings were cleared, and their complexity and vascular patterning analyzed as above. Images were taken using a Nikon Eclipse E600 microscope (Nikon Instruments Inc., Tokyo, Japan) and a Canon EOS D30 digital camera (Oita Canon Inc., Oita, Japan).

4.3. Computer Simulations of Leaf Venation and Growth

Computer simulations of PIN1 and auxin dynamics used the model and parameters described in [35], written within the VirtualLeaf software Version 1.0.2 package [42]. This dual-polarization model produces both CPs and inward-extending provascular tracks (see also [24,43]). Installation, files, and parameter values are discussed further in Supplementary Text S1.

The generation of the normal sequence of a primary (distally localized) auxin maximum with the subsequent formation of secondary (more proximal) auxin maxima was coupled to auxin-dependent cell growth. This generated a multi-lobed leaf shape (with a central terminal lobe and two lateral lobes), representing a compound leaf.

Auxin synthesis occurred from a precursor in the margin, A_{prec}. (See Introduction of [35]: the computational margin represents the epidermal and subepidermal cells in which auxin is synthesized and CPs form.) The earlier development of the terminal lobe was represented by a higher initial A_{prec} value than in the lateral lobes. A_{prec} increased in all cells to a maximum. Leaf growth was generated by increasing margin cell areas according to background and auxin-dependent rates, followed by cell division. PAT inhibitor treatment was simulated by decreasing the transmission of auxin through PIN1 (T parameter). Exogenous IAA treatment was simulated by increasing A_{prec} a given amount in all auxin-synthesizing cells at 2 h 47 min (computational units), when the lateral lobes become auxin-synthesizing.

Supplementary Materials: The following supporting information can be downloaded at: https://www.mdpi.com/article/10.3390/plants13182566/s1, Figure S1: Computer simulations of the effect of PAT inhibitor treatment on leaf margin complexity; Figure S2: Exogenous IAA affected leaflet formation in tomato (variety Glamour); Text S1: Computational Methods. References [44–46] are cited in the supplementary materials

Author Contributions: C.L.W. designed, performed, and supervised experiments. J.M. designed and supervised experiments. D.M.H. and C.L.W. designed computer simulations. D.M.H. performed computer simulations and did statistical analyses of experimental data. All authors contributed to writing the article. All authors have read and agreed to the published version of the manuscript.

Funding: This work was supported by a BCIT IRF grant to C.W., an NSERC Discovery grant to J.M., and an NSERC Discovery grant to D.H.

Data Availability Statement: Data are contained within the article and Supplementary Materials.

Acknowledgments: We thank Kristy Kyle, Konrad Blonski, Afsaneh Haghighi-Kia and Micky Yip for technical assistance, and Brandon Wong for help with literature searches. We also thank M. Ivachenko for the tomato seeds containing constructs from G. Hagen (*pDR5*::GUS) and J. Normanly (*pAtIAA2*::GUS).

Conflicts of Interest: The authors declare no conflict of interest.

References

1. Ash, A.; Ellis, B.; Hickey, L.J.; Johnson, K.; Wilf, P.; Wing, S. *Manual of Leaf Architecture–Morphological Description and Categorization of Dicotyledonous and Net-Veined Monocotyledonous Angiosperms. Leaf Architecture Working Group*; Smithsonian Institution: Washington, DC, USA, 1999.
2. Kerstetter, R.A.; Poethig, R.S. The Specification of Leaf Identity During Shoot Development. *Ann. Rev. Cell Dev. Biol.* **1998**, *14*, 373–398. [CrossRef] [PubMed]
3. Dengler, N.G. Comparison of Leaf Development in Normal (+/+), Entire (E/E), and Lanceolate (La/+) Plants of Tomato, Lycopersicon eculentum 'Ailsa Craig'. *Bot. Gaz.* **1984**, *145*, 66–77. [CrossRef]
4. Dengler, N.G.; Tsukaya, H. Leaf Morphogenesis in Dicotyledons: Current Issues. *Int. J. Plant Sci.* **2001**, *162*, 459–464. [CrossRef]
5. Hagemann, W.; Gleissberg, S. Organogenetic Capacity of Leaves: The Significance of Marginal Blastozones in Angiosperms. *Plant Syst. Evol.* **1996**, *199*, 121–152. [CrossRef]
6. Kaplan, D.R. Fundamental Concepts of Leaf Morphology and Morphogenesis: A Contribution to the Interpretation of Molecular Genetic Mutants. *Int. J. Plant Sci.* **2001**, *162*, 465–474. [CrossRef]
7. Holtan, H.E.; Hake, S. Quantitative Trait Locus Analysis of Leaf Dissection in Tomato Using Lycopersicon pennellii Segmental Introgression Lines. *Genetics* **2003**, *165*, 1541–1550. [CrossRef]
8. Maugarny-Calès, A.; Laufs, P. Getting Leaves into Shape: A Molecular, Cellular, Environmental and Evolutionary View. *Development* **2018**, *145*, dev161646. [CrossRef]
9. Kierzkowski, D.; Runions, A.; Vuolo, F.; Strauss, S.; Lymbouridou, R.; Routier-Kierzkowska, A.L.; Wilson-Sánchez, D.; Jenke, H.; Galinha, C.; Mosca, G.; et al. A Growth-Based Framework for Leaf Shape Development and Diversity. *Cell* **2019**, *177*, 1405–1418. [CrossRef]

10. Nikolov, L.A.; Runions, A.; Das Gupta, M.; Tsiantis, M. Leaf Development and Evolution. *Curr. Top. Dev. Biol.* **2019**, *31*, 109–139.
11. Van Volkenburgh, E. Leaf Expansion–An Integrating Plant Behaviour. *Plant Cell Environ.* **1999**, *22*, 1463–1473. [CrossRef]
12. Dengler, N.; Kang, J. Vascular Patterning and Leaf Shape. *Current Opinion in Plant Biol.* **2001**, *4*, 50–56. [CrossRef] [PubMed]
13. Runions, A.; Tsiantis, M.; Prusinkiewicz, P. A Common Developmental Program can Produce Diverse Leaf Shapes. *New Phytol.* **2017**, *216*, 401–418. [CrossRef] [PubMed]
14. Scarpella, E.; Marcos, D.; Friml, J.; Berleth, T. Control of Leaf Vascular Patterning by Polar Auxin Transport. *Genes Dev.* **2006**, *20*, 1015–1027. [CrossRef] [PubMed]
15. Wenzel, C.L.; Schuetz, M.; Yu, Q.; Mattsson, J. Dynamics of MONOPTEROS and PIN-FORMED 1 Expression during Leaf Vein Pattern Formation in Arabidopsis thaliana. *Plant J.* **2007**, *49*, 387–398. [CrossRef] [PubMed]
16. Okada, K.; Ueda, J.; Komaki, M.K.; Bell, C.J.; Shimura, Y. Requirement of the Auxin Polar Transport System in Early Stages of Arabidopsis Floral Bud Formation. *Plant Cell* **1991**, *3*, 677–684. [CrossRef]
17. Liu, C.-M.; Xi, Z.-H.; Chua, N.-H. Auxin Polar Transport is Essential for the Establishment of Bilateral Symmetry During Early Plant Embryogenesis. *Plant Cell* **1993**, *5*, 521–530. [CrossRef]
18. Benkova, E.; Michniewicz, M.; Sauer, M.; Teichmann, T.; Seifertova, D.; Jurgens, G.; Friml, J. Local, Efflux-Dependent Auxin Gradients as a Common Module for Plant Organ Formation. *Cell* **2003**, *115*, 591–602. [CrossRef]
19. Reinhardt, D.; Mandel, T.; Kuhlemeier, C. Auxin Regulates the Initiation and Radial Position of Plant Lateral Organs. *Plant Cell* **2000**, *12*, 507–518. [CrossRef]
20. Reinhardt, D.; Pesce, E.R.; Stieger, P.; Mandel, T.; Baltensperger, K.; Bennett, M.; Traas, J.; Friml, J.; Kuhlemeier, C. Regulation of Phyllotaxis by Polar Auxin Transport. *Nature* **2003**, *426*, 255–260. [CrossRef]
21. Scarpella, E.; Barkoulas, M.; Tsiantis, M. Control of Leaf and Vein Development by Auxin. *Cold Spr. Harb. Persp. Biol.* **2010**, *2*, a001511. [CrossRef]
22. Barkoulas, M.; Hay, A.; Kougioumoutzi, E.; Tsiantis, M. A Developmental Framework for Dissected Leaf Formation in the Arabidopsis Relative. *Cardamine Hirsute. Nat. Gen.* **2008**, *40*, 1136–1141. [CrossRef] [PubMed]
23. Koenig, D.; Bayer, E.; Kang, J.; Kuhlemeier, C.; Sinha, N. Auxin Patterns in Solanum lycopersicum Leaf Morphogenesis. *Development* **2009**, *136*, 2997–3006. [CrossRef] [PubMed]
24. Bayer, E.; Smith, R.; Mandel, T.; Nakayama, N.; Sauer, M.; Prusinkiewicz, P.; Kuhlemeier, C. Integration of Transport-Based Models for Phyllotaxis and Midvein Formation. *Genes Dev.* **2009**, *23*, 373–384. [CrossRef] [PubMed]
25. Mattsson, J.; Ckurshumova, W.; Berleth, T. Auxin Signaling in Arabidopsis Leaf Vascular Development. *Plant Physiol.* **2003**, *131*, 1327–1339. [CrossRef] [PubMed]
26. Mattsson, J.; Sung, Z.R.; Berleth, T. Responses of Plant Vascular Systems to Auxin Transport Inhibition. *Development* **1999**, *126*, 2979–2991. [CrossRef]
27. Yang, Z.; Xia, J.; Hong, J.; Zhang, C.; Wei, H.; Ying, W.; Sun, C.; Sun, L.; Mao, Y.; Gao, Y.; et al. Structural Insights into Auxin Recognition and Efflux by Arabidopsis PIN1. *Nature* **2022**, *609*, 611–618. [CrossRef]
28. Verna, C.; Janani Ravichandran, S.; Sawchuk, M.G.; Linh, N.M.; Scarpella, E. Coordination of Tissue Cell Polarity by Auxin Transport and Signaling. *eLife* **2019**, *8*, e51061. [CrossRef]
29. Avasarala, S.; Yang, J.; Caruso, J.L. Production of Phenocopies of the Lanceolate Mutant in Tomato Using Polar Auxin Transport Inhibitors. *J. Exp. Bot.* **1996**, *47*, 709–712. [CrossRef]
30. Wang, H.; Jones, B.; Li, Z.; Frasse, P.; Delanlande, C.; Regad, F.; Chaabouni, S.; Latche, A.; Pech, J.C.; Bouzayen, M. The Tomato Aux/IAA Transcription Factor IAA9 is Involved in Fruit Development and Leaf Morphogenesis. *Plant Physiol.* **2005**, *146*, 1759–1772. [CrossRef]
31. DeMason, D.A.; Chawla, R. Roles for Auxin during Morphogenesis of the Compound Leaves of Pea (Pisum sativum). *Planta* **2004**, *218*, 435–448. [CrossRef]
32. Zeng, L.; Zhang, N.; Zhang, Q.; Endress, P.K.; Huang, J.; Ma, H. Resolution of Deep Eudicot Phylogeny and Their Temporal Diversification using Nuclear Genes from Transcriptomic and Genomic Datasets. *New Phytol.* **2017**, *214*, 1338–1354. [CrossRef] [PubMed]
33. Bennett, T.; Liu, M.M.; Aoyama, T.; Bierfreund, N.M.; Braun, M.; Coudert, Y.; Dennis, R.J.; O'Connor, D.; Wang, X.Y.; White, C.D.; et al. Plasma Membrane-Targeted PIN Proteins Drive Shoot Development in a Moss. *Curr. Biol.* **2014**, *24*, 2776–2785. [CrossRef] [PubMed]
34. Bilsborough, G.D.; Runions, A.; Barkoulas, M.; Jenkins, H.W.; Hasson, A.; Galinha, C.; Laufs, P.; Hay, A.; Prusinkiewicz, P.; Tsiantis, M. Model for the Regulation of Arabidopsis thaliana Leaf Margin Development. *Proc. Natl. Acad. Sci. USA* **2011**, *108*, 3424–3429. [CrossRef] [PubMed]
35. Holloway, D.M.; Wenzel, C.L. Polar Auxin Transport Dynamics of Primary and Secondary Vein Patterning in Dicot Leaves. *Silico Plants* **2021**, *3*, diab030. [CrossRef]
36. Ravichandran, S.J.; Linh, N.M.; Scarpella, E. The Canalization Hypothesis—Challenges and Alternatives. *New Phytol.* **2020**, *227*, 1051–1059. [CrossRef]
37. Linh, N.M.; Scarpella, E. Leaf Vein Patterning is Regulated by the Aperture of Plasmodesmata Intercellular Channels. *PLoS Biol.* **2022**, *20*, e3001781.
38. Lincoln, C.; Britton, J.H.; Estelle, M. Growth and Development of the axr1 Mutants of Arabidopsis. *Plant Cell* **1990**, *2*, 1071–1080.

39. Ulmasov, T.; Murfett, J.; Hagen, G.; Guilfoyle, T.J. Aux/IAA Proteins Repress Expression of Reporter Genes Containing Natural and Highly Active Synthetic Auxin Response Elements. *Plant Cell* **1997**, *9*, 1963–1971.
40. Luschnig, C.; Gaxiola, R.A.; Grisafi, P.; Fink, G.R. EIR1, a Root-Specific Protein Involved in Auxin Transport, is Required for Gravitropism in Arabidopsis thaliana. *Genes Dev.* **1998**, *12*, 2175–2187. [CrossRef]
41. Park, S.H.; Morris, J.L.; Park, J.E.; Hirschi, K.D.; Smith, R.H. Efficient and Genotype-Independent Agrobacterium-Mediated Tomato Transformation. *J. Plant Physiol.* **2003**, *160*, 1253–1257. [CrossRef]
42. Merks, R.M.; Guravage, M.; Inzé, D.; Beemster, G.T. VirtualLeaf: An Open-Source Framework for Cell-Based Modeling of Plant Tissue Growth and Development. *Plant Physiol.* **2011**, *155*, 656–666. [CrossRef] [PubMed]
43. Cieslak, M.; Runions, A.; Prusinkiewicz, P. Auxin-Driven Patterning with Unidirectional Fluxes. *J. Exp. Bot.* **2015**, *66*, 5083–5102. [CrossRef] [PubMed]
44. Antonovici, C.C.; Peerdeman, G.Y.; Wolff, H.B.; Merks, R.M.H. Modeling Plant Tissue Development using VirtualLeaf. In *Methods in Molecular Biology*; Lucas, M., Ed.; Humana: New York, NY, USA, 2022; pp. 165–198. [CrossRef]
45. Merks, R.M.H.; Guravage, M.A. Building Simulation Models of Developing Plant Organs using VirtualLeaf. In *Methods in Molecular Biology*; Humana Press: Totowa, NJ, USA, 2012; pp. 333–352. [CrossRef]
46. Merks, R.M.; Van de Peer, Y.; Inzé, D.; Beemster, G.T. Canalization Without Flux Sensors: A Traveling-Wave Hypothesis. *Trends Plant Sci.* **2007**, *12*, 384–390. [CrossRef] [PubMed]

plants

Article

Impact of External Sources of Indole Acetic Acid and 2,3,5-Triiodobenzoic Acid on Alkaloid Production and Their Relationships with Primary Metabolism and Antioxidant Activity in *Annona emarginata* (Schltdl.) H. Rainer

Bruna Cavinatti Martin [1], Ivan De-la-Cruz-Chacón [2], Carolina Ovile Mimi [1,*], Carmen Silvia Fernandes Boaro [1], Felipe Girotto Campos [1], Inara Regiane Moreira-Coneglian [1] and Gisela Ferreira [1]

1 Department of Biodiversity and Biostatistics, Institute of Biosciences, São Paulo State University (UNESP), Prof. Dr. Antônio Celso Wagner Zanin Street, 250, Botucatu 18618-689, SP, Brazil; bruna.cavinatti@unesp.br (B.C.M.); carmen.boaro@unesp.br (C.S.F.B.); felipe.girotto@unesp.br (F.G.C.); inara.moreira@unesp.br (I.R.M.-C.); gisela.ferreira@unesp.br (G.F.)

2 Laboratorio de Fisiología y Química Vegetal, Instituto de Ciencias Biológicas, Universidad de Ciencias y Artes de Chiapas, Libramiento Norte-Poniente 1150, Tuxtla Gutiérrez 29039, Chiapas, Mexico; ivan.cruz@unicach.mx

* Correspondence: carolina.mimi@unesp.br

Abstract: *Annona emarginata* is a native Brazilian species capable of producing at least ten alkaloids of ecological, agronomic, and pharmacological importance. Some studies have explored the effect of external phytoregulators on the production of alkaloids, including the effect of auxins, which, like alkaloids, derive from the shikimic acid pathway. Thus, this study aimed to evaluate how indole acetic acid (IAA) and its inhibitor 2,3,5-triiodobenzoic acid (TIBA) impact the production of alkaloids and the primary metabolism of *A. emarginata*, which brings advances in the understanding of the mechanisms of alkaloid synthesis and can aid in the bioprospection of molecules of interest in Annonaceae. The design was completely randomized, with three treatments (control, IAA [10^{-6} M] and TIBA [10^{-6} M]) and five collection times (12, 36, 84, 156, and 324 h). The following variables were analyzed: total alkaloids, alkaloid profile, nitrate reductase activity, gas exchange in photosynthesis, chlorophyll *a* fluorescence, sugars, starch, and antioxidant activity. Of the twelve alkaloids analyzed, discretine and xylopine were not detected in the control plants; however, both were detected when IAA was applied (in roots and leaves) and xylopine (in roots) when the inhibitor was applied. The alkaloid asimilobine was not detected with the use of TIBA. Variations in alkaloid concentrations occurred in a punctual manner, without significant variations in photosynthesis and nitrate reductase activity, but with variations in the antioxidant system and sugar concentrations, mainly at 156 h, when the highest alkaloid concentrations were observed with the use of TIBA. It could be concluded that IAA is capable of selectively modulating the production of alkaloids in *A. emarginata*, either due to an external source or by the application of its inhibitor (TIBA).

Keywords: Annonaceae; IAA; TIBA; plant regulators; nitrogenous secondary metabolites

1. Introduction

Specialized metabolism differs in the different groups of species and is dependent on environmental conditions, being directly related to primary metabolism, which is common to all plants and provides precursor molecules for the development of specialized metabolism [1–3]. Among specialized metabolites, alkaloids stand out due to their role in the adaptation and survival of plants to various biotic and abiotic factors, constituting a nitrogen reserve and acting in plant defense [4]. In addition, they are bioactive compounds of medical, pharmaceutical, and agronomic interest due to their antifungal, bactericidal, antiprotozoal, cytotoxic, analgesic, and antiplatelet action [5–9].

Several abiotic factors, including hormonal balance, are capable of altering primary metabolism and triggering changes in specialized metabolism [10,11]. Among these plant regulators are auxins, which cause changes in primary metabolism, such as increased photosynthetic rates [12–14], and are associated with growth regulation, development, plant defense [15,16] and modulation of specialized metabolism [11,17–20].

Auxin and alkaloids compete for substrate because they share part of the same biosynthetic pathway, the shikimic acid pathway, mainly the tryptophan pathway, which produces indole alkaloids [12,21]. Thus, the application of auxins can interfere with the synthesis of alkaloids in a positive [11] or negative [22] manner, which may be related to the increase in the internal CO_2 concentration [23] and variations in the synthesis of sugars, which highlights the relationship between primary and specialized metabolism. Such metabolic modifications increase the production of reactive oxygen species, which implies the action of the antioxidant system for stress control [24].

The Annonaceae family is of interest in the study of specialized metabolism due to its broad biodiversity and chemodiversity. Alkaloids stand out, with approximately 934 of them having been documented in the family, distributed in 254 species [25]. In Brazil, the Annonaceae family is represented by approximately 389 species belonging to 32 genera [26], among them the genus *Annona*, with 80 species, 24 of which are endemic [27]. The most abundant alkaloids are benzylisoquinolines derived from amino acids phenylalanine and tyrosine; however, indole alkaloids derived from tryptophan have also been reported [25].

Annona emarginata (Schltdl.) H. Rainer 'Terra-Fria' morphotype is native to the South American continent [28], and in Brazil, it has importance in the recovery of degraded areas and in fruit growing, being used as rootstock for the commercial species atemoya (*Annona atemoya* Mabb.) [8,29]. Alkaloids such as liriodenine, anonaine, reticuline, and asimilobine have been identified in this species [8,11,30–32]. Some of these alkaloids are of pharmacological [5,6,9] and agricultural interest, with action in the control of phytopathogenic fungi [6,8].

Therefore, the diversity and importance of alkaloids in the Annonaceae family stimulate the understanding of factors that may impact their biosynthesis. In this context, considering that there are factors, such as auxins, that act in both primary and specialized metabolism, this work was carried out. The novelty is the study not only of the impact of auxin IAA but also of an auxin transport inhibitor (TIBA) on alkaloid biosynthesis and the possible relationships of these compounds in the availability of resources from primary metabolism [31], in addition to the action on the antioxidant system, indicating variations in the stress levels of processes. Thus, this study aimed to evaluate how indole acetic acid (IAA) and its transport inhibitor (2,3,5-triiodobenzoic acid—TIBA) impact the production of alkaloids and the relationships with primary metabolism (photosynthesis, sugar production), nitrate reductase, and antioxidant activity.

2. Results

Effect on alkaloid production. The highest concentrations of total alkaloids were detected in the roots of *Annona emarginata* regardless of treatment. In the leaves of plants that received IAA, an increase in total alkaloids was observed in relation to the control only in the last evaluation (324 h). In contrast, the application of TIBA, an auxin transport inhibitor, caused an increase in the concentrations of total alkaloids, mainly in the last three evaluations (84, 156, and 324 h). The roots of plants that received IAA or TIBA did not present differences in the amounts of total alkaloids in relation to control in most evaluation times (12, 36, 84, and 324 h). However, at 156 h after treatments, the highest concentration of total alkaloids was observed in roots with the use of TIBA (108.03 $\mu g\ g^{-1}$ dry mass), differing from control (78.91 $\mu g\ g^{-1}$ dry mass) and IAA (57.18 $\mu g\ g^{-1}$ dry mass) (Table 1).

Table 1. Concentration of total alkaloids (μg g^{-1} dry mass) in the leaves and roots of *Annona emarginata* in each treatment (Control, indole acetic acid—IAA and its transport inhibitor [2,3,5-triiodobenzoic acid—TIBA]), at different collection times (12 h, 36 h, 84 h, 156 h, and 324 h after the application of treatments).

	Total Alkaloids (μg g^{-1} Dry Mass)				
	12 h	**36 h**	**84 h**	**156 h**	**324 h**
			Leaves		
Control	2.69 ± 0.23 Cb*[1]	9.49 ± 1.46 Aa	5.81 ± 0.72 Bb	3.12 ± 0.8 Cb	0.917 ± 0.31 Dc
IAA	3.19 ± 0.94 Aab	2.83 ± 0.21 Ac	0.90 ± 0.08 Bc	3.22 ± 0.77 Ab	2.32 ± 0.43 Ab
TIBA	4.31 ± 1.63 Ba	7.43 ± 1.97 Bb	9.67 ± 2.01 Aba	10.99 ± 2.17 Aa	11.55 ± 3.04 Aa
Treatments *p*: 0.0001; f: 68.219; Hours: *p* 0.0001; f: 37.935; Treatments×Hours *p*: 0.0001; f: 74.589					
			Roots		
Control	66.07 ± 2.78 Aa	66.62 ± 3.27 Aab	76.87 ± 4.11 Aa	78.91 ± 3.65 Ab	77.44 ± 4.27 Aab
IAA	55.43 ± 3.05 Aa	37.73 ± 2.55 Bb	72.39 ± 4.14 Aa	57.18 ± 2.87 Ac	68.16 ± 3.32 Ab
TIBA	71.02 ± 3.67 Ba	70.83 ± 2.81 Ba	74.69 ± 5.66 Ba	108.03 ± 9.12 Aa	91.50 ± 6.96 ABa
Treatments *p*: 0.0001; f: 28.246; Hours: *p* 0.0001; f: 10.394; Treatments×Hours *p*: 0.0002; f: 3.455					

*[1] Results are presented as mean ± standard deviation. Means followed by the same uppercase letter in the row and lowercase letter in the column do not show significant differences according to the Tukey test.

The application of IAA and its transport inhibitor (TIBA) also caused changes in the presence and absence of the 12 investigated alkaloids in *Annona emarginata* (Figure 1), both in leaves and roots (Table 2).

Among the 12 alkaloids under investigation, it was not possible to detect the presence of any of them in the leaves of the control group. However, when IAA was applied, it was possible to detect lanuginosine, liriodenine, and xylopine. Lanuginosine, liriodenine, and *N*-methyllaurotetanine were detected with the application of TIBA, which indicates possible selectivity in relation to the synthesis of Xylopine and *N*-methylllaurotetanine depending on the treatment.

In the roots of control plants, alkaloids discretine and xylopine were not observed, but asimilobine, lanuginosine, laurotetanine, liriodenine, *N*-methyllaurotetanine, norglaucine, norpredicentine, oxoglaucine, reticuline, and xylopinine were observed. On the other hand, when IAA was applied, oxoglaucine and reticuline were not detected. When auxin transport was inhibited, oxoglaucine, reticuline, asimilobine, and discretine were not found.

Regarding the nitrate reductase activity, it was found that there was only a change due to the application of IAA and TIBA in relation to control at 324 h, which demonstrates that nitrogen assimilation was not altered with treatments over time nor at the time of greatest alkaloid synthesis (156 h); therefore, it is not possible to conclude that there was an increase in nitrogen availability for alkaloid synthesis (Table 3).

Effect on primary metabolism. Gas exchange and chlorophyll *a* fluorescence were impacted by treatments in a specific manner. When TIBA (IAA transport inhibitor) was applied at 12 h, 156 h, and 324 h, a reduction in transpiration was observed when compared to plants that received IAA, differing from control only at 324 h (Table 4). Some moments in which reductions in transpiration occurred coincided with greater production of total alkaloids in leaves (156 h and 324 h) and roots (156 h), in addition to lower stomatal conductance at 156 h. However, during these periods of higher concentrations of total alkaloids in plants treated with TIBA, no differences in relation to the potential quantum yield of photosystem II (Fv′/Fm′), specific quantum yield of PSII (yield), photochemical dissipation (qP), and net CO_2 assimilation rate (Anet) were observed.

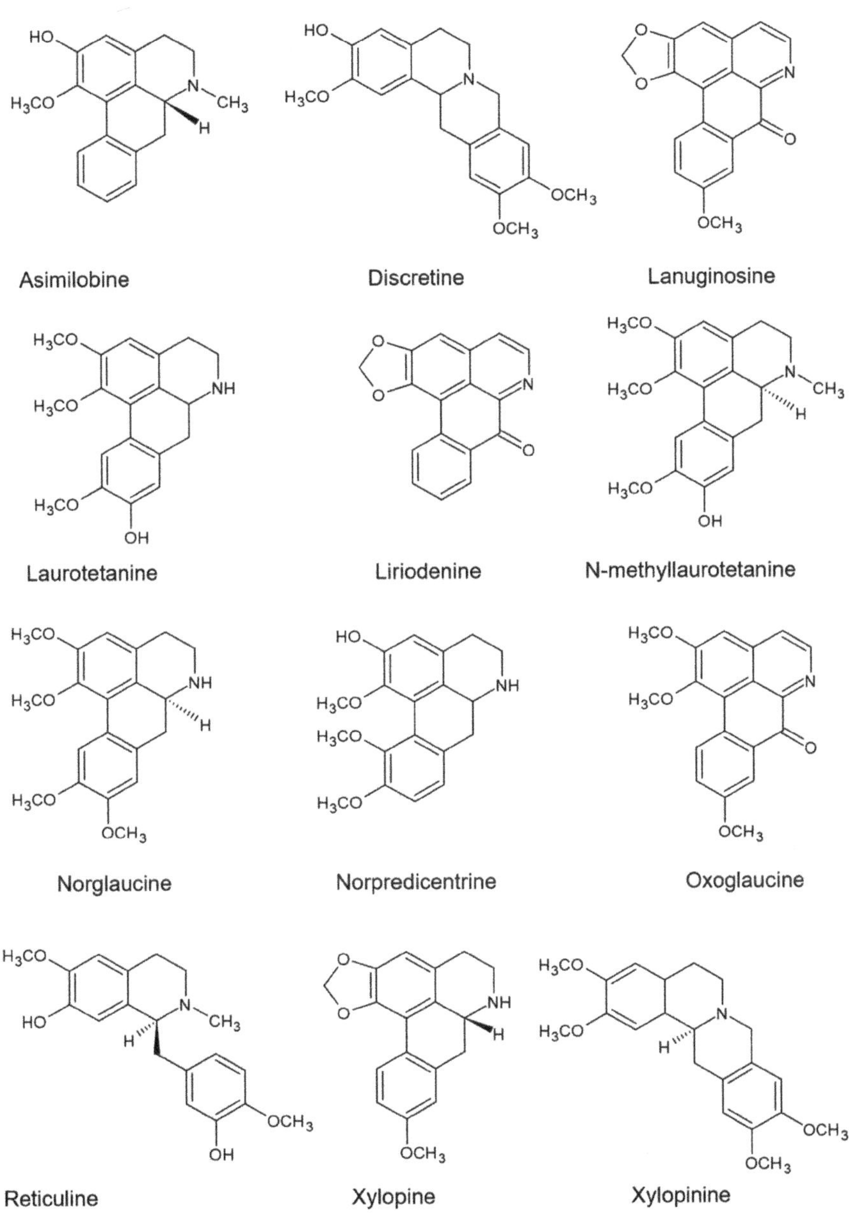

Figure 1. Alkaloids detected in *Annona emarginata* using high-performance liquid chromatograph (UHPLC).

Table 2. Presence (x) and absence (-) of alkaloids in leaves and roots of *Annona emarginata* after the application of treatments (Control, indole acetic acid IAA and its transport inhibitor (2,3,5-triiodobenzoic acid—TIBA)) at five collection times (12 h, 36 h, 84 h, 156 h, and 324 h after the application of treatments).

	Leaves (L)/Roots (R)				
	12 h (L/R)	**36 h (L/R)**	**84 h (L/R)**	**156 h (L/R)**	**324 h (L/R)**
		Control			
Asimilobine	-/-	-/x	-/x	-/-	-/-
Discretine	-/-	-/-	-/-	-/-	-/-
Lanuginosine	-/x	-/x	-/x	-/x	-/x
Laurotetanine	-/-	-/x	-/x	-/x	-/x
Liriodenine	-/x	-/x	-/x	-/x	-/x
N-methyllaurotetanine	-/x	-/x	-/x	-/x	-/x
Norglaucine	-/x	-/x	-/x	-/x	-/x
Norpredicentine	-/-	-/x	-/x	-/-	-/x
Oxoglaucine	-/-	-/x	-/-	-/-	-/-
Reticuline	-/x	-/x	-/x	-/x	-/x
Xylopine	-/-	-/-	-/-	-/-	-/-
Xylopinine	-/x	-/x	-/x	-/x	-/x
		IAA			
Asimilobine	-/-	-/-	-/-	-/x	-/-
Discretine	-/-	-/-	-/-	-/x	-/-
Lanuginosine	-/x	-/x	x/x	x/x	-/x
Laurotetanine	-/-	-/x	-/-	-/x	-/-
Liriodenine	-/x	x/x	x/x	x/x	-/x
N-methyllaurotetanine	-/x	-/x	-/x	-/x	-/x
Norglaucine	-/x	-/x	-/x	-/x	-/x
Norpredicentine	-/-	-/x	-/-	-/-	-/-
Oxoglaucine	-/-	-/-	-/-	-/-	-/-
Reticuline	-/-	-/-	-/-	-/-	-/-
Xylopine	-/-	x/x	-/x	-/-	-/x
Xylopinine	-/x	-/x	-/-	-/x	-/-
		TIBA			
Asimilobine	-/-	-/-	-/-	-/-	-/-
Discretine	-/-	-/-	-/-	-/-	-/-
Lanuginosine	-/x	-/x	x/x	x/x	-/x
Laurotetanine	-/-	-/x	-/-	-/x	-/x
Liriodenine	-/x	x/x	x/x	x/x	-/x
N-methyllaurotetanine	-/x	x/x	-/x	x/x	-/x
Norglaucine	-/x	-/x	-/x	-/x	-/x
Norpredicentine	-/-	-/-	-/-	-/-	-/x
Oxoglaucine	-/-	-/-	-/-	-/-	-/-
Reticuline	-/-	-/-	-/-	-/-	-/-
Xylopine	-/-	-/x	-/x	-/x	-/x
Xylopinine	-/-	-/x	-/-	-/x	-/-

"-" symbolizes the absence and "X" the presence of each alkaloid in the row by time in the column.

Table 3. Nitrate reductase activity (μg N-NO_2 g of MF^{-1} h^{-1}) in *Annona emarginata* leaves at each collection time after the application of treatments (Control, indole acetic acid IAA and its transport inhibitor [2,3,5-triiodobenzoic acid—TIBA]).

Nitrate Reductase (μg N-NO_2 g de MF^{-1} h^{-1})	12 h	36 h	84 h	156 h	324 h
Control	3.1665 ± 0.195 Ba*[1]	2.9165 ± 0.108 Ba	2.6959 ± 0.102 Ba	3.2196 ± 0.220 Ba	4.1528 ± 0.385 Aa
IAA	3.2182 ± 0.034 Aa	2.8571 ± 0.133 Aa	3.1928 ± 0.140 Aa	2.8261 ± 0.204 Aa	3.4460 ± 0.200 Ab
TIBA	3.4396 ± 0.244 Aa	3.0666 ± 0.148 Aa	3.0943 ± 0.152 Aa	3.0779 ± 0.109 Aa	3.2224 ± 0.112 Ab
Treatments *p*: 0.5742; f: 0.56; Hours: *p* 0.0002; f: 6.66; Treatments×Hours *p*: 0.014; f: 2.66					

*[1] Results are presented as mean ± standard deviation. Means followed by the same uppercase letter in the row and lowercase letter in the column do not present significant differences between them by the Tukey test.

Table 4. Gas exchange and chlorophyll *a* fluorescence evaluated after the application of treatments (IAA, TIBA, and control): potential quantum yield of photosystem II (Fv′/Fm′), specific quantum yield of PSII (yield), photochemical dissipation (qP), transpiration (mmol water vapor m^{-2} s^{-1}), net CO_2 assimilation rate [Anet] (μmol CO_2 m^{-2} s^{-1}), and stomatal conductance (gs) (mol m^{-2} s^{-1}).

Treatments	Fv′/Fm′	Yield	qP	Transpiration	Anet	Gs
			12 h			
Control	0.423 ± 0.089 AB*[1]	0.063 ± 0.008 A	0.156 ± 0.021 A	0.794 ± 0.102 AB	6.01 ± 1.97 AB	34.9 ± 5.9 AB
IAA	0.467 ± 0.076 A	0.081 ± 0.13 A	0.179 ± 0.067 A	1.281 ± 0.121 A	8.62 ± 2.01 A	60.8 ± 4.3 A
TIBA	0.317 ± 0.059 B	0.067 ± 0.007 A	0.2090.055 A	0.587 ± 0.097 B	3.99 ± 1.02 B	24.2 ± 4.6 B
	p: 0.0001 f: 22.019	*p*: 0.0001 f: 30.118	*p*: 0.0001 f: 20.342	*p*: 0.0002 f: 25.556	*p*: 0.0001 f: 20.488	*p*: 0.0002 f: 10.733
			36 h			
Control	0.434 ± 0.88 A	0.083 ± 0.01 A	0.194 ± 0.032 A	1.040 ± 0.2 A	8.28 ± 2.97 A	46.2 ± 6 A
IAA	0.365 ± 0.37 A	0.084 ± 0.008 A	0.225 ± 0.027 A	0.872 ± 0.186 A	7.94 ± 2.03 AB	39.5 ± 5.1 A
TIBA	0.358 ± 0.69 A	0.064 ± 0.005 A	0.187 ± 0.054 A	0.577 ± 0.099 A	5.24 ± 0.82 B	24.6 ± 4.5 A
	p: 0.0001 f: 13.441	*p*: 0.0003 f: 23.643	*p*: 0.0002 f: 24.559	*p*: 0.0005 f: 31.497	*p*: 0.0001 f: 22.362	*p*: 0.0005 f: 25.461
			156 h			
Control	0.509 ± 0.48 A	0.177 ± 0.025 A	0.352 ± 0.087 A	1.622 ± 0.182 AB	6.49 ± 1.8 A	63.7 ± 4.8 A
IAA	0.500 ± 0.67 A	0.180 ± 0.068 A	0.366 ± 0.075 A	1.819 ± 0.177 A	7.00 ± 1.81 A	72.5 ± 7.1 A
TIBA	0.444 ± 0.55 A	0.183 ± 0.045 A	0.411 ± 0.054 A	1.259 ± 0.098 B	6.78 ± 2 A	49.1 ± 6.7 B
	p: 0.0001 f: 10.231	*p*: 0.0001 f: 20.226	*p*: 0.0005 f: 26.768	*p*: 0.0002 f: 10.211	*p*: 0.0005 f: 20.389	*p*: 0.0001 f: 42.435
			324 h			
Control	0.450 ± 0.066 A	0.068 ± 0.012 A	0.159 ± 0.054 A	0.903 ± 0.134 A	7.59 ± 2.26 A	55.9 ± 5.4 A
IAA	0.474 ± 0.78 A	0.081 ± 0.009 A	0.172 ± 0.047 A	1.008 ± 0.122 A	9.24 ± 3.23 A	59.9 ± 4.9 A
TIBA	0.418 ± 0.098 A	0.071 ± 0.023 A	0.171 ± 0.074 A	0.694 ± 0.078 B	6.56 ± 1.87 A	45.6 ± 4.3 A
	p: 0.0005 f: 21.367	*p*: 0.0001 f: 13.793	*p*: 0.0002 f: 10.245	*p*: 0.0001 f: 18.742	*p*: 0.0003 f: 30.277	*p*: 0.0003 f: 61.223

*[1] Results are presented as mean ± standard deviation. Means followed by the same uppercase letter in the row and lowercase letter in the column do not show significant differences between them according to the Tukey test. IAA: indole acetic acid. TIBA (2,3,5-triiodobenzoic acid) is an inhibitor of IAA transport.

As part of the primary metabolism responses to IAA treatment, an increase in the concentration of total sugars was observed in leaves at 36 h and a reduction in roots, and when auxin transport was inhibited, a reduction in total sugars in leaves and an increase in roots was observed at 156 h, which coincided, at this time, with the highest concentrations of total alkaloids in roots and leaves. This increase in the concentration of total sugars in roots may indicate that there was greater translocation to this organ, the site of greatest alkaloid synthesis (Table 5).

Table 5. Concentration of total sugars (μg g^{-1} FW), reducing sugars (μg g^{-1} FW), starch (μg g^{-1} FW), and sucrose (μg g^{-1} FW) in the leaves and roots of *Annona emarginata* after the application of treatments with plant growth regulators at each collection time.

			Total sugars		
			Leaves		
Treatments	12 h	36 h	84 h	156 h	324 h
Control	1268.9 ± 322.1 Aa[*1]	1305.3 ± 245.2 Ab	1307.5 ± 245.5 Aa	1356.1 ± 343.8 Aa	1468.7 ± 321.5 Aa
IAA	1272.1 ± 298.3 Ba	1830.5 ± 356.6 Aa	1411.8 ± 365.4 Ba	1431.5 ± 278.9 Ba	1383.0 ± 289.2 Ba
TIBA	1191.3 ± 336.5 Ba	1279.5 ± 254.7 Ab	1486.2 ± 321.3 Aa	1206.5 ± 311.4 Bb	1568.7 ± 231.2 Aa
Treatments *p*: 0.0001; f: 31.22; Hours: *p* 0.0002; f: 23.11; Treatments×Hours *p*: 0.0002; f: 12.67					
			Roots		
Treatments	12 h	36 h	84 h	156 h	324 h
Control	269.8 ± 32.1 Cb	466.6 ± 52 Aa	333.7 ± 37.8 Ba	377.0 ± 44.5 Bb	362.7 ± 47.6 Ba
IAA	355.8 ± 21.7 Aa	349.7 ± 39.7 Ab	304.9 ± 40.5 Aa	379.3 ± 39.1 Ab	356.5 ± 36.7 Aa
TIBA	305.2 ± 45 Bab	426.1 ± 65.2 Aba	360.3 ± 32.7 Ba	478.3 ± 58.9 Aa	354.6 ± 34.1 ABa
Treatments *p*: 0.0001; f: 21.99; Hours: *p* 0.0001; f: 10.03; Treatments×Hours *p*: 0.0005; f: 32.45					
			Reducing sugars		
			Leaves		
Treatments	12 h	36 h	84 h	156 h	324 h
Control	1.15 ± 0.06 Aa	0.92 ± 0.07 Bb	0.94 ± 0.08 Ba	0.82 ± 0.1 Ba	0.79 ± 0.05 Bab
IAA	0.75 ± 0.02 Bb	1.06 ± 0.12 Aa	1.05 ± 0.09 Aa	0.77 ± 0.08 Ba	0.71 ± 0.07 Bb
TIBA	0.84 ± 0.03 Ab	0.84 ± 0.05 Ab	0.79 ± 0.03 Ab	0.85 ± 0.06 Aa	0.91 ± 0.1 Aa
Treatments *p*: 0.0001; f: 13.92; Hours: *p* 0.0001; f: 23.79; Treatments×Hours *p*: 0.0003; f: 12.86					
			Roots		
Treatments	12 h	36 h	84 h	156 h	324 h
Control	0.772 ± 0.12 Ab	0.780 ± 0.14 Ab	0.712 ± 0.23 Ab	0.728 ± 0.21 Ac	0.692 ± 0.14 Ab
IAA	1.191 ± 0.2 Aa	1.159 ± 0.21 Aba	1.085 ± 0.3 ABa	1.165 ± 0.32 ABa	1.06 ± 0.22 Ba
TIBA	0.531 ± 0.16 Bc	0.538 ± 0.1 Bc	0.598 ± 0.19 Bc	0.834 ± 0.31 Ab	0.582 ± 0.29 Bc
Treatments *p*: 0.0002; f: 25.36; Hours: *p* 0.0001; f: 17.64; Treatments×Hours *p*: 0.0001; f: 21.77					
			Starch		
			Leaves		
Treatments	12 h	36 h	84 h	156 h	324 h
Control	39.3 ± 5.3 Aba	43.4 ± 6.1 Aa	24.6 ± 2.1 Bb	35.4 ± 5.1 ABb	46.4 ± 4.8 Aa
IAA	31.6 ± 4.6 Bb	29.5 ± 4.2 Bb	46.2 ± 4.7 Aa	47.4 ± 6.2 Aa	22.5 ± 3.7 Cc
TIBA	22.1 ± 3.7 Cc	19.2 ± 2.6 Cc	28.4 ± 2.9 Bb	37.2 ± 3.3 Ab	31.3 ± 3.9 Bb
Treatments *p*: 0.0003; f: 27.37; Hours: *p* 0.0001; f: 10.55; Treatments×Hours *p*: 0.0002; f: 18.87					
			Roots		
Treatments	12 h	36 h	84 h	156 h	324 h
Control	7.61 ± 3.5 Aa	9.71 ± 5.2 Ab	7.43 ± 3.7 Aa	9.23 ± 3.6 Ab	8.78 ± 2.8 Aa
IAA	7.78 ± 4.1 Ba	14.87 ± 6.1 Aa	7.94 ± 3.2 Ba	14.23 ± 5.7 Aa	8.36 ± 3.1 Ba
TIBA	8.29 ± 3.9 Ba	13.49 ± 5.9 Aa	9.03 ± 4.1 Ba	9.41 ± 4.3 Bb	6.62 ± 2.2 Ca
Treatments *p*: 0.0002; f: 40.33; Hours: *p* 0.0001; f: 22.49; Treatments×Hours *p*: 0.0003; f: 12.28					

Table 5. *Cont.*

Sucrose					
Leaves					
Treatments	12 h	36 h	84 h	156 h	324 h
Control	33.57 ± 4.3 Aa	34.82 ± 3.8 Aa	34.31 ± 4.5 Aa	35.46 ± 3.1 Aa	33.59 ± 2.9 Aa
IAA	33.71 ± 3.9 Aa	33.33 ± 3.2 Aa	33.11 ± 5.1 Aa	34.70 ± 4.2 Aa	33.41 ± 3.5 Aa
TIBA	29.12 ± 2.8 Ab	29.48 ± 2.4 Ab	29.26 ± 1.9 Ab	29.75 ± 3 Ab	30.47 ± 2.6 Ab
Treatments *p*: 0.0005; f: 30.29; Hours: *p* 0.0005; f: 34.78; Treatments×Hours *p*: 0.0002; f: 9.54					
Roots					
Treatments	12 h	36 h	84 h	156 h	324 h
Control	25.7 ± 4.2 Ba	29.1 ± 2.9 Aa	26.7 ± 3.7 ABa	26.7 ± 4 ABa	24.0 ± 2.8 Ba
IAA	26.6 ± 4.8 Aa	27.2 ± 2.9 Aab	27.1 ± 2.8 Aa	26.6 ± 3.5 Aa	25.4 ± 2.2 Aa
TIBA	23.1 ± 3.9 Bb	25.3 ± 3.2 ABb	25.0 ± 3.1 ABa	26.3 ± 5.1 Aa	25.5 ± 3 Aba
Treatments *p*: 0.0001; f: 32.3; Hours: *p* 0.0001; f: 14.57; Treatments×Hours *p*: 0.0002; f: 19.71					

*[1] Results are presented as mean ± standard deviation. Means followed by the same capital letter in the row and lowercase letter in the column do not present significant differences between them by the Tukey test. IAA: indole acetic acid. TIBA (2,3,5-triiodobenzoic acid) IAA transport inhibitor.

Regarding sucrose in leaves, it was observed that the inhibition of IAA transport caused a reduction in the sucrose concentration in leaves at all collection times, both in relation to IAA application and in relation to control. However, in roots, the lowest sucrose concentrations with the use of TIBA were punctual in relation to IAA and control (12 h and 36 h) (Table 5).

Contrary to what was observed with sucrose in leaves, there was no response pattern regarding the reduction in the concentration of reducing sugars in plants with IAA transport inhibition. What was observed was a punctual reduction at 84 h and an increase in concentration at 324 h (when compared to plants that received IAA, but without differing from control), the time at which the highest concentrations of total alkaloids were observed in leaves treated with TIBA. However, it was possible to observe that the auxin inhibitor caused a reduction in the concentrations of reducing sugars in roots at all evaluation times, while IAA caused an increase in concentrations (Table 5).

Starch concentration in leaves at 84 h and 156 h was high after the application of IAA compared with plants treated with TIBA and control. At the other times (12 h, 36 h and 324 h), the auxin transport inhibitor caused a reduction in starch concentration in leaves. In roots, starch concentration was higher in plants treated with IAA only at 156 h (14.23 $\mu g\ g^{-1}$ FW) when compared to plants treated with the inhibitor (9.41 $\mu g\ g^{-1}$ FW) (Table 5). This lower value may be related to the increase in total sugars at 156 h and to the synthesis of alkaloids since there is a subsequent reduction in starch, total and reducing sugars in roots between 156 h and 324 h, and an increase in alkaloid concentration. Thus, the lowest sugar concentrations found in roots generally coincide with the periods of highest alkaloid concentrations, especially in roots, which may indicate that sugars were used as a source of energy and/or greater availability of carbon skeletons for the synthesis of alkaloids.

Effect on the antioxidant system. Hydrogen peroxide was not altered in leaves in relation to control, and in roots, IAA caused an increase in hydrogen peroxide concentrations at all evaluation times compared with TIBA inhibitor and control (Table 6).

Table 6. Analysis of antioxidant enzymes peroxidase (POD—μmol purpurogallin min^{-1} mg $prot^{-1}$), catalase (CAT—μKat $μg^{-1}$ protein) and superoxide dismutase (SOD—U $μg^{-1}$ protein) and reactive oxygen species: hydrogen peroxide (μmol H_2O_2 g^{-1} MF) and lipoperoxide (nmol g^{-1} MF) in leaves and roots of *Annona emarginata* after the application of treatments with regulators at each collection time.

Hydrogen Peroxide					
Leaves					
Treatments	12 h	36 h	84 h	156 h	324 h
Control	1.39 ± 0.29 Aa*[1]	1.37 ± 0.4 Aa	1.24 ± 0.3 Aab	1.19 ± 0.2 Aa	1.16 ± 0.2 Aa
IAA	1.21 ± 0.3 Aa	1.26 ± 0.3 Aa	1.30 ± 0.3 Aa	1.04 ± 0.3 Aa	1.08 ± 0.3 Aa
TIBA	1.28 ± 0.35 Aa	1.19 ± 0.4 ABa	1.03 ± 0.2 ABb	1.02 ± 0.2 ABa	0.97 ± 0.25 Ba
Treatments p: 0.0003; f: 27.322; Hours: p 0.0005; f: 13.224; Treatments × Hours p: 0.0001; f: 35.887					
Roots					
Treatments	12 h	36 h	84 h	156 h	324 h
Control	0.120 ± 0.03 Ab	0.134 ± 0.24 Ab	0.143 ± 0.23 Ab	0.138 ± 0.06 Ab	0.152 ± 0.024 Ab
IAA	0.218 ± 0.07 Aa	0.230 ± 0.56 Aa	0.417 ± 0.49 Aa	0.252 ± 0.4 Aa	0.174 ± 0.05 Aa
TIBA	0.088 ± 0.02 Ab	0.139 ± 0.45 Ab	0.090 ± 0.02 Ab	0.090 ± 0.002 Ab	0.084 ± 0.019 Ab
Treatments p: 0.0002; f: 21.134; Hours: p 0.0001; f: 29.624; Treatments × Hours p: 0.0002; f: 19.728					
Lipoperoxide					
Leaves					
Treatments	12 h	36 h	84 h	156 h	324 h
Control	11.4 ± 4.5 Aa	14.1 ± 5.6 Ab	9.9 ± 3 ABb	8.6 ± 2.7 Bb	8.3 ± 3.2 Bb
IAA	10.7 ± 3.9 Aa	12.7 ± 4.3 Ab	12.1 ± 4.2 Ab	10.5 ± 3 Aab	10.4 ± 3.8 Ab
TIBA	12.6 ± 5 Ca	22.8 ± 5.8 Aa	17.8 ± 5.1 Ba	12.7 ± 3.5 Ca	13.7 ± 4.2 Ca
Treatments p: 0.0001; f: 9.934; Hours: p 0.0001; f: 18.998; Treatments × Hours p: 0.0002; f: 30.609					
Roots					
Treatments	12 h	36 h	84 h	156 h	324 h
Control	6.33 ± 2 Bb	11.33 ± 3.1 Ab	9.44 ± 3.1 Ab	9.45 ± 3.2 Ac	8.80 ± 3 Aa
IAA	10.78 ± 3.2 Ca	19.43 ± 5 Ba	12.37 ± 4 Ca	27.60 ± 7.1 Aa	6.40 ± 1.7 Db
TIBA	7.72 ± 2.2 Bb	8.97 ± 2.9 Bc	8.93 ± 2.5 Bb	24.04 ± 7 Ab	10.23 ± 4.1 Ba
Treatments p: 0.0001; f: 18.293; Hours: p 0.0001; f: 14.576; Treatments × Hours p: 0.0002; f: 24.255					
Superoxide dismutase (SOD)					
Leaves					
Treatments	12 h	36 h	84 h	156 h	324 h
Control	176.0 ± 21 Ba	217.0 ± 32.1 Ba	490.6 ± 27.8 Aa	155.7 ± 13.6 Ba	119.3 ± 17.5 Ba
IAA	0.42 ± 0.12 Ab	0.16 ± 0.007 Ab	0.12 ± 0.003 Ab	0.06 ± 0.002 Ab	0.05 ± 0.001 Ab
TIBA	0.05 ± 0.007 Ab	0.04 ± 0.001 Ab	0.04 ± 0.009 Ab	0.04 ± 0.002 Ab	0.03 ± 0.001 Ab
Treatments p: 0.0001; f: 45.233; Hours: p 0.0003; f: 31.724; Treatments × Hours p: 0.0002; f: 20.776					
Roots					
Treatments	12 h	36 h	84 h	156 h	324 h
Control	39.8 ± 6.2 Ab	44.3 ± 6.1 Ab	40.3 ± 5.3 Ac	48.0 ± 6 Ab	48.7 ± 5.8 Ab
IAA	63.6 ± 8 Ba	76.1 ± 8.9 Aa	77.8 ± 7.2 Aa	76.9 ± 8.2 Aa	76.5 ± 8.6 Aa
TIBA	61.4 ± 8.5 Ba	68.7 ± 7.5 Ba	65.3 ± 7.1 Bb	80.8 ± 7.9 Aa	71.8 ± 8.1 ABa
Treatments p: 0.0001; f: 10.444; Hours: p 0.0001; f: 20.411; Treatments × Hours p: 0.0001; f: 5.981					

Table 6. *Cont.*

Catalase (CAT)					
Leaves					
Treatments	12 h	36 h	84 h	156 h	324 h
Control	0.0031 ± 0.001 Aa	0.0026 ± 0.001 Aa	0.0028 ± 0.0008 Aa	0.0011 ± 0.0001 Aa	0.0028 ± 0.0012 Aa
IAA	0.0014 ± 0.0007 Aa	0.0014 ± 0.001 Aa	0.0009 ± 0.0002 Aa	0.0008 ± 0.0002 Aa	0.0025 ± 0.0009 Aa
TIBA	0.0025 ± 0.001 Aa	0.0013 ± 0.001 Aa	0.0022 ± 0.001 Aa	0.0012 ± 0.0004 Aa	0.0019 ± 0.0007 Aa
Treatments *p*: 0.0005; f: 10.903; Hours: *p* 0.0001; f: 27.774; Treatments×Hours *p*: 0.0003; f: 12.356					
Roots					
Treatments	12 h	36 h	84 h	156 h	324 h
Control	0.00125 ± 0.00007 Aa	0.0010 ± 0.0002 Ab	0.0014 ± 0.0003 Aa	0.0011 ± 0.0005 Aa	0.0021 ± 0.0006 Aa
IAA	0.00249 ± 0.0012 Aa	0.0050 ± 0.001 Aa	0.0026 ± 0.0009 Aa	0.0024 ± 0.0008 Aa	0.0029 ± 0.0009 Aa
TIBA	0.00209 ± 0.0011 Aa	0.0015 ± 0.0003 Aab	0.0029 ± 0.0002 Aa	0.0015 ± 0.0003 Aa	0.0010 ± 0.0002 Aa
Treatments *p*: 0.0457; f: 3.25; Hours: *p* 0.8982; f: 0.27; Treatments×Hours *p*: 0.5585; f: 0.86					
Peroxidase					
Leaves					
Treatments	12 h	36 h	84 h	156 h	324 h
Control	0.420 ± 0.13 Ba	0.499 ± 0.2 ABa	0.275 ± 0.012 Ba	0.307 ± 0.15 Ba	1.460 ± 0.4 Aab
IAA	0.149 ± 0.05 Ba	0.870 ± 0.34 Ba	0.798 ± 0.3 Ba	0.525 ± 0.24 Ba	2.008 ± 0.6 Aa
TIBA	0.301 ± 0.23 Aa	0.373 ± 0.21 Aa	0.595 ± 0.22 Aa	0.335 ± 0.11 Aa	0.831 ± 0.3 Ab
Treatments *p*: 0.0001; f: 30.093; Hours: *p* 0.0001; f: 10.677; Treatments×Hours *p*: 0.0002; f: 21.589					
Roots					
Treatments	12 h	36 h	84 h	156 h	324 h
Control	0.222 ± 0.13 Ab	0.558 ± 0.2 Ab	0.481 ± 0.023 Ab	0.518 ± 0.16 Ab	0.739 ± 0.27 Ac
IAA	0.847 ± 0.24 Ba	1.439 ± 0.4 Aa	1.210 ± 0.48 ABa	1.149 ± 0.45 ABa	1.163 ± 0.5 Ab
TIBA	1.024 ± 0.42 Ba	0.840 ± 0.03 Bb	1.225 ± 0.5 Ba	1.108 ± 0.43 Ba	4.379 ± 1.1 Aa
Treatments *p*: 0.0001; f: 30.233; Hours: *p* 0.0001; f: 10.665; Treatments×Hours *p*: 0.0002; f: 10.801					

*[1] Results are presented as mean ± standard deviation. Means followed by the same capital letter in the row and lowercase in the column do not present significant differences between them according to the Tukey test. IAA: indole acetic acid. TIBA (2,3,5-triiodobenzoic acid) IAA transport inhibitor.

There was an increase in lipoperoxide concentrations in leaves at most evaluation times (except for 12 h) when auxin transport was inhibited. In contrast, in roots, the application of IAA caused an increase in lipoperoxide concentration at all evaluation times, except for 324 h, when a drastic reduction was observed, both in relation to TIBA and control. It is noteworthy that at 156 h, the highest lipoperoxide concentration was observed with IAA over time (27.60 nmol g^{-1} MF) in roots, as well as with the use of TIBA (24.04 nmol g^{-1} MF), with differences between both and control (9.45 nmol g^{-1} MF), and at this time, the highest concentration of total alkaloids was observed with the use of TIBA.

Regarding the activity of the antioxidant system, treatments caused a reduction in the superoxide dismutase (SOD) activity in leaves and an increase in roots. At 84 h, differences were observed in roots between treatments, with the highest activity due to the application of IAA (77.8 U μg^{-1} protein), followed by plants that received the transport inhibitor (65.3 U μg^{-1} protein) and control (40.3 U μg^{-1} protein).

No differences were observed in the catalase (CAT) activity in roots and leaves between treatments and control (Table 6).

Regarding the peroxidase (POD) activity in leaves, it was only possible to detect a change between treatments at 324 h, with greater activity due to the application of IAA, with no difference in relation to control. However, in roots, both treatments caused an increase in the POD activity in relation to control, except at 36 h, when only treatment with

IAA caused an increase in the POD activity and at 324 h, when treatment with TIBA caused the greatest increase (4.379), both in relation to IAA (1.163) and to control (0.739) (Table 6).

3. Discussion

The higher concentrations of total alkaloids found in *Annona emarginata* roots corroborate the reports for this and other species of the genus *Annona* that showed the same pattern of higher concentration of alkaloids in roots than in leaves [30,32,33].

Regarding the effect of IAA, the results of this experiment generally demonstrated that there was no significant change in the concentration of total alkaloids over time compared with the control group. However, at 156 h, the increase in total alkaloids with the application of IAA transport inhibitor (TIBA) and the decrease with IAA was noteworthy. These results differ from what was expected since, in a study with *Annona emarginata* and the use of indole butyric acid (IBA), Sousa et al. [11] observed a significant increase in total alkaloids in roots. Similarly, but with species from other botanical families, Mostafa and Abou Alham [34] found that in *Balanites aegyptiaca* leaves, the synthesis of alkaloids increased with the use of IAA. In *Catharanthus roseus*, the foliar application of IAA provides an increase in the production of the indole alkaloid vincristine [17].

On the other hand, Godjin et al. [22] and Pasquali et al. [35] detected a reduction in the concentration of indole alkaloids in the cell culture of *Daucus carota* and *C. Roseus* with the application of auxins 1-naphthaleneacetic acid (NAA), indole acetic acid (IAA) and 2,4-dichlorophenoxyacetic acid (2,4-D). Godjin et al. [22] and Pasquali et al. [35] justified the results due to the lower transcription of the tryptophan decarboxylase gene, which participates in the alkaloid biosynthesis route through the tryptophan pathway and also gives rise to IAA, and the level of these transcripts increased in medium without the auxins tested (IAA, NAA, and 2,4-D). Thus, the fact that indole acetic acid shares the tryptophan pathway in the production of alkaloids [12,21] may justify the lower concentration of alkaloids when there is a higher concentration of auxins. However, this information does not explain what occurs when there is an increase in alkaloids mediated by the application of IAA [17,34], and there is also a need for further studies to clarify the results when there is an increase in alkaloids in Annonaceae [11], which are largely benzylisoquinolinic alkaloids and, therefore, have tyrosine as precursor [25,36].

Plant regulators from other groups, such as cytokinins and gibberellins, have also been shown to modulate the synthesis of alkaloids in several *Annona* species. Sousa et al. [37] found that after soaking *Annona cacans* seeds in a solution containing GA_{4+7} + 6-Benzyladenine, the content of total alkaloids and liriodenine doubled in roots of seedlings and in cotyledonary leaves, the number of total alkaloids decreased, and the liriodenine concentration remained unchanged. Silva et al. [38] also observed changes in alkaloids during germination and in *Annona x atemoya* Mabb. cv. 'Gefner' seedlings with a reduction in the concentration of total alkaloids after treating seeds with GA_3 to overcome dormancy.

As observed in this experiment, changes in the detection of different alkaloids with the use of regulators were also observed in other studies, such as that of Sousa et al. [37], where the primary roots (5 cm) of *A. cacans* from seeds treated with a solution containing GA_{4+7} + 6-Benzyladenine showed a decrease in the proportion of alkaloids discretine and reticuline and a reduction in the content of total alkaloids, but the content of liriodenine was not altered. In another study by Sousa et al. [11], it was observed that in *A. emarginata* seedlings, the supply of abscisic acid resulted in the disappearance of the alkaloid laurotetanine when compared to control, while treatments did not cause changes in the content of total alkaloids.

Nitrogen is required for the synthesis of alkaloids since alkaloids are nitrogenous molecules of the specialized metabolism, which is related to the activity of the nitrate reductase enzyme [39]. This is the first enzyme to catalyze the reaction of nitrate to nitrite in the cytosol during the process of nitrate assimilation in organic compounds [14]. The nitrogen atoms of alkaloids come from amino acids, and in most cases, alkaloids maintain in their structure the carbon skeleton of the precursor amino acid [21,40]. In this context, the use of auxin did not result in increased activity of the nitrate reductase enzyme and, therefore, did not promote greater nitrogen availability that would result in greater alkaloid production, as expected. Furthermore, it is not possible to determine how much of the available nitrogen was destined for alkaloid synthesis [39] or for other metabolic processes, such as photosynthesis [41,42].

Regarding gas exchange and chlorophyll *a* fluorescence, IAA did not cause variations that could be related to lower photosynthetic efficiency, unlike what was observed by Sousa et al. [11] with the use of another auxin (indole butyric acid—IBA) also in *A. emarginata*. The authors found that IBA promoted a decrease in CO_2 assimilation that culminated in a reduction in the carboxylation efficiency of the Rubisco enzyme, which was not observed in this experiment. Furthermore, the punctual reduction in stomatal conductance at 156 h and in transpiration at 324 h occurred in response to the use of the IAA inhibitor but without variations in photosynthetic efficiency. Hormones such as auxin and cytokinin inhibit stomatal closure induced by IBA; however, in this experiment, auxin did not cause any variation in stomatal conductance compared with control.

Also, in relation to primary metabolism, variations in carbohydrate levels, especially reductions in concentrations when the highest concentrations of alkaloids (156 and 324 h) are observed with the auxin transport inhibition, are related to the greater availability of carbon skeletons and energy for the synthesis of compounds such as amino acids and, consequently, alkaloids, as suggested by Hikosaka et al. [43] and Nunes-Nessi et al. [44]. The higher concentrations of reducing sugars observed with IAA application corroborate reports by Ono et al. [44], who observed an increase in the concentrations of total and reducing sugars with the application of NAA and IBA.

When auxin transport was inhibited by TIBA, there was a reduction in sugars, possibly used for the greater synthesis of alkaloids, which corroborates Dudareva et al. [10], who reported that the biosynthesis of specialized metabolites depends on the availability of carbon (C), nitrogen (N), and energy provided by primary metabolism, and as a consequence, the availability of these building blocks affects the concentration of specialized metabolites, demonstrating a high degree of connectivity between primary and specialized metabolisms.

In order to avoid the harmful effect of reactive oxygen species generated by primary metabolism, an increase in enzymatic activity may occur, generating the accumulation of amino acids and the production of alkaloids, with an increase in nitrogen reserves, as proposed by Ghorbanpour [45]. In this context, although IAA has stimulated the highest hydrogen peroxide and lipoperoxide concentrations in most moments in roots, the antioxidant system (especially SOD and POD) acted similarly, both in plants that received IAA and in those that received TIBA and with greater antioxidant activity than that observed with control plants. In other words, there was an increase in the action of the antioxidant system due to the application of both IAA and TIBA and not specifically due to the greater stress caused by free radicals only with the use of IAA in order to guarantee the functionality of the system and avoid cellular damage, which was observed at all evaluation times, corroborating reports by Barbosa et al. [24]. However, the accumulation of amino acids proposed by Ghorbanpour [45] seems to have occurred only at 156 h, when the highest concentrations of total alkaloids were observed with the use of TIBA. At this time, greater stress and greater activity of the antioxidant system were expected; however, values were similar to those observed at other times, indicating that treatments were not effective in sufficiently stressing the system and increasing the synthesis of alkaloids or that the alkaloid synthesis resulted from substrates originating from sources not involved in the antioxidant system, without representing additional stress to the plant.

4. Materials and Methods

4.1. Plant Material and Experimental Area

Annona emarginata (Schltdl.) H. Rainer ('terra-fria' morphotype) seedlings were obtained from the "Seedling Production Center of São Bento do Sapucaí", CATI (Technical and Integrated Assistance Coordination), municipality of São Bento do Sapucaí—SP. Young 1.5-year-old plants were taken to a greenhouse at the Biodiversity and Biostatistics Department of the Unesp Biosciences Institute of Botucatu, SP (located at 22°53′25″ South, 48°27′19″ West, and an altitude of 800 m a.s.l.; Cwa climate), kept in bags from the original nursery, and irrigated with water daily (sprinkler irrigation) and weekly with 50% Hoagland nutrient solution (manual irrigation) for a period of ten months.

The greenhouse where the experiment was conducted consists of a metal arch structure covered with 150 μm transparent polyethylene (LDPE, anti-UV) agricultural plastic film and laminated 'Aluminet' shade (Polysack®, Nir Yitzhak, Israel) with an automated sprinkler irrigation system. The environmental conditions were determined, where leaf and air temperatures ranged from 24 °C to 29 °C, and relative humidity ranged from 45 to 62%.

The species was identified by Prof. Renato Melo-Silva, PhD, and samples were deposited at the "Herbário Irina Delanova Gemtchujnicov" BOTU herbarium, São Paulo State University, Campus of Botucatu—SP, under code 33120.

4.2. *Experimental Design*

The study was conducted in a completely randomized experimental design with a 3 × 5 factorial scheme with five replicates of one plant per treatment, three treatments (auxin, auxin transport inhibitor, and control), and five collection times (12 h, 36 h, 84 h, 156 h, and 324 h after the last application of treatments).

Indole acetic acid (IAA) at a concentration of 10^{-6} M was used as the source of auxin, and 2,3,5-triiodobenzoic acid (TIBA) at a concentration of 10^{-6} M was used as an auxin transport inhibitor. These IAA and TIBA concentrations were defined according to a previous experiment carried out by Sousa et al. [11]. IAA and TIBA were purchased from Sigma-Aldrich® Brasil Ltda, Sao Paulo, Brazil.

4.3. *Application of Indole Acetic Acid and TIBA Transport Inhibitor*

Three foliar applications were performed at 48 h intervals using a pressurized carbon dioxide backpack sprayer with a fan nozzle and pressure of 276 bar. The non-ionic Haiten® (Ningbo, China) adhesive was used as the spreading agent at a rate of 1 mL/10 L of solution, as recommended by the manufacturer, and was also used with water in the control treatment.

4.4. *Quantitative and Qualitative Analysis of Alkaloid Extracts*

For the extraction of alkaloids, leaf and root material from the five replicates of each treatment (auxin, auxin inhibitor, and control) was used at each collection time (12 h, 36 h, 84 h, 156 h, and 324 h after the last application of treatments). To obtain the total alkaloid extracts, leaf and root samples (1.5 g) were dried in an oven with forced ventilation (40 °C) until reaching constant dry mass and then ground using the mechanical mill. Extraction was performed using the acid-base method [6,46]. Alkaloid extracts were kept in the dark until the total alkaloid content was determined using a UV-visible spectrophotometer (single beam—model UV-M51—BEL Engineering®, Monza, Italy) at 254 nm. Liriodenine was used with a calibration curve of 1–100 mg mL^{-1} to prepare the standard curve ($y = 0.0881x - 0.0112$, $R^2 = 0.9949$) since it is one of the most widely distributed alkaloids among species of the Annonaceae family [47]. Liriodenine standard was provided by Iván De-la-Cruz-Chacón (coauthor). The methods for the isolation and data identification of liriodenine were reported in De-la-Cruz-Chacón and González-Esquinca [46].

The presence of alkaloids was detected using a high-performance liquid chromatograph (UHPLCfocused ThermoFisher-Scientific, Waltham, MA, USA) with gradient pump and UV-VIS detector using a C18 reversed-phase column (150 × 4.6 mm and particle diameter of 5 μm) at 30 °C. The mobile phase was water (with trifluoroacetic acid pH 3.5) and methanol at a ratio of 30:70, with a flow rate of 1 mL/min. Detection was performed in UV at 280 nm [46]. The identification of alkaloids asimilobine, discretine, lanuginosine, laurotetanine, liriodenine, *N*-methyllaurotetanine, norglaucine, norpredicentine, oxoglaucine, reticuline, xylopine, and xylopinine was also carried out by the comparison to standards provided by Emmanoel Vilaça Costa and Jackson Roberto Guedes da Silva Almeida (spectrometric data were previously reported in Sousa et al. [37]).

4.5. Nitrate Reductase Activity

The activity of the nitrate reductase enzyme was determined according to the methodology proposed by Mulder et al. [48], which is based on the production of NO_2^- during the incubation of 1 g of leaf cuttings in the presence of NO_3^-. Leaf samples (200 mg) from the five replicates of each treatment (auxin, auxin inhibitor, and control) were used to perform the analysis at each collection time (12 h, 36 h, 84 h, 156 h, and 324 h after the last application of treatments) in duplicate.

4.6. Gas Exchange and Chlorophyll a Fluorescence

Evaluations were performed at 12 h, 36 h, 156 h, and 324 h after the last application of plant growth regulators in the period from 09:00 a.m. to 11:00 a.m. on the third or fourth fully expanded leaves in five replicates per treatment, except for 84 h after the application of treatments, as it was raining, which did not allow the equipment to function properly. An open photosynthesis system equipment with CO_2 and water vapor analyzer by infrared radiation (Infra Red Gas Analyzer—IRGA, model GSF 3000- Heinz Walz GmbH, Effeltrich, Germany) was used. The gas exchange variables analyzed were net CO_2 assimilation rate (A, µmol CO_2 m^{-2} s^{-1}), transpiration rate (E water vapor m^{-2} s^{-1}), stomatal conductance (gs, mol m^{-2} s^{-1}), and internal CO_2 concentration in the substomatal chamber (Ci, µmol m^{-2} s^{-1}), adopting the average natural light conditions (DFFFA) of each evaluation time incident inside the cultivation environment, on average 867 PAR µmol m^{-2} s^{-1}. From these data, it was possible to calculate the carboxylation efficiency (A/Ci) (µmol CO_2 m^{-2} s^{-1}/ µmol m^{-2} s^{-1}). Regarding fluorescence, the potential quantum yield of photosystem II (Fv/Fm), the electron transport rate (ETR), and the photochemical extinction coefficient (qP) were analyzed [49].

4.7. Carbohydrate Concentration

Total soluble sugars were extracted from leaf (100 mg) and root (100 mg) material obtained from five replicates of each treatment (auxin, auxin inhibitor, and control) at each collection time (12 h, 36 h, 84 h, 156 h, and 324 h after the last application of treatments), according to Garcia et al. [50], and to determine the concentration of total soluble sugars, the method proposed by to Morris [51] was used. Starch was extracted according to Clegg [52] and quantified according to Yemm and Folkes [53]. For reducing sugars, the methodology established by Miller [54] was used, and for sucrose, the methodology was determined by Passos [55].

4.8. Antioxidant Activity and Reactive Oxygen Species

Leaf and root samples (100 mg) from the five replicates of each treatment (auxin, auxin inhibitor, and control) at each collection time (12 h, 36 h, 84 h, 156 h, and 324 h after the last application of treatments) were immediately immersed in liquid nitrogen and subsequently macerated in a mortar with the aid of liquid nitrogen and stored in a freezer at −20 °C. The methodology proposed by Kar and Misshra [56] was followed to obtain the enzymatic extract. To determine the specific activity of antioxidant enzymes, it was necessary to evaluate the protein concentration using the Bradford methodology [57]. Catalase activity (CAT, EC 1.11.1.6) was determined according to Peixoto et al. [58], peroxidase activity (POD, EC. 1.11.1.7) according to Teisseire and Guy [59], and superoxide dismutase activity (SOD, EC 1.15.1.1) according to the Beauchamp and Fridovich technique [60].

Regarding reactive oxygen species, hydrogen peroxide content was determined according to the method of Alexieva et al. [61], and lipid peroxidation was determined according to the methodology proposed by Heath and Paker [62].

4.9. Data Analysis

Data were submitted to normality and homogeneity of variance tests. The results of quantification analyses of indole acetic acid, gas exchange, and chlorophyll *a* fluorescence were submitted to one-way ANOVA analysis of variance, and the means were compared using Tukey's test ($p \leq 0.05$).

The results of quantification analyses of total alkaloids and liriodenine, the activity of the nitrate reductase enzyme, carbohydrate concentration (total sugars, reducing sugars, starch, and sucrose), and the activity of antioxidant enzymes (SOD, POD, and CAT) were submitted to two-way ANOVA, and means were compared using Tukey's test ($p \leq 0.05$).

5. Conclusions

Indole acetic acid promotes selective variations in alkaloid production in the leaves and roots of young *Annona emarginata* plants. In leaves, with the use of IAA, it was possible to detect lanuginosine, liriodenine, and xylopine, and when the IAA transport was inhibited by TIBA, the synthesis of xylopine was inhibited, and the production of *N*-methyllaurotetanine was induced, while lanuginosine and liriodenine remained present. In roots, 10 alkaloids were detected, and IAA promoted the appearance of discretine and xylopine (which did not appear in control) and inhibited the synthesis of oxoglaucine and reticuline. With the application of TIBA (auxin transport inhibition), asimilobine, discretine, oxoglaucine, and reticuline were not observed, but xylopine (absent in control) and the other alkaloids analyzed were observed. IAA did not cause significant variations in photosynthesis over a long period of evaluation in a way that impacts the production of alkaloids. However, when the IAA transport was inhibited by TIBA, there was a punctual increase in the production of sugars and starch, which resulted in higher alkaloid concentrations. Greater stress was observed with the application of IAA and TIBA and greater activity of the antioxidant system; however, no increase in stress was observed with the increase in the production of total alkaloids. The activity of nitrate reductase also did not increase with the application of IAA and IBA, demonstrating that nitrogen redistribution occurred in order to increase alkaloid production.

Author Contributions: Conceptualization, B.C.M. and G.F.; formal analysis, B.C.M., I.D.-l.-C.-C., C.O.M., F.G.C. and G.F.; investigation, B.C.M.; resources, C.S.F.B. and G.F.; writing—original draft preparation, B.C.M. and G.F.; writing—review and editing, B.C.M., I.D.-l.-C.-C., C.O.M., F.G.C., I.R.M.-C., C.S.F.B. and G.F.; supervision, G.F.; funding acquisition, G.F. All authors have read and agreed to the published version of the manuscript.

Funding: This research was funded by Coordenação de Aperfeiçoamento de Pessoal de Nível Superior—Brasil (CAPES)—Finance Code 001, CAPESPrint—Unesp and National Council for Scientific and Technological Development (CNPq) (140073/2022-1).

Data Availability Statement: The original contributions presented in the study are included in the article, further inquiries can be directed to the corresponding author.

Acknowledgments: To Seedling Production Center of São Bento do Sapucaí, CATI (Technical and Integrated Assistance Coordination), municipality of São Bento do Sapucaí—São Paulo, for the donation of young *Annona emarginata* plants ("araticum de terra-fria" morphotype).

Conflicts of Interest: The authors declare no conflicts of interest.

References

1. Wasternack, C. A Plant's Balance of Growth and Defense—Revisited. *New Phytol.* **2017**, *215*, 1291–1294. [CrossRef] [PubMed]
2. Gobbo-Neto, L.; Lopes, N.P. Plantas Medicinais: Fatores de Influência No Conteúdo de Metabólitos Secundários. *Quim. Nova* **2007**, *30*, 374–381. [CrossRef]
3. Pichersky, E.; Lewinsohn, E. Convergent Evolution in Plant Specialized Metabolism. *Annu. Rev. Plant Biol.* **2011**, *62*, 549–566. [CrossRef] [PubMed]
4. Bhambhani, S.; Kondhare, K.R.; Giri, A.P. Diversity in Chemical Structures and Biological Properties of Plant Alkaloids. *Molecules* **2021**, *26*, 3374. [CrossRef] [PubMed]

5. Vinche, A.D.L.; De-La-Cruz-Chacón, I.; González-Esquinca, A.R.; Silva, J.d.F.d.; Ferreira, G.; Santos, D.C.d.; Garces, H.G.; Oliveira, D.V.M.d.; Marçon, C.; Cavalcante, R.d.S.; et al. Antifungal Activity of Liriodenine on Agents of Systemic Mycoses, with Emphasis on the Genus *Paracoccidioides*. *J. Venom. Anim. Toxins Incl. Trop. Dis.* **2020**, *26*, e20200023. [CrossRef]
6. De-La-Cruz-Chacón, I.; González-Esquinca, A.R.; Fefer, P.G.; Garcia, L.F.J. Liriodenine, Early Antimicrobial Defence in *Annona diversifolia*. *Z. Für Naturforschung C* **2011**, *66*, 377–384. [CrossRef]
7. Khan, T.-M.; Gul, N.S.; Lu, X.; Wei, J.-H.; Liu, Y.-C.; Sun, H.; Liang, H.; Orvig, C.; Chen, Z.-F. In Vitro and in Vivo Anti-Tumor Activity of Two Gold(III) Complexes with Isoquinoline Derivatives as Ligands. *Eur. J. Med. Chem.* **2019**, *163*, 333–343. [CrossRef]
8. Silva, G.C.; Dutra, L.M.; Silva Almeida, J.R.G.; Silva, F.M.A.; Harakava, R.; Honório, A.B.M.; De-la-Cruz-Chacón, I.; Martínez-Vázquez, M.; Ferreira, G. Alkaloid Screening of *Annona emarginata* (Schltdl.) H. Rainer Rootstocks to Increase Fungal Tolerance in *Annona atemoya* Mabb. Crops Revealed by MS and NMR Chemical Profiling. *Ind. Crops Prod.* **2024**, *212*, 118335. [CrossRef]
9. Li, H.-T.; Wu, H.-M.; Chen, H.-L.; Liu, C.-M.; Chen, C.-Y. The Pharmacological Activities of (−)-Anonaine. *Molecules* **2013**, *18*, 8257–8263. [CrossRef]
10. Dudareva, N.; Klempien, A.; Muhlemann, J.K.; Kaplan, I. Biosynthesis, Function and Metabolic Engineering of Plant Volatile Organic Compounds. *New Phytol.* **2013**, *198*, 16–32. [CrossRef]
11. Sousa, M.C.; De-la-Cruz-Chacón, I.; Campos, F.G.; Vieira, M.A.R.; Corrêa, P.L.C.; Marques, M.O.M.; Boaro, C.S.F.; Ferreira, G. Plant Growth Regulators Induce Differential Responses on Primary and Specialized Metabolism of *Annona emarginata* (Annonaceae). *Ind. Crops Prod.* **2022**, *189*, 115789. [CrossRef]
12. Ljung, K. Auxin Metabolism and Homeostasis during Plant Development. *Development* **2013**, *140*, 943–950. [CrossRef] [PubMed]
13. Casanova-Sáez, R.; Mateo-Bonmatí, E.; Ljung, K. Auxin Metabolism in Plants. *Cold Spring Harb. Perspect. Med.* **2021**, *13*, a039867. [CrossRef] [PubMed]
14. Joshi, G.; Shukla, A.; Shukla, A. Synergistic Response of Auxin and Ethylene on Physiology of *Jatropha curcas* L. *Braz. J. Plant Physiol.* **2011**, *23*, 66–77. [CrossRef]
15. Koornneef, A.; Pieterse, C.M.J. Cross Talk in Defense Signaling. *Plant Physiol.* **2008**, *146*, 839–844. [CrossRef]
16. Huot, B.; Yao, J.; Montgomery, B.L.; He, S.Y. Growth–Defense Tradeoffs in Plants: A Balancing Act to Optimize Fitness. *Mol. Plant* **2014**, *7*, 1267–1287. [CrossRef]
17. Muthulakshmi, S.; Pandiyarajan, V. Influence of IAA on the Vincristine Content of *Catharanthus roseus* (L). G. Don. *Asian J. Plant Sci. Res.* **2013**, *3*, 81–87.
18. Murcia, G.; Fontana, A.; Pontin, M.; Baraldi, R.; Bertazza, G.; Piccoli, P.N. ABA and GA3 Regulate the Synthesis of Primary and Secondary Metabolites Related to Alleviation from Biotic and Abiotic Stresses in Grapevine. *Phytochemistry* **2017**, *135*, 34–52. [CrossRef]
19. Jamwal, K.; Bhattacharya, S.; Puri, S. Plant Growth Regulator Mediated Consequences of Secondary Metabolites in Medicinal Plants. *J. Appl. Res. Med. Aromat. Plants* **2017**, *9*, 26–38. [CrossRef]
20. Parmoon, G.; Ebadi, A.; Jahanbakhsh, S.; Hashemi, M.; Moosavi, S.A. Effect of Exogenous Application of Several Plant Growth Regulators on Photosynthetic Pigments of Fennel Plants. *Not. Sci. Biol.* **2018**, *10*, 508–515. [CrossRef]
21. Maeda, H.; Dudareva, N. The Shikimate Pathway and Aromatic Amino Acid Biosynthesis in Plants. *Annu. Rev. Plant Biol.* **2012**, *63*, 73–105. [CrossRef] [PubMed]
22. Goddijn, O.J.M.; de Kam, R.J.; Zanetti, A.; Schilperoort, R.A.; Hoge, J.H.C. Auxin Rapidly Down-Regulates Transcription of the Tryptophan Decarboxylase Gene from *Catharanthus roseus*. *Plant Mol. Biol.* **1992**, *18*, 1113–1120. [CrossRef] [PubMed]
23. Singh, A.; Pandey, B.; Kumari, S.; Agrawal, M. Nitrogen Availability Modulates CO_2-Induced Responses of *Catharanthus roseus*: Biomass Allocation, Carbohydrates and Alkaloids Profile. *J. Appl. Res. Med. Aromat. Plants* **2015**, *2*, 160–167. [CrossRef]
24. Barbosa, M.R.; Silva, M.M.d.A.; Willadino, L.; Ulisses, C.; Camara, T.R. Geração e Desintoxicação Enzimática de Espécies Reativas de Oxigênio Em Plantas. *Ciência Rural.* **2014**, *44*, 453–460. [CrossRef]
25. Lúcio, A.S.S.C.; Almeida, J.R.G.d.S.; Da-Cunha, E.V.L.; Tavares, J.F.; Barbosa Filho, J.M. Alkaloids of the Annonaceae: Occurrence and a Compilation of Their Biological Activities. *Alkaloids Chem. Biol.* **2015**, *74*, 233–409. [CrossRef]
26. Flora e Funga do Brasil. Annonaceae. Available online: https://floradobrasil.jbrj.gov.br/FB110219 (accessed on 18 April 2024).
27. Mendes-Silva, I.; Lopes, J.C.; Silva, L.V.; Bazante, M.L. Annona in Flora Do Brasil 2020. Available online: https://floradobrasil.jbrj.gov.br/FB110235 (accessed on 18 April 2024).
28. Maas, P.J.M.; Kamer, H.M.d.; Junikka, L.; Mello-silva, R.D.; Rainer, H. Annonnaceae from Central-Eastern Brazil. *Rodriguésia* **2001**, *52*, 65–98. [CrossRef]
29. Baron, D.; Esteves Amaro, A.C.; Pina, A.; Ferreira, G. An Overview of Grafting Re-Establishment in Woody Fruit Species. *Sci. Hortic.* **2019**, *243*, 84–91. [CrossRef]
30. Mimi, C.O.; De-la-Cruz-Chacón, I.; Sousa, M.C.; Vieira, M.A.P.; Marques, M.O.M.; Ferreira, G.; Boaro, C.S.F. Chemophenetics as a Tool for Distinguishing Morphotypes of *Annona emarginata* (Schltdl.) H. Rainer. *Chem. Biodivers.* **2021**, *18*, e2100544. [CrossRef]
31. González-Esquinca, A.R.; De-La-Cruz-Chacón, I.; Castro-Moreno, M.; Orozco-Castillo, J.A.; Riley- Saldaña, C.A. Alkaloids and Acetogenins in Annonaceae Development: Biological Considerations. *Rev. Bras. Frutic.* **2014**, *36*, 01–16. [CrossRef]
32. De-la-Cruz-Chacón, I.; Riley-Saldaña, C.A.; Arrollo-Gómez, S.; Sancristóbal-Domínguez, T.J.; Castro-Moreno, M.; González-Esquinca, A.R. Spatio-Temporal Variation of Alkaloids in *Annona purpurea* and the Associated Influence on Their Antifungal Activity. *Chem. Biodivers.* **2019**, *16*, e1800284. [CrossRef]

33. Corrêa, P.L.C.; De-la-Cruz-Chacón, I.; Sousa, M.C.; Vieira, M.A.R.; Campos, F.G.; Marques, M.O.M.; Boaro, C.S.F.; Ferreira, G. Effect of Nitrogen Sources on Photosynthesis and Biosynthesis of Alkaloids and Leaf Volatile Compounds in *Annona sylvatica* A. St.-Hil. *J. Soil. Sci. Plant Nutr.* **2022**, *22*, 956–970. [CrossRef]
34. Mostafa, G.G.; Abou Alham, M.F. Effect of Gibberellic Acid and Indole 3-Acetic Acid on Improving Growth and Accumulation of Phytochemical Composition in Balanites Aegyptiaca Plants. *Am. J. Plant Physiol.* **2010**, *6*, 36–43. [CrossRef]
35. Pasquali, G.; Goddijn, O.J.M.; de Waal, A.; Verpoorte, R.; Schilperoort, R.A.; Hoge, J.H.C.; Memelink, J. Coordinated Regulation of Two Indole Alkaloid Biosynthetic Genes from *Catharanthus roseus* by Auxin and Elicitors. *Plant Mol. Biol.* **1992**, *18*, 1121–1131. [CrossRef] [PubMed]
36. De-La-Cruz Chacón, I.; González-Esquinca, A.R.; Riley-Saldaña; Anabí, C. Biosíntesis de Alcaloides Bencilisoquinolínicos. *Univ. Sci.* **2012**, *17*, 189–202. [CrossRef]
37. Sousa, M.C.; Bronzatto, A.C.; González-Esquinca, A.R.; Campos, F.G.; Dalanhol, S.J.; Boaro, C.S.F.; Martins, A.L.; da Silva Almeida, J.R.G.; Costa, E.V.; De-la-Cruz-Chacón, I.; et al. The Production of Alkaloids in *Annona cacans* Seedlings Is Affected by the Application of GA 4+7 + 6-Benzyladenine. *Biochem. Syst. Ecol.* **2019**, *84*, 47–51. [CrossRef]
38. da Silva, G.C.; de-la-Cruz-Chacón, I.; Honório, A.B.M.; Martin, B.C.; Sousa, M.C.; Campos, F.G.; Boaro, C.S.F.; Ferreira, G. Temperature and GA3 as Modulating Factors in the Biosynthesis of Alkaloids during Imbibition and Early Development of *Annona x atemoya* Mabb. Cv. 'Gefner' Seedlings. *Horticulturae* **2022**, *8*, 766. [CrossRef]
39. Mengel, K.; Kirkby, E.A.; Kosegarten, H.; Appel, T. *Principles of Plant Nutrition*, 5th ed.; Mengel, K., Kirkby, E.A., Kosegarten, H., Appel, T., Eds.; Springer: Dordrecht, The Netherlands, 2001; ISBN 978-1-4020-0008-9.
40. Dewick, P.M. Secondary Metabolism: The Building Blocks and Construction Mechanisms. In *Medicinal Natural Products;* John Wiley & Sons, Ltd.: Chichester, UK, 2002; Volume 2, pp. 7–34. ISBN 0471496405.
41. Dodd, A.N.; Salathia, N.; Hall, A.; Kévei, E.; Tóth, R.; Nagy, F.; Hibberd, J.M.; Millar, A.J.; Webb, A.A.R. Plant Circadian Clocks Increase Photosynthesis, Growth, Survival, and Competitive Advantage. *Science (1979)* **2005**, *309*, 630–633. [CrossRef]
42. Nunes-Nesi, A.; Fernie, A.R.; Stitt, M. Metabolic and Signaling Aspects Underpinning the Regulation of Plant Carbon Nitrogen Interactions. *Mol. Plant* **2010**, *3*, 973–996. [CrossRef]
43. Hikosaka, K. Interspecific Difference in the Photosynthesis?Nitrogen Relationship: Patterns, Physiological Causes, and Ecological Importance. *J. Plant Res.* **2004**, *117*, 481–494. [CrossRef]
44. Ono, E.O.; Rodrigues, J.D.; Pinho, S.Z. De Efeito de Auxinas e Boro No Enraizamento de Estacas Caulinares de Kiwi Retiradas Em Diferentes Épocas. *Pesqui. Agropecu. Bras.* **1998**, *33*, 221–223.
45. Ghorbanpour, M. Role of Plant Growth Promoting Rhizobacteria on Antioxidant Enzyme Activities and Tropane Alkaloids Production of *Hyoscyamus niger* under Water Deficit Stress. *Turk. J. Biol.* **2013**, *37*, 14. [CrossRef]
46. De-La-Cruz-Chacón, I.; González-Esquinca, A.R. Liriodenine Alkaloid in *Annona diversifolia* during Early Development. *Nat. Prod. Res.* **2012**, *26*, 42–49. [CrossRef]
47. Février, A.; Ferreira, M.E.; Fournet, A.; Yaluff, G.; Inchausti, A.; de Arias, A.R.; Hocquemiller, R.; Waechter, A.-I. Acetogenins and Other Compounds from *Rollinia emarginata* and Their Antiprotozoal Activities. *Planta Med.* **1999**, *65*, 047–049. [CrossRef] [PubMed]
48. Mulder, E.G.; Boxma, R.; Van Veen, W.L. The Effect of Molybdenum and Nitrogen Deficiencies on Nitrate Reduction in Plant Tissues *. *Plant Soil.* **1959**, *4*, 335–355. [CrossRef]
49. Demmig-Adams, B.; Adams III, W.W.; Barker, D.H.; Logan, B.A.; Bowling, D.R.; Verhoeven, A.S. Using Chlorophyll Fluorescence to Assess the Fraction of Absorbed Light Allocated to Thermal Dissipation of Excess Excitation. *Physiol. Plant* **2008**, *98*, 253–264. [CrossRef]
50. Garcia, I.S.; Souza, A.; Barbedo, C.J.; Dietrich, S.M.C.; Figueiredo-Ribeiro, R.C.L. Changes in Soluble Carbohydrates during Storage of *Caesalpinia echinata* LAM. (Brazilwood) Seeds, an Endangered Leguminous Tree from the Brazilian Atlantic Forest. *Braz. J. Biol.* **2006**, *66*, 739–745. [CrossRef]
51. Morris, D.L. Quantitative Determination of Carbohydrates with Dreywood's Anthrone Reagent. *Science (1979)* **1948**, *107*, 254–255. [CrossRef]
52. Clegg, K.M. The Application of the Anthrone Reagent to the Estimation of Starch in Cereals. *J. Sci. Food Agric.* **1956**, *7*, 40–44. [CrossRef]
53. Yemm, E.W.; Folkes, B.F. The Regulation of Respiration during the Assimilation of Nitrogen in *Torulopsis utilis*. *Biochem. J.* **1954**, *57*, 495–508. [CrossRef]
54. Miller, G.L. Use of Dinitrosalicylic Acid Reagent for Determination of Reducing Sugar. *Anal. Chem.* **1959**, *31*, 426–428. [CrossRef]
55. Passos, L.P. *Métodos Analíticos e Labora Toriais em Fisiologia Vegetal;* EMBRAPA-CNPGL: Coronel Pacheco, Brazil, 1996; ISBN 8585748087.
56. Kar, M.; Mishra, D. Catalase, Peroxidase, and Polyphenoloxidase Activities during Rice Leaf Senescence. *Plant Physiol.* **1976**, *57*, 315–319. [CrossRef] [PubMed]
57. Bradford, M. A Rapid and Sensitive Method for the Quantitation of Microgram Quantities of Protein Utilizing the Principle of Protein-Dye Binding. *Anal. Biochem.* **1976**, *72*, 248–254. [CrossRef] [PubMed]
58. Peixoto, P.H.P.; Cambraia, J.; Sant'Anna, R.; Mosquim, P.R.; Moreira, M.A. Aluminum Effects on Lipid Peroxidation and on the Activities of Enzymes of Oxidative Metabolism in *Sorghum*. *Braz. J. Plant Physiol.* **1999**, *11*, 137–143.

59. Teisseire, H.; Guy, V. Copper-Induced Changes in Antioxidant Enzymes Activities in Fronds of Duckweed (*Lemna minor*). *Plant Sci.* **2000**, *153*, 65–72. [CrossRef]
60. Beauchamp, C.; Fridovich, I. Superoxide Dismutase: Improved Assays and an Assay Applicable to Acrylamide Gels. *Anal. Biochem.* **1971**, *44*, 276–287. [CrossRef] [PubMed]
61. Alexieva, V.; Sergiev, I.; Mapelli, S.; Karanov, E. The Effect of Drought and Ultraviolet Radiation on Growth and Stress Markers in Pea and Wheat. *Plant Cell Environ.* **2001**, *24*, 1337–1344. [CrossRef]
62. Heath, R.L.; Packer, L. Photoperoxidation in Isolated Chloroplasts. *Arch. Biochem. Biophys.* **1968**, *125*, 189–198. [CrossRef]

plants

MDPI

Article

Allelic Expression Dynamics of Regulatory Factors During Embryogenic Callus Induction in ABB Banana (*Musa* spp. cv. Bengal, ABB Group)

Xiaobing Zhao [1], Yiting Zhuang [2], Wangyang Xie [2], Yixin Yang [2], Jingyu Pu [2], Zhengyang Fan [1], Yukun Chen [1], Yuling Lin [1,*] and Zhongxiong Lai [1,*]

[1] Institute of Horticultural Biotechnology, Fujian Agriculture and Forestry University, Fuzhou 350002, China; zxb1344@126.com (X.Z.); ffzzyy789@163.com (Z.F.); cyk68@163.com (Y.C.)

[2] Center for Genomics and Biotechnology, Fujian Provincial Key Laboratory of Haixia Applied Plant Systems Biology, Key Laboratory of Genetics, Breeding and Multiple Utilization of Crops, Ministry of Education, Fujian Agriculture and Forestry University, Fuzhou 350002, China; zhuang_yiting@163.com (Y.Z.); xwy971211@163.com (W.X.); yyx1231209@163.com (Y.Y.); pjy13990877535@126.com (J.P.)

* Correspondence: buliang84@163.com (Y.L.); laizx01@163.com (Z.L.)

Abstract: The regulatory mechanisms underlying embryogenic callus (EC) formation in polyploid bananas remain unexplored, posing challenges for genetic transformation and biotechnological applications. Here, we conducted transcriptome sequencing on cultured explants, non-embryogenic callus, EC, and browning callus in the ABB cultivar 'MJ' (*Musa* spp. cv. Bengal). Our analysis of differentially expressed genes (DEGs) revealed significant enrichment in plant hormones, MAPK, and zeatin biosynthesis pathways. Notably, most genes in the MJ variety exhibited balanced expression of the A and B alleles, but A-specific allele expression was dominant in the key signaling pathways, whereas B-specific allele expression was very rare during EC induction. In the auxin signaling pathway, six A-specific *MJARF* genes were markedly downregulated, underscoring their critical roles in the negative regulation of callus formation. Additionally, six A-specific *MJEIN3* alleles were found to play negative regulatory roles in ethylene signaling during EC development. We also identified phenylpropanoids responsible for enzymatic browning. Furthermore, the expression patterns of transcription factors in bananas exhibited specific expression modes, highlighting the unique mechanisms of callus formation. This study enhanced our understanding of the regulatory roles of these alleles in EC induction and offers new insights into the utilization of alleles to improve the efficiency of somatic embryogenesis in bananas.

Keywords: embryogenic callus; specific allele; transcriptional regulation; browning

1. Introduction

Bananas (*Musa* spp.) are perennial, herbaceous, monocotyledonous plants belonging to the Musaceae family within the order Zingiberales. As one of the most produced fruits globally, bananas are a staple food for millions of people. The ABB group, which consists of starchy cooking bananas, plays a crucial role in food security, particularly in Southeast Asia [1]. Almost all cultivated banana varieties originated from the intra- and interspecific hybridization of two wild species in the Eumusa section, namely *Musa acuminata* (AA) and *Musa balbisiana* (BB), as well as long-term evolutionary variations, resulting in different genomic types such as diploids (AA, AB, and BB), triploids (AAA, AAB, ABB), and tetraploids (AAAA, AAAB, AABB, ABBB) [2]. Most commercial bananas are triploids, making it challenging to improve them using traditional breeding methods. To enhance the diversity, disease resistance, and cold tolerance of banana varieties, breeding bananas using biotechnological approaches is inevitable. The establishment of a robust genetic transformation receptor system is the foundation of plant genetic engineering [3]. A deeper understanding of the regulatory mechanisms governing banana embryogenic callus (EC) will significantly enhance genetic transformation efficiency.

The receptor system for exogenous genes in bananas is much more complex than the general tissue culture and reproduction systems used for banana production. Embryogenic cell suspension (ECS) lines have been successfully established in various banana varieties (lines) with multiple genotypes (AA, AB, AAB, ABB, AAA, and AAAA) [4,5]. The embryogenic system has been successfully used for Banana *Agrobacterium* transformation [6]; however, the transformation efficiency remains relatively low. Genotypes are the most crucial factor affecting efficiency. In this sense, somatic embryogenesis (SE) has been successfully achieved in several banana genotypes, including AA, BB, AB, AAA, ABB, and AAB) [3]. Currently, three varieties of *Musa* AAB, namely 'French Plantain', 'Mysore', and 'Silk', have been reported with SE induction efficiencies of 2%, 3%, and 7%, respectively. However, the *Musa* AAA 'Grande Naine' variety exhibited the highest induction rate among all the tested varieties, reaching 37% [7]. The reason for the highest induction rate in AAA varieties, and whether this was due to gene duplication in the A genome, has not yet been reported. Despite the existing research on SE induction in bananas, the process still faces challenges of low and unstable induction rates. The low induction efficiency of EC is still the key to hindering the genetic transformation of bananas [3]. Therefore, unveiling the key genes and molecular mechanisms underlying somatic embryogenic callus induction has become a focal point of research on the banana embryogenic regeneration system.

The use of plant genetic transformation systems to investigate gene functions has become a widely adopted research approach. Recent reports have suggested that the overexpression of certain key genes can significantly enhance the ability of plants to induce callus differentiation and regeneration, thereby improving the efficiency of genetic transformation. Overexpression of the maize *GOLDEN2* gene in rice and maize callus tissues promotes callus differentiation, consequently enhancing genetic transformation efficiency [8]. Similarly, the *TaWOX5* gene could significantly enhance wheat embryogenic calli and plant regeneration. Utilization of the *TaWOX5* gene markedly improves genetic transformation efficiency and reduces genotype dependence in most wheat varieties [9]. Furthermore, the *TaLAX1* gene enhances wheat malt regeneration by activating cytokinin synthesis and auxin transport-related gene expression, thereby enhancing wheat genetic transformation and gene-editing efficiency [10]. In *Arabidopsis*, ectopic expression of *LEC2* promotes SE, whereas, in cassava, *MeLEC2* overexpression induces embryogenesis, making it a valuable tool for transformation recovery [11,12]. Furthermore, *HB52* and *CRF3* overexpression in *Arabidopsis* aids callus formation without requiring exogenous auxins, whereas *DoLEA43* overexpression enhances callus induction, such as *WIND1* expression [13,14]. All these genes, which can significantly enhance callus differentiation and regeneration capabilities, have been successfully applied in genetic transformation following in-depth studies on the mechanisms of embryogenic calli and callus differentiation. This suggests that if we can identify the key genes involved in the induction of embryogenic calli in bananas and thoroughly elucidate their regulatory mechanisms, it may be possible to apply this knowledge to enhance the efficiency of banana genetic transformation.

There has been limited research on the regulatory mechanisms of EC induction and SE in bananas, and the main regulatory genes have not yet been identified. Understanding somatic cell development at the molecular level can be facilitated by transcriptome and proteome analyses [15]. Various differentially expressed proteins and genes involved in banana SE have been identified using multi-omics approaches, particularly transcriptomics and proteomics [16–18]. Studies suggest that embryogenic cells (ECs) are associated with an elevated accumulation of reactive oxygen species (ROS)-scavenging proteins, heat shock proteins (HSPs), and proteins involved in growth regulation [16,17]. Calcium signaling pathways and plant growth regulators, including indole-3-acetic acid (IAA), benzylaminopurine (BAP), and kinetin, play critical roles in the development and germination of banana somatic embryos [3].

The hormone signaling pathway, an important carrier of regulatory substances in organisms, plays an important role in callus formation [19]. Some studies have found that exogenous auxin can activate the expression of *LECs* and then activate the expression of auxin synthesis-related genes *YUC2* and *YUC4* and promote the development of *Arabidopsis* somatic embryo [20]. The LATERAL ORGAN BOUNDARIES DOMAIN (LBD)/asymmetric leaf 2-like transcription factor is involved in the regulation of callus formation in *Arabidopsis*. Auxin response factor ARF can activate the expression of the downstream LBD family of transcription factors LBD16, LBD17, LBD18, and LBD29 [21]. The molecular mechanism by which cytokinins promote callus formation remains to be further studied; however, B-ARRs, key transcription factors of the cytokinin signaling pathway, are closely related to callus formation [22]. Studies on *Arabidopsis thaliana* showed that overexpression of *ARR4* significantly promoted bud formation in the presence of cytokinin, while overexpression of *ARR8* inhibited bud formation and callus greening [23]. Overexpression of *ARR1* in a medium containing cytokinins promotes the formation of *Arabidopsis* calli [24]. Mitogen-activated protein kinase (MAPK) is an important signal transmitter from the cell surface to the nucleus and plays a crucial regulatory role in plant growth and development [25,26]. Transcriptome sequencing of different calli from two *Eucalyptus* species revealed that most of the genes were involved in signaling pathways such as plant hormones and MAPK [27]. In *Hevea brasiliensis*, differentially expressed genes (DEGs) were mainly enriched in MAPK and plant hormone signaling, biosynthesis, and metabolic pathways of secondary metabolites [28]. Numerous genes associated with SE have been identified through transcriptome analyses and callus induction, including ABA-induced genes [29], AGL 15 [30], calmodulin [31], DcAGP [32], LEC [33], and SERK [34]. Further studies suggest that the transcription factors BBM, WUS, VIVIVIPAROUS1, and LEC may play important roles in banana SE, particularly BBM2 and WUS2, in both embryogenic and non-embryogenic cell suspensions [18]. To date, the regulatory network for EC induction in triploid bananas of the ABB genome group has not been established, and the contributions of different alleles during the induction process remain unclear. The "Bengal Banana" (*Musa* spp. cv. Bengal, ABB Group, referred to as MJ banana) is part of the cooking banana series and was introduced from Bangladesh to China. Its primary phenotypic traits include a robust plant structure, large fruit size, and strong resistance to both diseases and drought [35]. Investigating the EC regulation mechanism of the MJ cultivar and establishing its callus induction system is crucial for developing transgenic banana varieties that are cold-tolerance, disease-resistant, and of high quality.

In this study, we developed an EC induction system for the MJ cultivar and uncovered a framework for the molecular events of callus formation in the ABB banana MJ. Furthermore, we elucidated the functional roles of the A and B alleles, providing a critical foundation for a deeper understanding of the regulatory mechanisms underlying EC formation in triploid bananas.

2. Results

2.1. Morphologies of Calli and the Crucial Role of the Growth Regulator 2,4-D in Promoting EC Induction

When cultured on previously reported callus induction media with varying hormone ratios [36,37], the MJ cultivar (ABB genotype) exhibited the earliest initiation of callus growth. Therefore, we selected this cultivar as the primary experimental material. We developed several distinct culture media to induce callus formation and observed that immature flowers of the MJ cultivar could be induced to form callus on any of the four media tested (Supplementary Table S1). After a 5-month induction period, explants from immature male flowers showed the highest callus induction rate on B1 medium with 0.5 mg/L 2,4-D, resulting in the emergence of crystalline small granular callus on the explant surfaces (Figure 1a–d, Supplementary Table S1). The callus induction rate on the B2 medium with 1 mg/L 2,4-D was similar, with most explants producing larger yellowish granular calli (Figure 1e) and small crystalline granular calli. However, some explants on B2 medium produced velvety white calli. The induction rate on B3 medium with 2 mg/L 2,4-D was significantly decreased compared to that on B1 and B2, showing a mix of crystalline small granular calli and yellowish larger granular calli (Supplementary Table S1). For the B4 medium with 4 mg/L 2,4-D, the induction rate of the explants was 28.1% (Supplementary Table S1). The callus state was suboptimal, with a low number of induced calli, exhibiting velvety white calli and very few crystalline small granular calli with a more pronounced brown coloration. From these results, we concluded that the concentration of 2,4-D is crucial for callus induction rate and quality. Embryogenic calli of Banana MJ were successfully induced from the B2 medium after 9 months of induction, displaying a soft, yellow, moist, and loose morphology (Figure 1f). Therefore, the B2 medium (basal medium supplemented with 1 mg/L 2,4-D, 1 mg/L NAA, and 1 mg/L IAA) was determined to be the optimal medium for subsequent experiments.

To further analyze the morphology of callus cells at different developmental stages, we examined the status of various callus cell forms in paraffin sections. The non-embryogenic calli appeared white, loose and lacked embryo-like structures on the surface (Figure 1g). The cells of the non-embryogenic calli were highly vacuolated, contained fewer starch granules, and exhibited abnormal shapes and sizes (Figure 1g–h). Larger yellowish granular calli were another special non-embryogenic callus (Figure 1e), suggesting that they may be intermediates in the process of EC formation or SE. In contrast, the embryogenic calli of MJ were pale yellow and delicate. These cells were small, spherical, and had centrally located nuclei. Both the nucleus and the cytoplasm were deeply stained and displayed characteristics typical of ECs. Darkly stained cell aggregates confirmed the presence of dense cytoplasm with abundant starch granules in the EC (Figure 1i). Embryogenic cells were spherical and multiplied at a high rate.

2.2. Comparison of Transcriptional Profiling During Formation of EC in ABB Banana

To identify the key DEGs and essential pathways involved during banana EC induction, we selected samples from the five stages for global transcriptome analysis. These stages included immature male flowers of bananas cultivated on B2 medium for one week (0M), flowers cultivated on B2 medium for 5 months (5M) and 12-month callus (12M), embryogenic calli (EC), and browning embryogenic calli (BC) (Figure 2a–e). The published genome of *M. acuminata* (AA genome) of DH-Pahang and *Musa balbisiana* (BB genome) of DH-PKW were used as reference genomes for transcriptome data analysis [38,39]. The correlation between the biological replicates for each stage was calculated based on the FPKM results. Pearson's correlation analysis indicated a high linear correlation among different samples (Supplementary Figure S1a). Furthermore, the PCA of the 15 samples (with three biological replicates for each sample) showed clear distinctions at different stages and a compact distribution of biological replicates (Supplementary Figure S1b), indicating that the data quality was highly reproducible, allowing for further analysis.

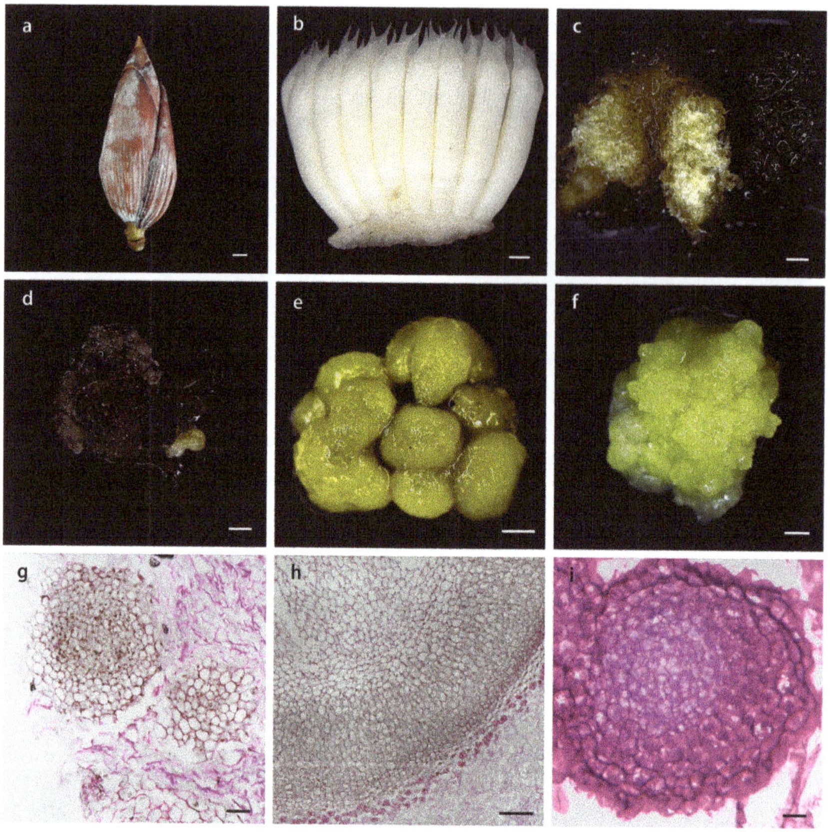

Figure 1. The process of embryogenic callus (EC) induction and morphological observation in banana MJ. (**a**) MJ male inflorescence. (**b**) The internal flower combs of MJ banana. (**c**,**d**) Crystalline small granular callus on B2 medium for 12 months. (**e**) Yellowish large granular callus cultivated on B2 medium for 8 months; (**f**) EC cultivated on B2 medium for 14 months. (**g**,**h**) Paraffin section of banana non-embryogenic callus as c and e, respectively. (**i**) Paraffin section of banana EC as (**f**). Scale bar = 2 cm (**a**), scale bar = 1 mm (**b**–**f**), scale bar = 20 μm (**g**–**i**).

We identified 6440, 5053, 4335, and 1787 upregulated DEGs and 6634, 6005, 4143, and 1018 downregulated DEGs across four comparisons: 5M versus 0M, 12M versus 5M, EC versus 12 M, and BC versus EC, respectively, using the AA genome as a reference (Figure 2f–g). Similar patterns of upregulated and downregulated DEGs were observed when the BB genome was used as a reference (Figure 2f). Additionally, using the AA genome as a reference, we identified 4756, 2159, 1937, and 299 specific DEGs across the four groups, respectively (Figure 2g). Similarly, using the BB genome as a reference, we detected 4286, 2026, 1696, and 256 DEGs in the respective groups (Supplementary Figure S2). The number of specifically expressed genes progressively decreased across the four stages, suggesting that gene expression was significantly influenced during the early phases of explant differentiation. To investigate the specific expression of the A and B alleles in MJ bananas further, we identified homologous and specific genes using scripts based on mapping of the AA and

BB genomes. We identified 28,969 homologous genes, 7936 A-specific genes, and 2772 B-specific genes in the MJ bananas. The number of A-specific genes was 2.8 times greater than that of B-specific genes, indicating the significant predominance of A-specific genes in the MJ ABB genome.

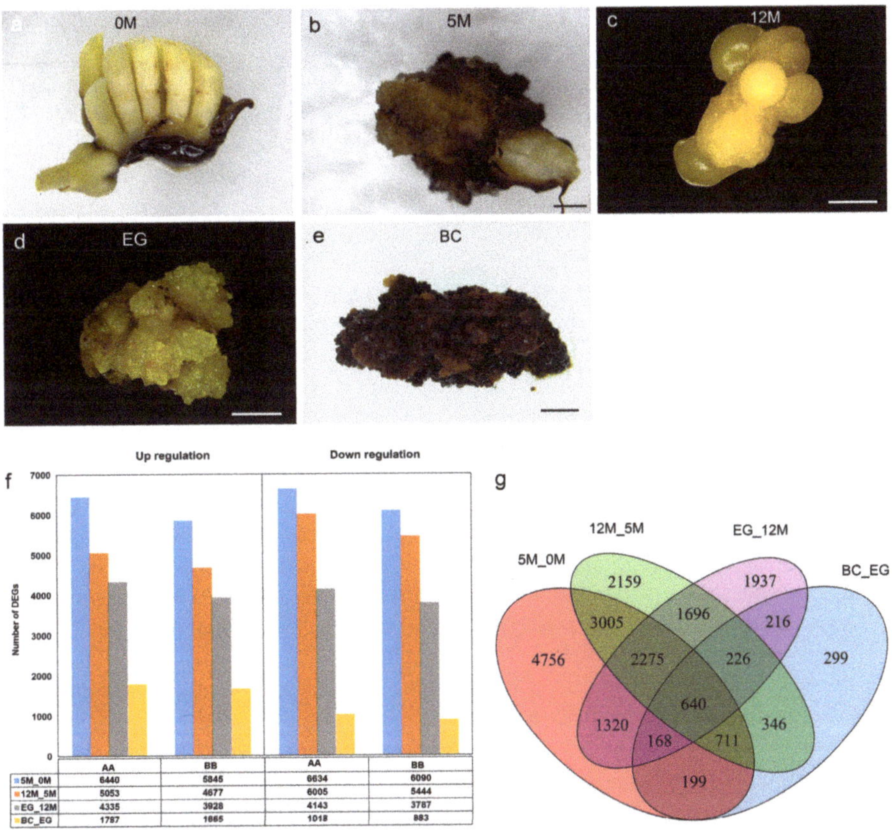

Figure 2. Overview of differentially expressed genes among five samples during the stages of EC induction in banana MJ. (**a**) MJ flower buds. (**b**) Explants induced on B2 medium for 5 months (5M). (**c**) Explants induced on B2 medium for 12 months (12 M). (**d**) EC obtained after 14 months of induction on the B2 medium. (**e**) Browned EC (BC). (**f**) The number of DEGs up- or downregulated during embryogenic callus formation. (**g**) Venn diagram showing overlap and specific DEGs in five samples. Scale bar = 2 mm (**a**–**e**).

2.3. *Global Analysis of DEGs Expressed in Calli During the EC Induction of ABB Banana*

Subsequently, GO classification and KEGG pathway analyses were conducted for the DEGs across the four comparisons. In the comparison between the 0M and 5M callus, DEGs were significantly enriched in biological processes such as response to stimulus, response to stress, embryogenic root development, and protein phosphorylation (Supplementary Figure S3a).

Additionally, in the comparison between the 12M and 5M callus, DEGs were enriched in biological processes, including response to stimulus, response to stress, oxidation–reduction, and DNA metabolic processes (Supplementary Figure S3b). As the 12M callus developed into an EC, the DEGs were primarily enriched in plant organ development, response to hormones, response to organic substances, and response to acidic chemicals

(Supplementary Figure S3c). When comparing browning calli and embryogenic calli, enrichment was observed for phenylpropanoid metabolic processes, phenylpropanoid biosynthetic processes, secondary metabolic processes, lignin metabolic processes, and flavonoid metabolic processes (Supplementary Figure S3d).

Furthermore, significant differences were identified in the KEGG pathways of DEGs during the formation of embryogenic calli in MJ at various stages. In the comparison between 12M and 5M callus, DEGs were enriched in pathways such as plant hormone signal transduction, tyrosine metabolism, fatty acid degradation, ABC transporters, and starch and sucrose metabolism (Supplementary Figure S4a). Following the development of the 5-month callus, the enriched pathways included plant hormone signal transduction, phenylpropanoid biosynthesis, glutathione metabolism, starch, and sucrose metabolism, and the MAPK signaling pathway (Supplementary Figure S4b). Notably, aside from metabolism and secondary metabolism, the hormone signaling pathway emerged as the most significantly enriched pathway, underscoring the pivotal role of hormones at this stage. Additionally, KEGG enrichment analysis revealed a distinct MAPK signaling pathway in plants. Previous evidence suggests that MAPK plays a crucial role in physiological processes such as the regulation of plant growth, development, and stress resistance [40].

In the comparison between 12M callus and embryogenic calli, plant hormone signal transduction and MAPK signaling pathways were the two most enriched pathways. Concurrently, significant enrichment was observed in metabolites such as glutathione metabolism, tropane, piperidine, and pyridine alkaloid biosynthesis; phenylpropanoid biosynthesis; and starch and sucrose metabolism (Supplementary Figure S4c). When comparing browning calli to normal calli, pathways including phenylpropanoid biosynthesis, tropane, piperidine, and pyridine alkaloid biosynthesis, phenylalanine, tyrosine, and tryptophan biosynthesis, glutathione metabolism, and flavonoid biosynthesis were predominantly enriched (Supplementary Figure S4d), suggesting that the accumulation of metabolites primarily begins at the EC stage. Subsequently, we focused on the changes in the expression levels of genes significantly enriched in these pathways to elucidate the cellular changes occurring in plant cells.

Next, we concentrated on the plant hormone signaling pathways, MAPK signaling pathways, and zeatin biosynthesis signaling pathways to uncover the critical roles of A-specific and B-specific alleles in the banana MJ EC induction process. Consequently, we identified 48, 32, and 5 genes within the A-specific alleles in three signaling pathways. Interestingly, we identified only two specific B alleles expressed in the hormone signaling pathway, whereas no specific B allele was identified in the MAPK or zeatin biosynthesis signaling pathways (Supplementary Tables S2 and S3). We selected seven specific genes for identify chromosomal position and specificity determination, encompassing six A-specific genes: *Macma4_04_g11940 (MJ_GID2)*, *Macma4_06_g14450 (MJ_ARF)*, *Macma4_07_g28170 (MJ_IAA)*, *Macma4_02_g00180 (MJ_CALM)* involved in the MAPK signaling pathway, *Macma4_08_g22960 (MJ_EIN3)*, and *Macma4_04_g24700 (MJ_CISZOG)* associated with the zeatin biosynthesis pathway, in addition to one B-specific gene, *Mba10_g12780 (MJ_PYL)*. Through a comprehensive synteny analysis of the genes flanking these seven specific loci, we observed that the interval lengths of these A-specific genes and their adjacent syntenic genes were significantly greater than those of the syntenic genes in the BB genome (Figure 3). The results showed that the emergence of these specific genes was attributable to gene loss in the BB genome or insertional structural variations. This finding not only corroborates the precision of identifying specific genes but also reveals potential mechanisms underlying the formation of these specific genes.

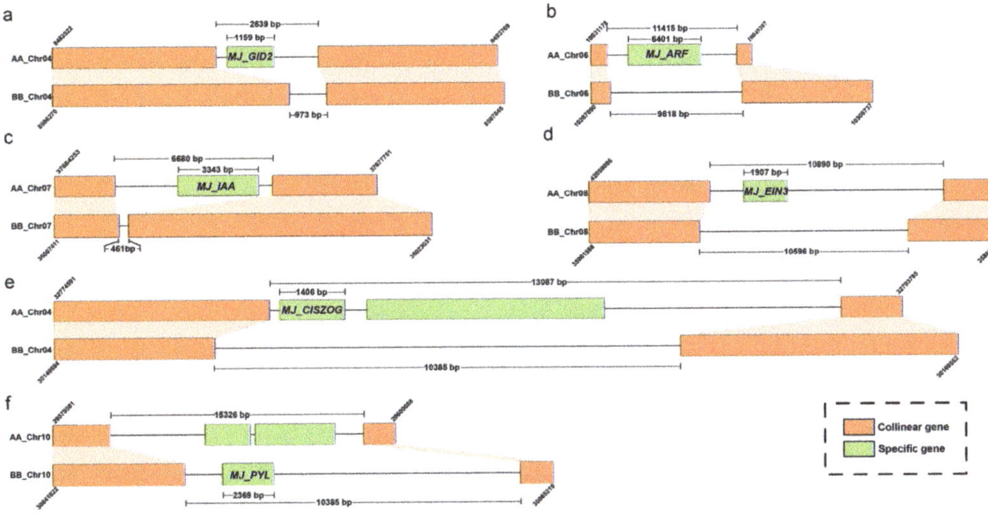

Figure 3. Visualization of the Specific Gene Locations Identified in the AA and BB Genomes. Specific genes within the AA and BB genomes were identified using BLASTn, and the positions of homologous genes flanking these specific genes in both genomes are displayed. This figure demonstrates the accuracy of specific gene identification and indicates that these specific genes arise due to insertion and deletion mutations. (**a**–**d**) Panels illustrate specific genes in the hormone signaling pathways, including *Macma4_04_g11940 (MJ_GID2), Macma4_06_g14450 (MJ_ ARF), Macma4_07_g28170 (MJ_IAA)* in the A genome compared to the B genome, *Macma4_02_g00180 (MJ_CALM)* in the MAPK signaling pathway, *Macma4_08_g22960 (MJ_EIN3)*, and *Macma4_04_g24700 (MJ_CISZOG)* in the zeatin biosynthesis pathway (**e**), as well as the specific gene *Mba10_g12780 (MJ_PYL)* in the BB genome (**f**).

2.4. The Dominant Expression Changes of Genes Involved in Auxin Signaling During EC Formation

To elucidate the regulatory network involved in EC dedifferentiation, we analyzed the DEGs enriched in hormonal signaling pathways across the four stages of callus development. Specifically, 234, 210, 183, and 58 genes were enriched in the comparisons of 5M vs. 0M, 12M vs. 5M, EC vs. 12M, and BC vs. EC, respectively. A Venn diagram was constructed to identify the stage-specific genes (Figure 2g, Supplementary Figure S2), highlighting the significant involvement of hormone-related genes in the differentiation potential of embryogenic calli. Our initial focus was on the auxin and cytokinin signaling pathways during callus induction.

The primary inactivation route for the natural auxin indole-3-acetic acid (IAA) involves the GH3-ILR1-DAO pathway. Initially, GH3 IAA amidosynthetases transform IAA into IAA-amino acid conjugates. These conjugates, specifically IAA-aspartate (IAA-Asp) and IAA-glutamate (IAA-Glu) act as storage forms of IAA. They can be reconverted to IAA through the action of ILR1/ILL amidohydrolases [41]. We identified 17 A allele GH3 genes and 19 B allele GH3 genes. Among the A alleles, *MJGH3.8_1*, *MJGH3.8_2*, *MJGH3.1_3*, and *MJGH3.8_5* were significantly upregulated during the 5M stage of MJ, whereas *MJGH3.1_1*, *MJGH3.1_5*, and *MJGH3.8_7* showed notable induction at the 12M stage, with a marked increase in expression in the EC (Figure 4a). These findings suggest that following callus induction, *MJGH3* expression is induced by auxin in the culture medium, leading to the production of more IAA-amino acid conjugates within the cells. This indicated that GH3 plays a crucial role in maintaining auxin concentrations and ensuring cell division, thereby promoting callus induction.

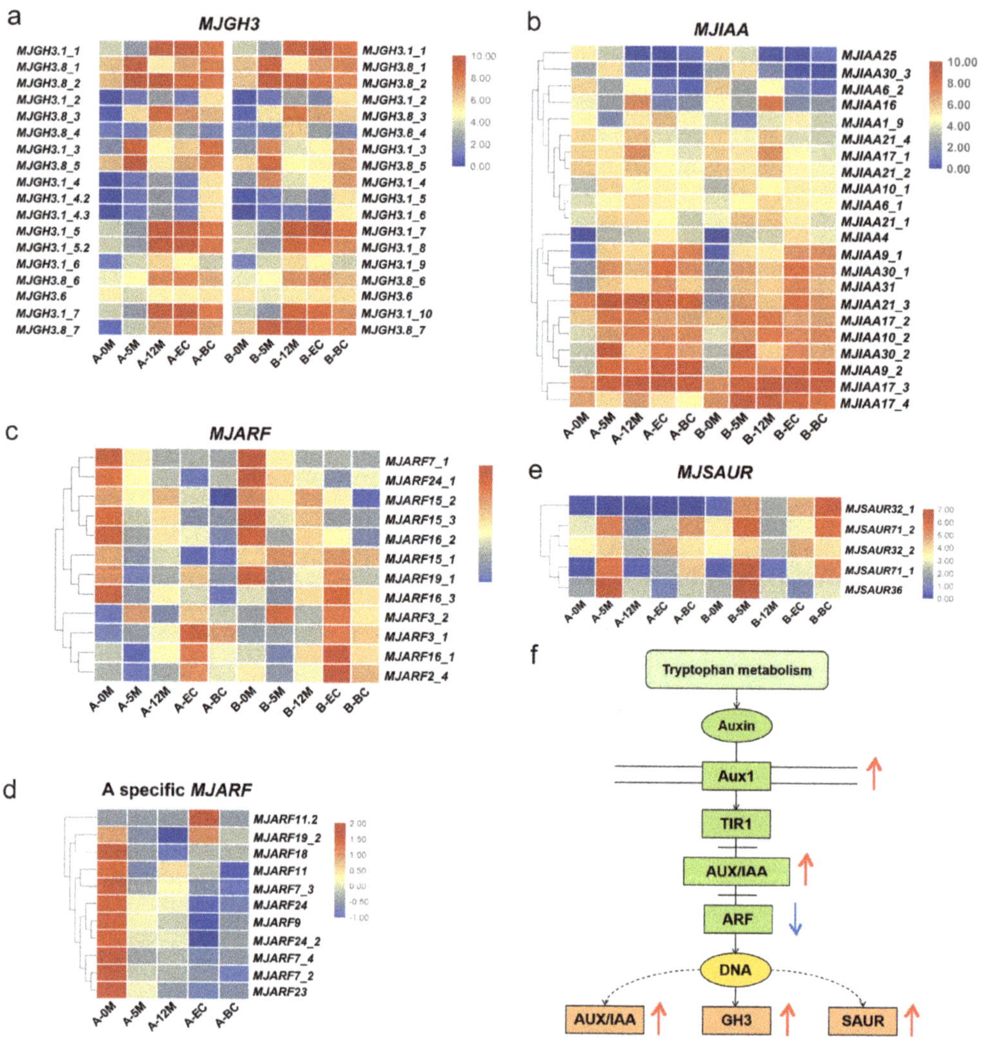

Figure 4. Expression of genes involved in the auxin pathway during embryogenic callus formation. The expression levels were visualized by using OmicStudio tools at https://www.omicstudio.cn/tool (accessed on 5 December 2023). Numbers beneath the heat map indicate the relative expression intensities and the higher expression intensities are indicated by more reddish colors. Identification of A and B alleles in the ABB genome was conducted using allele tables, where A-5M denotes the gene expression levels of the A allele at the 5M stage, and B-5M denotes those of the B allele at the same stage. (**a–c**,**e**) The expression levels of *MJGH3s*, *MJIAAs*, *MJARFs*, and *MJSAUR*, which are associated with the auxin pathway. (**d**) A heat map showing A-specific allele expression levels of *MJARF*. (**f**) The diagram illustrates the upregulation and downregulation of components in the auxin signaling pathway.

In the auxin signaling pathway, Aux/IAA plays an inhibitory role in the regulation of auxin-responsive gene expression. Aux/IAA proteins form dimers with the Auxin Response Factors (ARFs), thereby inhibiting their transcriptional regulatory functions of ARFs. ARF transcription factors can bind to the cis-regulatory elements of auxin-inducible genes and regulate their transcriptional expression [42]. We identified 27 differentially expressed members of the IAA family in both A and B alleles during MJ callus induction. Notably, *MJIAA21_2*, *MJIAA30_3*, *MJIAA16*, and *MJIAA21_4* were notably induced at the 5M stage of callus initiation, with B alleles showing a consistent expression pattern (Figure 4b). This suggests that IAA transcriptional repressors were significantly upregulated after callus induction, initiating transcriptional repression. Additionally, 12 differentially expressed ARF genes were identified. Interestingly, these alleles exhibited low expression levels during the MJ callus induction stage, except for one gene, *MJARF24_1*, which was highly expressed in MJ explants but showed a significant decrease in expression at the 5M stage (Figure 4c,f). Notably, we identified 11 specific A alleles of *MJARF* genes, all of which displayed a significant downward trend during callus induction (Figure 4d,f). These findings indicate that, during banana callus induction, the accumulation of IAA family members leads to an increase in bound ARFs, resulting in decreased ARF expression, thereby regulating callus proliferation and differentiation.

Small auxin-up RNA (SAUR) is a rapidly responsive gene to auxin [43]. A total of five *MJSAUR* genes that exhibited differential expression during the callus induction process with MJ treatment were identified. Notably, almost all these genes showed a significant increase in expression at the 5M stage (Figure 4e,f). This suggests that *MJSAUR* plays a crucial role in the early response to callus induction.

Cytokinin signaling relies primarily on downstream auxin response regulators (ARRs) to convey signals. Interestingly, we observed minimal differential expressions of cytokinin-related genes during callus induction in bananas. Only three ARR-B genes were identified as A-specific alleles. However, they exhibited very low expression levels in both MJ explants and callus induction stages (Supplementary Table S2). Additionally, cytokinin dehydrogenase (CKX) inactivates cytokinins by oxidatively removing their active side chains, rendering them biologically inactive. We identified one specific A allele of *MJCKX* with no differential expression; the B allele was not expressed during callus formation (Supplementary Tables S2 and S3).

2.5. BSK and BZR1 Are Involved in the Regulation of EC Induction

Brassinosteroids (BRs) constitute a class of plant steroid hormones playing pivotal roles in plant growth and development [44,45]. Studies have demonstrated that supplementation with BRs at suitable concentrations enhances SE formation in various plant species, including plumules [46], *Coffea arabica* [47], and *cotton* [48]. In our study, we identified five DEGs enriched in brassinosteroid (BR) synthesis and signal transduction pathways, comprising two BR-signaling kinase genes (BSK) and three brassinosteroid-resistant 1/2 genes (BZR1_2). The expression abundance of the *MJBSK1* gene peaked during the 0M period, followed by initial downregulation and subsequent gradual upregulation after the 12M period. Conversely, the expression of the *MJBSK2* gene was upregulated during the 5M period, reaching its peak at the 12M period and then gradually decreasing during EC development, with an interesting upregulation observed in the browning EC (Figure 5a). For the *MJBZR1_1* genes, expression abundance peaked in the EC, whereas *MJBZR1_2* expression peaked at the 12M stage (Figure 5a). The expression patterns of the B allele of *MJBSK1* and *MJBZR1* were highly similar to those of the A allele, suggesting the conservation of *BSK* and *BZR1* in the functions of both A and B alleles (Figure 5a). These results indicated that *MJBZR1_1* and *MJBZR1_2* play positive roles in inducing EC formation.

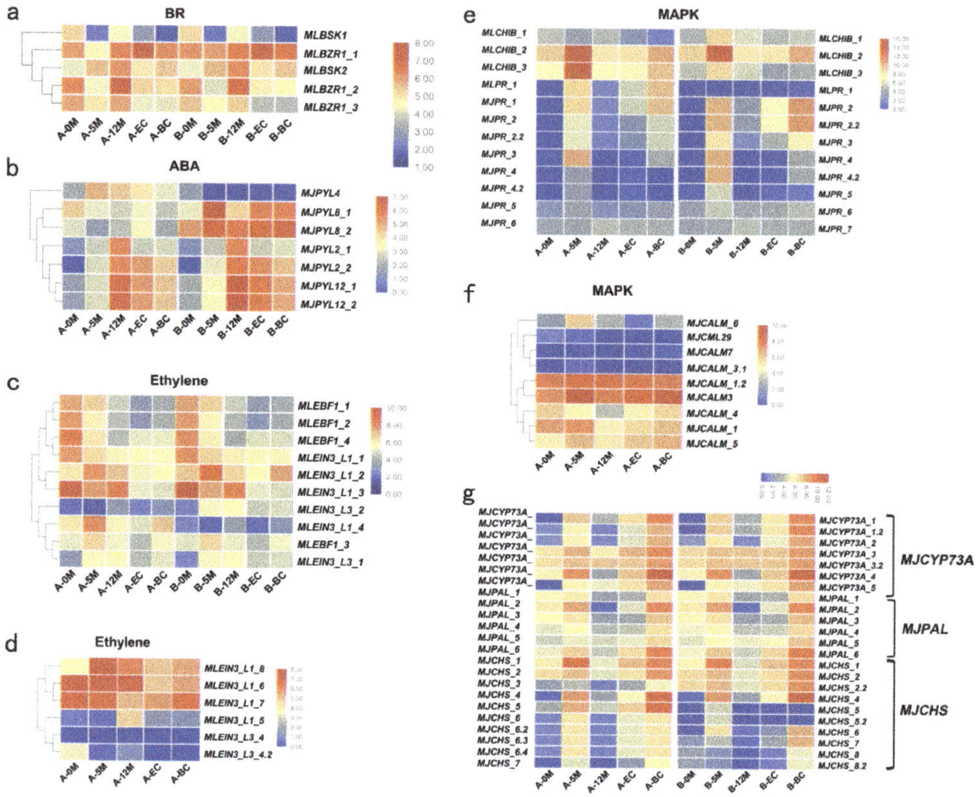

Figure 5. Expression of genes involved in the hormone pathway pathway during embryogenic callus formation. Numbers beneath the heat map indicate the relative expression intensities and the higher expression intensities are indicated by more reddish colors. Identification of A and B alleles in the ABB genome was conducted using allele tables, where A-5M denotes the gene expression levels of the A allele at the 5M stage, and B-5M denotes those of the B allele at the same stage. (**a**) A heat map illustrating gene expression levels in the BR signaling pathway. (**b**) A heat map for gene expression levels in the ABA signaling pathway. (**c**) A heat map depicting gene expression levels in the ethylene signaling pathway. (**d**) A heat map showing A-specific allele expression levels within the ethylene signaling pathway. (**e**) A heat map of gene expression levels in the MAPK signaling pathway. (**f**) A heat map of A-specific allele gene expression levels in the MAPK signaling pathway. (**g**) The expression levels of *MJCYP73As*, *MJPALs*, and *MJCHSs*, which are associated with embryogenic callus browning.

2.6. Abscisic Acid Pathway Gene PYL Is Very Important to Respond EC Induction

Abscisic acid (ABA) is a classical plant growth inhibitor that regulates various physiological processes, including embryo maturation, seed dormancy, germination, cell division, and elongation [49]. ABA also inhibits the germination of mature embryos and their conversion into plantlets [50]. ABA is perceived by the intracellular PYR/PYL/RCAR receptors, which are encoded by *PYL* genes [51,52]. When ABA binds to PYL receptors, it triggers a signaling cascade that activates downstream responses [53]. In our study, we identified eight DEGs enriched in the abscisic acid receptor PYR/PYL family. Four *MJPYL* genes (*MJPYL2_1*, *MJPYL12_1*, *MJPYL12_2*, and *MJPYL2_2*) were highly expressed during EC induction. However, the B alleles of *MJPYL8_1* and *MJPYL8_2* were significantly upregulated in Stage 2 (five months after callus induction in explants), whereas the expression of the A alleles showed no significant changes (Figure 5b). This indicated that the B alleles of these two alleles were specifically expressed. Interestingly, we also identified one B allele of *MJPYL* that was specifically expressed and exhibited the highest expression levels in the embryogenic calli (Supplementary Table S3). This suggests that the B-specific allele of *MJPYL* is crucial for EC formation in MJ bananas.

2.7. Genes Involved in Ethylene Signaling Negatively Regulate the Formation of EC

Ethylene regulates organogenesis and SE, and its role has been extensively studied [54]. EBF1 (EIN3-Binding F-box 1) and EIN3 (Ethylene Insensitive 3) are key components of the ethylene signaling pathway, where EBF1 negatively regulates EIN3 by promoting its degradation. In our study, three out of four *MJEBF1* genes showed high expression in the explants, but all were downregulated after callus induction, with similar trends observed for both the A and B alleles (Figure 5c). Furthermore, six *MJEIN3* genes exhibited similar expression patterns in both the A and B alleles. Among them, *MJEIN3_L1_1* and *MJEIN3_L1_3* showed the highest expression in the explants, with their expression levels gradually decreasing after callus induction. However, *MJEIN3_L1_2* and *MJEIN3_L1_4* displayed different patterns, with their expression levels significantly upregulated at Stage 2 (five months after callus induction), followed by a decrease (Figure 5c). Interestingly, we identified six *MJEIN3* genes that were specifically expressed in the A allele. Among these, *MJEIN3_L1_5*, *MJEIN3_L3_4*, and *MJEIN3_L3_4.2* exhibited very low expression levels, whereas *MJEIN3_L1_6*, *MJEIN3_L1_7*, and *MJEIN3_L1_8* exhibited allele-specific expression (Figure 5d). Specifically, *MJEIN3_L1_6* and *MJEIN3_L1_7* alleles negatively regulated callus formation, whereas *MJEIN3_L1_8* was significantly highly expressed only at stage 2, and its expression decreased at other stages (Figure 5d). These findings indicate that *MJEBF1* and *MJEIN3* serve as negative regulators of callus induction. Notably, the expression of specific *MJEIN3* A alleles appears to play a crucial role in negative regulation during embryogenic development. This suggests a complex regulatory mechanism by which these alleles modulate key developmental processes, potentially influencing the efficiency and outcomes of callus formation.

2.8. Specific A Alleles Involved in the MAPK Signaling Pathway Positively Regulates the Development of EC

In plants, MAPK signaling pathways play crucial regulatory roles in various biological processes, including responses to biotic and abiotic stresses, as well as hormone and developmental signaling pathways [55]. BraMAPK3 demonstrates a rapid response to treatments involving salt, heat, waterlogging, wounding, 6-BA, and NAA, suggesting that BraMAPK3 may be a key regulator of abiotic stress responses [56]. Additionally, miRNAs target genes within the MAPK signaling pathway to regulate the accumulation of functional metabolites in longans [57].

We investigated the MAPK signaling pathway and identified 128 and 103 DEGs in comparisons between 12M vs. 5M and between EC vs. 12M, respectively. Among these, 68 DEGs were common to both comparisons, whereas 60 of 128 and 35 of 103 DEGs were unique to each comparison. CHIB basic endochitinase B (CHIB) and pathogenesis-related protein 1(PR1) are both involved in plant defense mechanisms and can be linked to the MAPK signaling pathway, which is crucial for transmitting stress and defense signals in plants [58]. Specifically, we identified three *CHIB*, six *PR1* for the A allele, and three *CHIB* and seven *PR1* genes for the B allele that were differentially expressed during callus formation in Banana MJ (Figure 5e).

Notably, the three *CHIB* genes for the A and B alleles were significantly upregulated at the 5-month callus stage but were downregulated at other stages, particularly *MJCHIB2* and *MJCHIB3*, indicating their dominance in banana callus formation. Plant pathogenesis-related (PR) proteins are crucial components of plant defense mechanisms and are primarily associated with responses to biotic and abiotic stress [59,60]. For the A allele of PR1, all *MJPR* were highly expressed in the 5-month callus and browning calli, except for the B alleles of *MJPR_1* and *MJPR_6* (Figure 5e). This indicates that the *MJPR1* gene exhibits a similar effect in both the A and B alleles.

Interestingly, our analysis revealed 32 genes associated with the MAPK signaling pathway within specific A alleles, including notable genes such as calmodulin (CALM), *MKK4_5, MKK9, MPK1,* and transcription factor MYC2 (Supplementary Table S2). Among these nine members of the *MJCALM* gene family, *MJCALM_1, MJCALM_1.2,* and *MJCALM3* (Figure 5f) showed remarkably high expression levels during callus formation. Of special interest, *MJCALM3* was significantly induced at the 5M stage, underscoring its pivotal role in callus induction. These observations imply that the 32 genes involved in the MAPK signaling pathway may have a substantial impact on callus induction.

2.9. Pheylalanine and Flavonoids Are the Main Phenolic Compounds Responsible for Enzymatic Browning

Tissue culture is an effective method for rapidly breeding seedlings and enhancing production efficiency. However, explant browning poses a key limitation to successful tissue culture. Enzymatic browning is the primary cause of explant browning in plant tissue culture. In banana tissue culture, browning of embryogenic calli is a significant hurdle in banana callus transformation. Through a comparative analysis between embryogenic calli and browning calli, we identified 49 DEGs significantly enriched in phenylpropanoid biosynthesis, 18 DEGs in the Tropane, piperidine, and pyridine alkaloid biosynthesis pathways, 17 DEGs in ubiquinone and other terpenoid–quinone biosynthesis pathway and 21DEGs in the Flavonoid Biosynthesis pathway. Cinnamate 4-hydroxylase (C4H; CYP73A) is a cytochrome P450 monooxygenase located in the endoplasmic reticulum of plants. This enzyme utilizes NADPH-cytochrome P450 reductase as an electron donor and hydroxylates cinnamic acid to produce 4-coumaric acid during phenylpropanoid metabolism. Notably, all six *MaCYP73A* genes and their alleles exhibited markedly upregulated expression levels in the browning calli, with particular emphasis on *MJCYP73A_5* (Figure 5g). In the phenylalanine metabolism pathway, genes encoding phenylalanine ammonia-lyase (PAL), peroxidase, and cinnamyl-alcohol dehydrogenase (CAD) displayed lower expression during the formation stage of the EC but significantly higher expression in browning calli. Notably, the inhibition of PAL activity has been linked to a reduction in phenolic compound biosynthesis, and our analysis identified six *MaPAL* genes and alleles that showed elevated expression in the browning callus compared to the EC (Figure 5g). CHS (chalcone synthase) is a key enzyme in the biosynthetic pathway of flavonoids. We identified nine MaCHS DEGs, four of which were specific genes lacking the B allele. All of these CHS genes exhibited elevated expression levels in the browning callus, with particular emphasis on the specific gene *MJCHS_5* (Figure 5g). When explants are cut, the cells are damaged, leading to the release of phenols that undergo enzymatic oxidation to form quinones, resulting in enzymatic browning [61].

2.10. Specific Expression of Transcription Factors During the EC Formation Process

Transcription factors (TFs) are master regulators that play crucial roles in various biological processes by modulating a wide array of downstream targets. Our RNA-seq data revealed that the bHLH, MYB, ERF, NAC, and WRKY families were the top five TF families identified in the DEG involved in the four comparisons (Figure 6a,b). Notably, one-third of the transcription factors were significantly downregulated upon initiation of EC induction, indicating their negative regulatory role in this process. In contrast, the remaining two-thirds of the transcription factors exhibited varying degrees of upregulation during both the 5-month and 12-month stages of EC development, suggesting at positive regulatory function for the induction of embryogenic calli (Figure 6c).

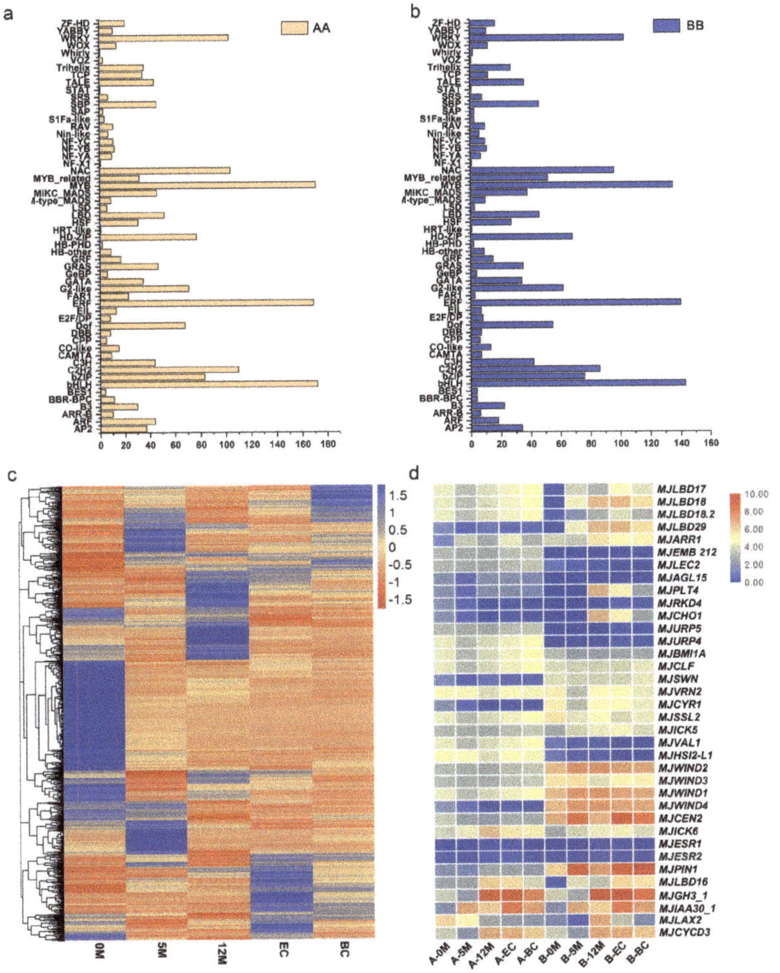

Figure 6. Expression levels of differentially expressed transcription factors and homologous genes previously reported to be involved in callus induction. (**a**,**b**) Statistics on the number of transcription factors encoded by differentially expressed genes identified during the banana embryogenic callus induction process, using AA and BB as reference genomes, respectively. (**c**) Expression levels of transcription factors encoded by differentially expressed genes during the embryogenic callus induction process using AA as reference. (**d**) Expression levels of homologous genes in banana MJ corresponding to genes reported in *Arabidopsis*, which are essential in the callus induction process. Genes highlighted in blue indicate lower expression levels of transcription factors (TFs), while those highlighted in red represent TFs with higher expression levels.

Over the past decade, studies of mutants with abnormal callus induction and development have revealed that callus induction is governed by intricate regulatory mechanisms [62]. Several key molecules involved in callus development regulation have been identified in *Arabidopsis*, many of which are crucial transcription factors. We investigated homologous genes in bananas and examined the expression of their A and B alleles. Interestingly, we found that nearly 80% of these homologous genes exhibited very low or no significant expression during EC induction in bananas. However, some previously reported transcription factors, such as *MJLBD* and *MJWIND1*, were significantly induced. Specifically, *MJLBD17* and *MJLBD18* were markedly induced in the BB genome, whereas both A and B alleles of the *MJLBD16* gene are induced during banana development. Additionally, *MJWIND1*, *MJWIND2*, *MJWIND3*, and *MJWIND4*, and the auxin transport carrier genes *MJPIN* and *MJLAX2* showed significant induction only in the B allele during EC induction. *CYCD3*, previously reported as a cell cycle-related gene, also showed significantly high expression of both the A and B alleles during the callus induction stage in bananas (Figure 6d). To validate the accuracy of the transcriptome data, we selected six genes (*MJGH3_1*, *MJIAA30_1*, *MHLAX2*, *MJCYCD3*, *MJWIND2*, *MJAGL15*) for quantitative analyses. The observed expression patterns were consistent with those of the RNA-Seq, confirming the reliability of the data (Figure 7, Supplementary Table S4). These findings suggest that the expression of these genes, which are known to play critical roles in callus induction in *Arabidopsis*, plays an allele-specific role in bananas.

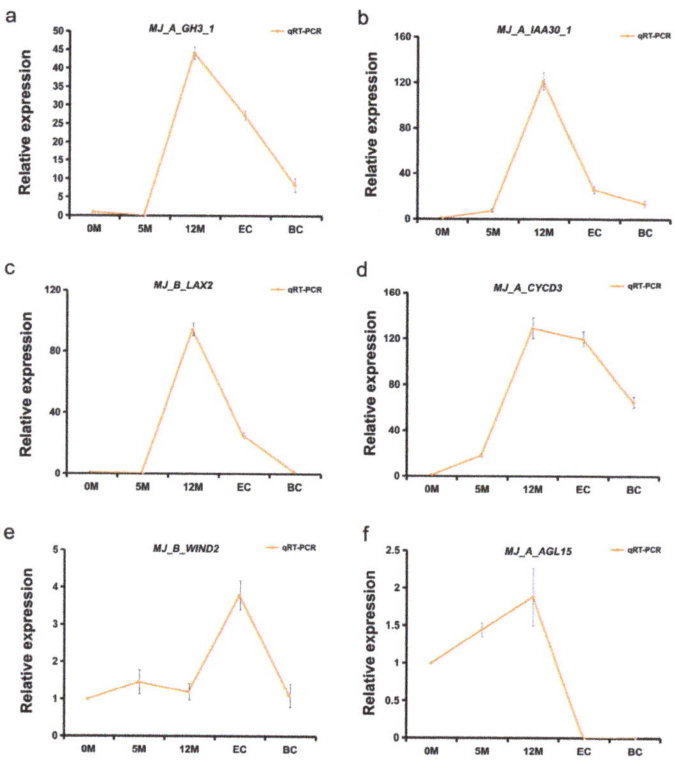

Figure 7. Transcript levels of four genes as determined by qRT-PCR. (**a**–**f**) Illustrate the expression of the A allele of *MJGH3* (*MJ_A_GH3.1*), *MJ_A_IAA30_1*, *MJ_A_CYCD3*, the B allele of *MJLAX2* (*MJ_B_LAX2*) and *MJ_B_WIND2*, respectively. The data presented are the means ± standard deviation (S.D.) from three biological replicates.

3. Discussion

The induction of embryogenic calli in bananas is challenging. In this study, we successfully induced embryogenic calli in the ABB genotype of the banana MJ variety under various test conditions, and histological analysis confirmed the excellent state of the EC. Through transcriptome analysis during the formation of embryogenic calli, we discovered that auxin signaling pathway genes played a dominant role in the induction of

MJ embryogenic calli, followed by the involvement of the MAPK signaling pathway, ABA, and ethylene. This indicates that the induction of banana callus is a process of coordinated hormonal signaling regulation, with callus induction and explant defense responses being synergistically controlled. Surprisingly, we demonstrated the minimal involvement of cytokinin signaling molecules in the establishment of embryogenic calli.

In our study of the banana ABB genotype cultivar MJ, we focused on the dominant and specific expression of the A and B alleles. Interestingly, we found that the homologous genes of the A and B alleles in MJ exhibited very similar expression patterns in the auxin, ABA, ethylene, and MAPK signaling pathways, with no homologous genes showing dominant expression levels (Figure 8). However, we identified some A-specific genes expressed in these signaling pathways, including the specific expression of the ARF gene family in the AA genome, which negatively regulates EC induction (Figure 8). *MJEIN3* in the ethylene signaling pathway also acts as a negative regulator, whereas *MJCALM* in the MAPK signaling pathway acts as a positive regulator (Figure 8). This suggests that these specific A genes have unique effects on embryogenic calli in bananas. In the B allele, we identified only two PYL genes in the ABA signaling pathway, and only one was specifically expressed in embryogenic calli, indicating that ABA signaling molecules in B-specific genes promote EC expression to some extent (Figure 8). In addition, our analysis of embryogenic calli and browned embryogenic calli revealed that browning was primarily caused by the accumulation of phenylpropanoids and flavonoids (Figure 8). Browning can be effectively suppressed by inhibiting the synthesis of these phenolic compounds, thereby enhancing the activity of banana embryogenic calli and improving their genetic transformation.

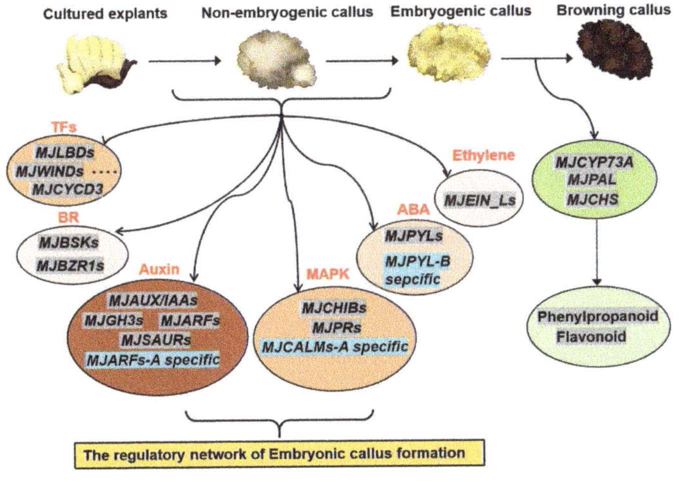

Figure 8. A schematic representation of gene expression regulation during the induction of embryogenic callus from explants in the banana cultivar MJ. The depth of the orange color in the oval signifies the importance of the involved signaling pathways, with darker hues indicating a higher likelihood of involvement. Genes highlighted in gray denote the presence of both A and B alleles, while those in blue indicate alleles specific to A or B. The auxin signaling pathway genes are primarily responsible for the induction of embryonic callus in bananas, with significant contributions from the MAPK signaling pathway, as well as the ABA, Ethylene, and BR pathways. Importantly, transcription factors such as LBD, WIND, and CYCD3, previously identified as key players in callus induction, are also integral to the formation of embryogenic callus in bananas, establishing a comprehensive regulatory network. Furthermore, our findings indicate that the predominant cause of browning in banana embryogenic callus is the marked upregulation of genes involved in the phenylalanine and flavonoid biosynthesis pathways, resulting in the accumulation of these compounds and potentially leading to browning.

3.1. Auxin Master Regulator in the Induction of EC in ABB Banana

Auxins and cytokinins are primary hormones that induce callus formation. Recent studies have highlighted auxin as a well-known inducer of lateral root formation in *Arabidopsis*. In the auxin signaling pathway, enzymes encoded by GH3 genes primarily function to conjugate auxins with amino acids, thereby regulating auxin activity and levels. This regulatory mechanism helps maintain the auxin balance during various developmental stages and environmental conditions [41]. Studies have shown that within a few hours of SE induction, IAA concentration decreases while CcGH3 levels increase, resulting in reduced auxin levels [63]. These findings suggest that endogenous auxin accumulation is crucial for altering cell fate and establishing embryogenesis. Furthermore, in our studies, during the induction of embryogenic calli in bananas, we observed significant changes in *MJGH3* gene expression. Genes such as *MJGH3.8_1*, *MJGH3.1_3*, and *MJGH3.8_5* were highly expressed at the 5M stage and during both the 12M and EC stages, whereas genes such as *MJGH3.1_5* were significantly expressed during the 12M and EC stages (Figure 4a). This indicates that auxin levels in MJ banana callus induction are regulated by multiple GH3 genes, with endogenous auxin decreasing after 5M induction, promoting EC formation through precise regulation by multiple GH3 genes, differing from previous studies.

Previous studies have indicated that nearly 86 transcription factors related to auxin synthesis and signaling are differentially expressed from the dedifferentiation process to the SE development stage. In embryogenic calli, *GH3.17*, *PIN3*, *AUX.IAA*, *YUC*, *AUX/LIKE1*, and *SAUR* show highly expressed; however, their expression decreases during embryo proliferation. However, their expression levels decline as embryo proliferation progresses [64]. We found that *MJIAAs* were also significantly upregulated after callus induction (Figure 4b), suggesting that the expression changes in IAA genes during callus induction are relatively conserved. In our study, *MJSAURs* were significantly induced at the 5M stage (Figure 4e), indicating their involvement in EC formation in bananas. Our study also found significant upregulation of *MJPIN1* at the onset of the 5M stage (Figure 6d), indicating a significant correlation between PIN1 and banana EC formation, which is consistent with previous studies.

Various members of the LBD family also referred to as the ASYMMETRIC LEAVES2-LIKE transcription factors, such as LBD16, LBD17, LBD18, and LBD29, play a role in callus formation on explants cultured on callus induction medium (CIM) by interacting with ARF7 and ARF19 [65,66]. Our results indicate that *MJLBD16* was significantly induced at the 12-month stage of callus induction (Figure 6d), suggesting that its role is also conserved in monocotyledonous bananas, potentially acting as a key transcription factor that activates the expression of downstream genes related to callus induction.

ARRs play a pivotal role in the induction of callus through cytokinin signaling. These transcription factors are activated via a multistep phosphorelay process, leading to the activation of various target genes [22]. One notable target of type-B ARRs is CYCD3, which is crucial for reinitiating the cell cycle. The expression of *CYCD3* is significantly increased within an hour of cytokinin exposure, and its overexpression can promote callus formation even without external cytokinin application [67]. Although our callus-induction medium did not include exogenous cytokinins, we observed a significant increase in *MJCYCD3* expression at the 12M stage of callus induction (Figure 6d). We did not detect significantly high expression of type-B ARRs during the callus induction stage, suggesting that MJ callus formation may be promoted by the activation of *MJCYCD3* expression through endogenous cytokinins. The AP2/ERF transcription factors, including ESR (also referred to as DORNRÖSCHEN [DRN]), ESR1, and ESR2, are potential candidates involved in cytokinin-driven callus formation. Notably, the overexpression of either ESR1 or ESR2 can trigger callus formation even without the addition of external plant hormones [68,69].

However, in bananas, we found that *MJESR1* was not expressed during the callus formation, indicating that the underlying mechanism in bananas is significantly different from that in papaya.

3.2. The PYL Genes Play a Specific Promotive Role in the B Genome During Banana Callus Formation

In *Ormosia henryi*, the levels of IAA, cytokinins, gibberellins, and ABA significantly increase during the development of embryogenic calli (EC). The expression patterns of DEGs involved in ABA biosynthesis and signal transduction, such as ZEP, ABA2, AAO3, CYP97A3, PYL, and ABF, are consistent with endogenous hormone levels observed during SE [70]. As receptors of ABA, PYL proteins play crucial roles in ABA-mediated seed germination, with ABI5 modulating seed germination through the feedback regulation of PYR/PYL/RCAR ABA receptor gene expression [71]. In banana tissues, the expression levels of most *PYL* genes peaked during the 5M and 12M stages and gradually decreased during the EC and BC stages (Figure 5b). However, previous studies indicated that ABA represses callus formation, highlighting its negative effects on cell proliferation. In our study, we did not observe the upregulation of *PYL* during SE in banana tissues. Notably, within the hormone signaling pathways, we identified only two B-specific genes, the *MJPYL* genes, whose expression began to increase gradually from the 5M stage, reaching the highest levels in embryogenic calli (Supplementary Table S3). ABA concentration in plants increases upon exposure to salt stress. ABA binds to its receptor proteins (PYR/PYL/RCAR), thereby transmitting the ABA signal to downstream transcription factors that regulate the expression of stress-responsive genes. These findings suggest that the *PYL* gene, a downstream receptor of ABA, is significantly upregulated during banana callus formation, indicating a potential promotive role in callus development rather than merely responding to stress signals. The specific inducible expression of the two B alleles of *MJPYL* suggests that they may play a crucial role in promoting callus formation via the ABA signaling pathway. This represents a polyploid-specific expression pattern that is significantly different from that of ordinary diploids.

3.3. Elucidating the Primary Browning Compounds Contributes to Mitigating Browning During the Process of EC Induction

Most studies on callus responses to browning have concentrated on physiological indicators, with limited research on the transcriptomic responses of banana calli to browning [72]. In the present study, we analyzed the genetic differences between embryogenic and browning calli in bananas. We identified DEGs that were enriched in pathways such as flavonoid biosynthesis, phenylalanine metabolism, the biosynthesis of phenylalanine, tyrosine, and tryptophan, and phenylpropanoid biosynthesis. These findings suggest that phenolic compounds, which are ubiquitous secondary metabolites in plants, serve as primary substrates for enzymatic browning. These substances can be categorized as flavonoids, phenolic acids, and tannins.

Browning results from a physiological enzymatic process where phenolic production, influenced by sugars, is essential. Preventing browning can be achieved by inhibiting flavonoid production. CHS is responsible for catalyzing the first critical step in the flavonoid biosynthesis pathway [73,74]. The expression of the CHS genes was upregulated during the browning of banana explants, which accelerated the browning process (Figure 5g). Phenolic secondary metabolites significantly influence the quality of plant-derived foods by affecting their appearance, flavor, and health-promoting properties. Various factors affect the contents of these metabolites, influencing their stability, biosynthesis, and degradation. Notably, PAL is a key enzyme in their biosynthesis, inducible by diverse stress and environmental conditions [75]. PAL and peroxidase (POD) act as catalysts for polyphenol biosynthesis, leading to wound-induced browning [76]. The PAL gene family is crucial for plant growth, development, and abiotic stress response and has been identified in various plants [77]. Browning of banana fruit during ripening results from the release of pre-existing PPO enzyme, which is synthesised very early in fruit development [78]. Genome-wide analysis of the polyphenol oxidase gene family in bananas reveals that MaPPO1 and MaPPO6 are major contributors to fruit browning during fruit ripening [79]. Phenylpropanoid compounds serve multiple functions in plant defense mechanisms, acting as both preformed and inducible barriers-physical and chemical-against infections. They also function as signaling molecules that trigger defense gene activation both locally and systemically [80]. Cinnamate 4-hydroxylase (C4H) is a crucial gene within the phenylpropanoid pathway, playing a role in the regulation of flavonoid and lignin biosynthesis in plants. Additionally, cinnamyl alcohol dehydrogenase (CAD) facilitates the final step in monolignol biosynthesis by converting cinnamyl aldehydes into alcohols, utilizing NADPH as a cofactor [81]. 4-Coumarate: coenzyme A ligase (4CL) is a critical enzyme in the phenylpropanoid biosynthetic pathway [82]. The upregulation of these genes in MJ ultimately leads to flavonoid biosynthesis and the accumulation of phenylpropanoids, with the phenolic substances responsible for enzymatic browning originating primarily from the phenylpropanoid pathway (Figure 5g). These phenolic compounds cause browning in MJ embryogenic calli. By identifying the main metabolites responsible for browning in bananas, we can effectively inhibit the production of these compounds during banana tissue culture using substances that suppress their synthesis, thereby enhancing callus induction rates and transformation efficiency. Previous studies on plant tissue cultures of *Taxus chinensis* and *Glycyrrhiza inflata* identified flavonoid compounds from the phenylpropanoid pathway as the primary contributors to browning, with myricetin and quercetin being the main substrates of the browning reaction. The inhibition of flavonoid biosynthesis can prevent browning, and the use of gibberellic acid (GA3) as an inhibitor of flavonoid biosynthesis has been shown to effectively control browning [73], further confirming the role of flavonoids in inducing browning. Research has shown that incorporating ascorbic acid into basal media can significantly reduce phenolic secretion, enhancing culture quality and boosting the survival rates of okra cotyledonary node explants (*Abelmoschus esculentus* L.) [83]. Additional research has demonstrated that activated charcoal in the culture medium can mitigate the negative effects of leached phenolics on the regeneration of *Celastrus paniculatus* and *C. orchioides* [84]. Based on these insights, we suggest that testing various combinations of GA3, ascorbic acid, activated charcoal, and other additives during callus induction in the ABB genotype MJ cultivar could inhibit embryogenic calli browning, thereby enhancing the efficiency of banana genetic transformation.

4. Materials and Methods

4.1. Callus Induction

The MJ variety of bananas (*Musa* spp. cv. Bengal, ABB Group), with a genotype of ABB, was introduced from Bangladesh and cultivated on the campus of Fujian Agriculture and Forestry University. Young male inflorescences were sourced from MJ banana plants exhibiting normal flowering patterns. They were used as explant material and initially rinsed under clean running water. The outer bracts and flowers were removed, leaving an inner inflorescence approximately 4 cm in size. These were then transferred to a laminar flow hood, where the inflorescences were rinsed three times with sterile distilled water. Following this, the inflorescences were immersed in 75% ethanol for 5 min with occasional stirring. The ethanol was discarded, and the inflorescences were rinsed 3–5 times with sterile distilled water. The inflorescence is then trimmed to approximately 1.5 cm, and immature male flowers are selected for inoculation. Eight banana cultivars were collected to establish an embryogenic callus induction system. The culture medium was prepared using MS basal medium and single distilled water, incorporating chemicals such as MS basal medium salts, 2,4-D, NAA, and IAA, as outlined in Supplementary Table S1. The concentrations of NAA and IAA were 1 mg/L in the culture medium, and the concentrations of 2,4-D in media B1 to B4 were 0.5, 1, 2, and 4 mg/L, respectively. The callus induction rate statistics were analyzed using IBM SPSS Statistics v.26.0.

4.2. Histological Observation of Callus

Embryogenic and non-embryogenic calli were fixed and preserved using FAA fixative, composed of 50 mL of 95% ethanol, 5 mL of glacial acetic acid, 10 mL of 37% formaldehyde, and 35 mL of double-distilled water. A vacuum was applied until bubbles appeared on the surface, maintained for 10 min, and then gradually released, allowing the tissue to settle at the bottom of the tube. Post-vacuum, the FAA fixative was replenished as necessary, and samples were stored at 4 °C overnight. Dehydration was subsequently performed by transitioning to absolute ethanol, starting with 50% ethanol, each concentration applied for 60 min, followed by an overnight treatment in absolute ethanol. Sequential treatments with 75% ethanol and 25% xylene, 50% ethanol and 50% xylene, 25% ethanol and 75% xylene, and finally 100% xylene were conducted, each for 60 min, with an overnight treatment in 100% xylene. Samples were then treated at 60 °C with 25%, 50%, and 75% paraffin–xylene solutions, each for 3–5 h, and finally with 100% paraffin for one week, changing the paraffin twice daily. For embedding, the embedding box was placed on an ice–water mixture, and melted paraffin was quickly poured into it. Once the paraffin slightly solidified at the bottom, the material was positioned in the paraffin using preheated tweezers and allowed to cool. The embedded tissue was sectioned into 8 μm slices using a microtome (LEICA, RM2145, Wetzlar, Germany), and the sections were floated on water on glass slides and incubated overnight at 37 °C. The samples were deparaffinized with 100% xylene and treated twice for 20 min each. Rehydration was performed sequentially with a 50% ethanol/xylene mixture, followed by 100%, 95%, 80%, and 70% ethanol, each for 2–5 min. The samples were stained with 1% periodic acid solution for 15–20 min, followed by Schiff reagent for 10–15 min, rinsed with running water for 5–10 min, stained with 0.5% hematoxylin for 2 min, and rinsed again with running water for 5 min. A few drops of mounting resin were applied to the slide, which was then covered with a coverslip. The morphology, internal structure, and cell division of the callus tissues at various stages were observed using a Nikon Ni-U research-grade upright microscope.

4.3. RNA Extraction and Library Conduction

Total RNA extraction utilized the TaKaRa MiniBEST Plant RNA Extraction kit (Takara Biomedical Technology (Beijing), Beijing, China). The integrity and quality of the RNA were assessed through 1.0% agarose gel electrophoresis and NanoDrop 2000 (Thermo Fisher Scientific, Waltham, MA, USA). RNA-Seq libraries were constructed using the NEBNext® Ultra™ RNA Library Prep Kit for Illumina® (New England Biolabs (NEB), Ipswich, MA, USA). Prior to sequencing on the Illumina NovaSeq platform at Novogene (Beijing, China), the concentrations of the prepared libraries were quantified using the Qubit® 2.0 fluorimeter (Qubit dsDNA HS Assay Kit, Thermo Fisher Scientific, Waltham, MA, USA).

4.4. Differential Gene Expression Analysis and Function Enrichment Analysis

Banana explants were subjected to flower culture, resulting in the generation of callus at various developmental stages, including EC and browning EC. Explants cultured on B2 for seven days were as the beginning control (MJ_0M), and after 5 months (MJ_5M) were selected for further culture, and those collected after 12 months (MJ_12M) were also cultured. The EC (MJ_EC) and browning EC (MJ_BC) obtained were then subjected to transcriptome sequencing with three replicates for each sample. Illumina sequencing was employed to generate the sequencing data, followed by quality control filtering using Fastp software (v.0.23.2, https://github.com/OpenGene/fastp (accessed on 8 July 2023). Assembly and annotation of the *M. acuminata* (AA genome) of DH-Pahang and *Musa balbisiana* (BB genome) of DH-PKW were downloaded from NCBI as reference genomes for the analysis of transcriptome data. Comparative analysis was performed using Hisat2 software [85] (v.2.1.0, https://daehwankimlab.github.io/hisat2/download/ (accessed on 15 July 2023), and feature counts were quantified using featureCounts [86] (v.2.0.2, https://github.com/ShiLab-Bioinformatics/subread (accessed on 20 July 2023) to calculate TMM (Trimmed Mean of M-values normalized data) value. Differential expression analysis was conducted using the Trinity software (v.2.11.0, https://github.com/trinityrnaseq/trinityrnaseq/wiki/Installing-Trinity, accessed on 20 July 2023), and the DESeq2 (v.1.42.0) package in R (v.4.3.1). DEGs were identified based on log2FoldChange (|log2FoldChange| > 1) and padj (<0.05). Gene Ontology (GO) enrichment analysis and Kyoto Encyclopedia of Genes and Genomes (KEGG) pathway enrichment analysis for the DEGs were conducted using the EggNOG Database (http://eggnog5.embl.de/#/app/home (accessed on 10 November 2023) and OmicShare (https://www.omicshare.com/tools/ (accessed on 15 November 2023). The visualization of the gene expression heatmap was also performed using the OmicShare platform(https://www.omicshare.com/tools/ (accessed on 15 November 2023). The prediction of the functional roles of the DEGs utilized the nr database; details regarding its setup can be found at https://www.matrixscience.com/help/seq_db_setup_nr.html (accessed on 10 December 2023).

4.5. Identification of Genomic Alleles in AA and BB

The collinear genes between AA and BB genome were identified using Mcscan (https://github.com/tanghaibao/jcvi/wiki/MCscan-(Python-version) (accessed on 2 August 2024), complemented by the reciprocalBLAST_plus.pl script with identity set to 0.6 and coverage set to 0.2. The remaining genes were added to the list of contributing genes through homologous alignment, resulting in a comprehensive collinearity table for AA and BB. Genes specific to AA and BB, with no matches, were further confirmed as specific genes by aligning them to the AA and BB genomes, respectively, using the GMAP software (http://research-pub.gene.com/gmap/, accessed on 20 October 2024) [87]. Additionally, seven specific genes were selected, and their adjacent genes were detected using BLAST (https://blast.ncbi.nlm.nih.gov/Blast.cgi, accessed on 20 October 2024). The positions of these specific genes were visualized using R (v.4.3.1), illustrating whether the variations were due to segment insertions or deletions. Structural variation analysis was conducted to confirm the variant genes.

4.6. Quantitative Analysis

The relative expression changes of six candidate genes (*MJ_A_GH3_1*, *MJ_A_IAA30_1*, *MJ_B_LAX2*, *MJ_A_CYCD3*, *MJ_B_WIND2*, and *MJ_A_AGL15*) in banana callus at five different stages were analyzed using real-time quantitative PCR (qPCR) technology. The Banana *RPS2* gene was used as the reference gene [88]. Sangon Biotech synthesised specific primers for each transcript. Extracting RNA from the material at different periods using Tiangen's RNAprep Pure Plant Plus Kit (Polysaccharides and Polyphenolics-rich). The extracted RNA was then reverse transcribed into cDNA using the EvoM-MLV reverse transcription premix kit (Accurate Biotechnology, Changsha, Hunan, China) according to the manufacturer's instructions. Materials from five different developmental stages (0M, 5M, 12M, EG, BC) were cultured on fresh media for 10 days before sampling. The five samples were collected around 7 p.m. and immediately frozen in liquid nitrogen.

All samples were stored at −80 °C. Once all the materials were collected, we proceeded with qPCR analysis. The RT-qPCR was performed using the TB Green Premix Ex Taq II kit (Takara Biomedical Technology (Beijing), China). The amplification program was as follows: 95 °C for 2 min, 95 °C for 5 s, 60 °C for 30 s, cycle 40, followed by a melting curve analysis from 65–95 °C. Relative transcript levels were calculated with the ΔΔCt method. The primer sequences are listed in Supplementary Table S4.

5. Conclusions

In this study, we performed a dynamic expression analysis of the EC induction process in the banana cultivar MJ. Our findings reveal that this process is mainly regulated by the plant hormone, MAPK, and zeatin biosynthesis pathways. Notably, while homologous genes did not exhibit dominant expression between the A and B alleles, there was a marked predominance of A-specific allele expression within these pathways, with B-specific allele expression being notably scarce during EC induction. This suggests that A-specific alleles may play a pivotal role in the induction of banana EC. Furthermore, we discovered that the browning of banana callus could be attributed to the substantial upregulation of key genes in the phenylpropanoid and flavonoid metabolic pathways, resulting in the accumulation of these metabolites. Additionally, we identified several key transcription factors known to be highly expressed during the induction of banana EC. Our results provide valuable insights into the allele-specific expression effects in banana EC induction and serve as important references for enhancing the banana embryogenic callus induction system.

Supplementary Materials: The following supporting information can be downloaded at: https://www.mdpi.com/article/10.3390/plants14050761/s1. Supplementary Table S1: Effects of 2,4-D on embryonic callus induction of Banana ABB cultivar MJ. Supplementary Table S2: Annotation of specific A Alleles enriched in plant hormone signal transduction, MAPK signaling pathways and zeatin biosynthesis. Supplementary Table S3: Annotation of specific A Alleles enriched in plant hormone signal transduction, MAPK signaling pathways and zeatin biosynthesis. Supplementary Table S4: Primers for RT-PCR are used to verify gene expression levels. Supplementary Figure S1: The Pearson's correlation analysis and Principal Component Analysis (PCA) were conducted on 15 samples. Each with three biological replicates, using the AA genome as a reference. Supplementary Figure S2: Venn diagram showing overlap and specific DEGs between four samples at different stages using BB genome as reference. 0M: immature male flower cultivated on B2 medium for one week as cultured explant; 5M: flower cultivated on B2 medium for 5 months; 10M: 10-month callus; EC:embryogenic callus. BC: Browning embryogenic callus. Supplementary Figure S3: Gene Ontology (GO) analysis of differentially expressed genes across four groups. (a) The 20 most significantly enriched GO biological process categories among DEGs in the comparison of 5M versus 0M. (b) The top 20 enriched GO pathways identified in the 12M versus 5M comparison, sorted by the number of DEGs with a p value ≤ 0.05. (c) The 20 most enriched GO categories among DEGs in the EC versus 12M comparison. (d) The top 20 GO pathway categories in the BC versus EC DEGs. Supplementary Figure S4: KEGG analysis of differentially expressed genes across four groups. (a) The 20 most significantly enriched GO biological process categories among DEGs in the comparison of 5M versus 0M. (b) The top 20 enriched GO pathways identified in the 12M versus 5M comparison, sorted by the number of DEGs with a p value ≤ 0.05. (c) The 20 most enriched GO categories among DEGs in the EC versus 12M comparison. (d) The top 20 GO pathway categories in the BC versus EC DEGs.

Author Contributions: Conceptualization and writing—original draft preparation, X.Z.; validation, Y.Z., W.X. and Y.Y.; formal analysis, Z.F.; investigation, X.Z., Y.Z. and J.P.; data curation, writing—review and editing, Z.L., Y.L. and Y.C.; supervision, project administration, and funding acquisition, Z.L. All authors have read and agreed to the published version of the manuscript.

Funding: This work was supported by the National Modern Agricultural Industrial Technology System (Banana) Special Fund (GARS-31-15) from Zhongxiong Lai, the National Key Research and

Development Program of China (2019YFD1000900) from Zhongxiong Lai, and the Special Fund for Scientific and Technological Innovation of Fujian Agricultural and Forestry University (KFb22021XA) from Zhongxiong Lai.

Data Availability Statement: The transcriptome data of banana callus at various developmental stages have been uploaded to the NCBI Sequence Read Archive under the accession number PRJNA1212789.

Conflicts of Interest: The authors declare no conflict of interest.

References

1. Cenci, A.; Sardos, J.; Hueber, Y.; Martin, G.; Breton, C.; Roux, N.; Swennen, R.; Carpentier, S.C.; Rouard, M. Unravelling the complex story of intergenomic recombination in ABB allotriploid bananas. *Ann. Bot.* **2021**, *127*, 7–20. [CrossRef] [PubMed]
2. Simmonds, N. *The Evolution of the Bananas*; Longmans: London, UK, 1962; pp. 101–131.
3. Wang, J.; Gan, S.; Zheng, Y.; Jin, Z.; Cheng, Y.; Liu, J. Banana somatic embryogenesis and biotechnological application. *Trop. Plants* **2022**, *1*, 12. [CrossRef]
4. Novak, F.J.; Afza, R.; Van Duren, M.; Perea-Dallos, M.; Conger, B.V.; Tang, X. Somatic embryogensis and plant regeneration in suspension cultures of dessert (AA and AAA) and cooking (ABB) bananas (*Musa* spp.). *Biotechnology* **1989**, *7*, 154–159.
5. Remakanthan, A.M.T.; Soniya, E.V. Somatic embryogenesis in banana (*Musa acuminata* AAA cv. Grand Naine): Effect of explant and culture conditions. *Vitr. Cell. Dev. Biol. Plant* **2014**, *50*, 127–136.
6. Huang, X.; Huang, X.L.; Xiao, W.; Zhao, J.T.; Dai, X.M.; Chen, Y.F.; Li, X.J. Highly efficient Agrobacterium-mediated transformation of embryogenic cell suspensions of *Musa acuminata* cv. Mas (AA) via a liquid co-cultivation system. *Plant Cell Rep.* **2007**, *26*, 1755–1762. [CrossRef] [PubMed]
7. Escalant, J.V.; Teisson, C.; Cote, F. Amplified somatic embryogenesis from male flowers of triploid banana and plantain cultivars (*Musa* spp.). *In Vitro-Plant* **1994**, *30*, 181–186. [CrossRef]
8. Li, X.; Yu, S.; Cheng, Z.; Chang, X.; Yun, Y.; Jiang, M.; Chen, X.; Wen, X.; Li, H.; Zhu, W.; et al. Origin and evolution of the triploid cultivated banana genome. *Nat. Genet.* **2024**, *56*, 136–142. [CrossRef]
9. Wang, K.; Shi, L.; Liang, X.; Zhao, P.; Wang, W.; Liu, J.; Chang, Y.; Hiei, Y.; Yanagihara, C.; Du, L.; et al. The gene TaWOX5 overcomes genotype dependency in wheat genetic transformation. *Nat. Plants* **2022**, *8*, 110–117. [CrossRef]
10. Yu, Y.; Yu, H.; Peng, J.; Yao, W.J.; Wang, Y.P.; Zhang, F.L.; Wang, S.R.; Zhao, Y.; Zhao, X.Y.; Zhang, X.S.; et al. Enhancing wheat regeneration and genetic transformation through overexpression of TaLAX1. *Plant Commun.* **2023**, *28*, 100738. [CrossRef]
11. Brand, A.; Quimbaya, M.; Tohme, J.; Chavarriaga-Aguirre, P. Arabidopsis LEC1 and LEC2 Orthologous Genes Are Key Regulators of Somatic Embryogenesis in Cassava. *Front. Plant Sci.* **2019**, *10*, 673. [CrossRef]
12. Guo, F.; Liu, C.; Xia, H.; Bi, Y.; Zhao, C.; Zhao, S.; Hou, L.; Li, F.; Wang, X. Induced expression of AtLEC1 and AtLEC2 differentially promotes somatic embryogenesis in transgenic tobacco plants. *PLoS ONE* **2013**, *8*, e71714. [CrossRef] [PubMed]
13. He, C.; Liu, X.; Teixeira da Silva, J.A.; Wang, H.; Peng, T.; Zhang, M.; Si, C.; Yu, Z.; Tan, J.; Zhang, J.; et al. Characterization of LEA genes in Dendrobium officinale and one Gene in induction of callus. *J. Plant Physiol.* **2021**, *259*, 30. [CrossRef]
14. Xu, K.; Liu, J.; Fan, M.; Xin, W.; Hu, Y.; Xu, C. A genome-wide transcriptome profiling reveals the early molecular events during callus initiation in Arabidopsis multiple organs. *Genomics* **2012**, *100*, 116–124. [CrossRef]
15. Marsoni, M.; Bracale, M.; Espen, L.; Prinsi, B.; Negri, A.S.; Vannini, C. Proteomic analysis of somatic embryogenesis in *Vitis vinifera*. *Plant Cell Rep.* **2008**, *27*, 347–356. [CrossRef]
16. Kumaravel, M.; Uma, S.; Backiyarani, S.; Saraswathi, M.S.; Vaganan, M.M.; Muthusamy, M.; Sajith, K.P. Differential proteome analysis during early somatic embryogenesis in *Musa* spp. AAA cv. Grand Naine. *Plant Cell Rep.* **2017**, *36*, 163–178. [CrossRef]
17. Kumaravel, M.; Uma, S.; Backiyarani, S.; Saraswathi, M.S. Proteomic analysis of somatic embryo development in *Musa* spp. cv. Grand Naine (AAA). *Sci. Rep.* **2020**, *10*, 020–61005. [CrossRef] [PubMed]
18. Shivani; Awasthi, P.; Sharma, V.; Kaur, N.; Pandey, P.; Tiwari, S. Genome-wide analysis of transcription factors during somatic embryogenesis in banana (*Musa* spp.) cv. Grand Naine. *PLoS ONE* **2017**, *12*, e0182242. [CrossRef]
19. Su, Y.H.; Zhang, X.S. The hormonal control of regeneration in plants. *Curr. Top. Dev. Biol.* **2014**, *108*, 35–69. [PubMed]
20. Braybrook, S.A.; Harada, J.J. LECs go crazy in embryo development. *Trends Plant Sci.* **2008**, *13*, 624–630. [CrossRef]
21. Fan, M.; Xu, C.; Xu, K.; Hu, Y. LATERAL ORGAN BOUNDARIES DOMAIN transcription factors direct callus formation in Arabidopsis regeneration. *Cell Res.* **2012**, *22*, 1169–1180. [CrossRef] [PubMed]
22. Hwang, I.; Sheen, J.; Müller, B. Cytokinin signaling networks. *Annu. Rev. Plant Biol.* **2012**, *63*, 353–380. [CrossRef] [PubMed]
23. Osakabe, Y.; Miyata, S.; Urao, T.; Seki, M.; Shinozaki, K.; Yamaguchi-Shinozaki, K. Overexpression of Arabidopsis response regulators, ARR4/ATRR1/IBC7 and ARR8/ATRR3, alters cytokinin responses differentially in the shoot and in callus formation. *Biochem. Biophys. Res. Commun.* **2002**, *293*, 806–815. [CrossRef] [PubMed]

24. Sakai, H.; Honma, T.; Aoyama, T.; Sato, S.; Kato, T.; Tabata, S.; Oka, A. ARR1, a transcription factor for genes immediately responsive to cytokinins. *Science* **2001**, *294*, 1519–1521. [CrossRef] [PubMed]
25. Yang, L.; Zhang, Y.; Guan, R.; Li, S.; Xu, X.; Zhang, S.; Xu, J. Co-regulation of indole glucosinolates and camalexin biosynthesis by CPK5/CPK6 and MPK3/MPK6 signaling pathways. *J. Integr. Plant Biol.* **2020**, *62*, 1780–1796. [CrossRef]
26. Guo, K.; Sui, Y.; Li, Z.; Huang, Y.; Zhang, H.; Wang, W. Colonization of Trichoderma viride Tv-1511 in peppermint (*Mentha* × *piperita* L.) roots promotes essential oil production by triggering ROS-mediated MAPK activation. *Plant Physiol. Biochem.* **2020**, *151*, 705–718. [CrossRef]
27. Zhang, Y.; Li, J.; Li, C.; Chen, S.; Tang, Q.; Xiao, Y.; Zhong, L.; Chen, Y.; Chen, B. Gene expression programs during callus development in tissue culture of two *Eucalyptus* species. *BMC Plant Biol.* **2022**, *22*, 1. [CrossRef]
28. Li, L.; Sun, X.; Yu, W.; Gui, M.; Qiu, Y.; Tang, M.; Tian, H.; Liang, G. Comparative transcriptome analysis of high- and low-embryogenic Hevea brasiliensis genotypes reveals involvement of phytohormones in somatic embryogenesis. *BMC Plant Biol.* **2023**, *23*, 489. [CrossRef]
29. Yang, X.; Zhang, X. Regulation of Somatic Embryogenesis in Higher Plants. *Crit. Rev. Plant Sci.* **2010**, *29*, 36–57. [CrossRef]
30. Zheng, Q.; Zheng, Y.; Ji, H.; Burnie, W.; Perry, S.E. Gene Regulation by the AGL15 Transcription Factor Reveals Hormone Interactions in Somatic Embryogenesis. *Plant Physiol.* **2016**, *172*, 2374–2387. [CrossRef]
31. Mahalakshmi, A.; Singla, B.; Khurana, J.P.; Khurana, P. Role of calcium–calmodulin in auxin-induced somatic embryogenesis in leaf base cultures of wheat (*Triticum aestivum* var. HD 2329). *Plant Cell Tissue Organ Cult.* **2007**, *88*, 167–174. [CrossRef]
32. Baldwin, T.C.; Domingo, C.; Schindler, T.; Seetharaman, G.; Stacey, N.; Roberts, K. DcAGP1, a secreted arabinogalactan protein, is related to a family of basic proline-rich proteins. *Plant Mol. Biol.* **2001**, *45*, 421–435. [CrossRef] [PubMed]
33. Niu, D.; He, Y. LEAFY COTYLEDONs: Old genes with new roles beyond seed development. *F1000Research* **2019**, *27*, 2144. [CrossRef] [PubMed]
34. Kumar, V.; Van Staden, J. Multi-tasking of SERK-like kinases in plant embryogenesis, growth, and development: Current advances and biotechnological applications. *Acta Physiol. Plant.* **2019**, *41*, 31. [CrossRef]
35. Lin, J.Y. Preliminary Research on the Cold and Disease Resistance of Bengal Plantain. *J. Zhongkai Agrotech. Coll.* **1994**, *7*, 85–87.
36. Jekayinoluwa, T.; Tripathi, J.N.; Obiero, G.; Muge, E.; Tripathi, L. Phytochemical Analysis and Establishment of Embryogenic Cell Suspension and Agrobacterium-mediated Transformation for Farmer Preferred Cultivars of West African Plantain (*Musa* spp.). *Plants* **2020**, *9*, 789. [CrossRef]
37. Kulkarni, V.M.; Bapat, V.A. Somatic embryogenesis and plant regeneration from cell suspension cultures of Rajeli (AAB), an endangered banana cultivar. *J. Plant Biochem. Biotechnol.* **2012**, *22*, 132–137. [CrossRef]
38. Wang, Z.; Miao, H.; Liu, J.; Xu, B.; Yao, X.; Xu, C.; Zhao, S.; Fang, X.; Jia, C.; Wang, J.; et al. *Musa balbisiana* genome reveals subgenome evolution and functional divergence. *Nat. Plants* **2019**, *5*, 810–821. [CrossRef] [PubMed]
39. D'Hont, A.; Denoeud, F.; Aury, J.-M.; Baurens, F.-C.; Carreel, F.; Garsmeur, O.; Noel, B.; Bocs, S.; Droc, G.; Rouard, M.; et al. The banana (*Musa acuminata*) genome and the evolution of monocotyledonous plants. *Nature* **2012**, *488*, 213–217. [CrossRef]
40. Zhang, M.; Zhang, S. Mitogen-activated protein kinase cascades in plant signaling. *J. Integr. Plant Biol.* **2022**, *64*, 301–341. [CrossRef]
41. Hayashi, K.-i.; Arai, K.; Aoi, Y.; Tanaka, Y.; Hira, H.; Guo, R.; Hu, Y.; Ge, C.; Zhao, Y.; Kasahara, H.; et al. The main oxidative inactivation pathway of the plant hormone auxin. *Nat. Commun.* **2021**, *12*, 789. [CrossRef]
42. Li, S.B.; Xie, Z.Z.; Hu, C.G.; Zhang, J.Z. A Review of Auxin Response Factors (ARFs) in Plants. *Front. Plant Sci.* **2016**, *7*, 47. [CrossRef] [PubMed]
43. Spartz, A.K.; Lee, S.H.; Wenger, J.P.; Gonzalez, N.; Itoh, H.; Inzé, D.; Peer, W.A.; Murphy, A.S.; Overvoorde, P.J.; Gray, W.M. The SAUR19 subfamily of SMALL AUXIN UP RNA genes promote cell expansion. *Plant J.* **2012**, *70*, 978–990. [CrossRef] [PubMed]
44. Xie, Z.; Nolan, T.; Jiang, H.; Tang, B.; Zhang, M.; Li, Z.; Yin, Y. The AP2/ERF Transcription Factor TINY Modulates Brassinosteroid-Regulated Plant Growth and Drought Responses in Arabidopsis. *Plant Cell* **2019**, *31*, 1788–1806. [CrossRef]
45. Chen, C.; He, B.; Liu, X.; Ma, X.; Liu, Y.; Yao, H.Y.; Zhang, P.; Yin, J.; Wei, X.; Koh, H.J.; et al. Pyrophosphate-fructose 6-phosphate 1-phosphotransferase (PFP1) regulates starch biosynthesis and seed development via heterotetramer formation in rice (*Oryza sativa* L.). *Plant Biotechnol. J.* **2020**, *18*, 83–95. [CrossRef]
46. Azpeitia, A.; Chan, J.L.; Sáenz, L.; Oropeza, C. Effect of 22(S),23(S)-homobrassinolide on somatic embryogenesis in plumule explants of *Cocos nucifera* (L.) culturedin vitro. *J. Hortic. Sci. Biotechnol.* **2015**, *78*, 591–596. [CrossRef]
47. Chone, R.M.S.; Rocha, D.I.; Monte-Bello, C.C.; Pinheiro, H.P.; Dornelas, M.C.; Haddad, C.R.B.; Almeida, J.A.S. Brassinosteroid increases the cytokinin efficiency to induce direct somatic embryogenesis in leaf explants of *Coffea arabica* L. (Rubiaceae). *Plant Cell Tissue Organ Cult. (PCTOC)* **2018**, *135*, 63–67. [CrossRef]
48. Aydin, Y.; Talas-Ogras, T.; Ipekçi-Altas, Z.; Gözükirmizi, N. Effects of brassinosteroid on cotton regeneration via somatic embryogenesis. *Biologia* **2006**, *61*, 289–293. [CrossRef]
49. Finkelstein, R.R.; Rock, C.D. Abscisic Acid Biosynthesis and Response. *Arab. Book* **2002**, *1*, e0058. [CrossRef]

50. Ruduś, I.; Kępczyńska, E.; Kępczyński, J. Comparative efficiency of abscisic acid and metyl jasmonate for indirect somatic embryogenesis in *Medicago sativa* L. *Plant Growth Regul.* **2006**, *48*, 1–11. [CrossRef]
51. Ma, Y.; Szostkiewicz, I.; Korte, A.; Moes, D.; Yang, Y.; Christmann, A.; Grill, E. Regulators of PP2C phosphatase activity function as abscisic acid sensors. *Science* **2009**, *324*, 1064–1068. [CrossRef]
52. Park, S.Y.; Fung, P.; Nishimura, N.; Jensen, D.R.; Fujii, H.; Zhao, Y.; Lumba, S.; Santiago, J.; Rodrigues, A.; Chow, T.F.; et al. Abscisic acid inhibits type 2C protein phosphatases via the PYR/PYL family of START proteins. *Science* **2009**, *324*, 1068–1071. [CrossRef]
53. Fidler, J.; Graska, J.; Gietler, M.; Nykiel, M.; Prabucka, B.; Rybarczyk-Płońska, A.; Muszyńska, E.; Morkunas, I.; Labudda, M. PYR/PYL/RCAR Receptors Play a Vital Role in the Abscisic-Acid-Dependent Responses of Plants to External or Internal Stimuli. *Cells* **2022**, *11*, 1352. [CrossRef]
54. Neves, M.; Correia, S.; Cavaleiro, C.; Canhoto, J. Modulation of Organogenesis and Somatic Embryogenesis by Ethylene: An Overview. *Plants* **2021**, *10*, 1208. [CrossRef]
55. Colcombet, J.; Hirt, H. Arabidopsis MAPKs: A complex signalling network involved in multiple biological processes. *Biochem. J.* **2008**, *413*, 217–226. [CrossRef]
56. Lu, K.; Guo, W.; Lu, J.; Yu, H.; Qu, C.; Tang, Z.; Li, J.; Chai, Y.; Liang, Y. Genome-Wide Survey and Expression Profile Analysis of the Mitogen-Activated Protein Kinase (MAPK) Gene Family in Brassica rapa. *PLoS ONE* **2015**, *10*, e0132051. [CrossRef]
57. Li, H.; Chen, X.; Wang, Y.; Yao, D.; Lin, Y.; Lai, Z. Exploration of the effect of blue light on microRNAs involved in the accumulation of functional metabolites of longan embryonic calli through RNA-sequencing. *J. Sci. Food Agric.* **2019**, *99*, 1533–1547. [CrossRef]
58. Farooq, M.A.; Zeeshan Ul Haq, M.; Zhang, L.; Wu, S.; Mushtaq, N.; Tahir, H.; Wang, Z. Transcriptomic Insights into Salt Stress Response in Two Pepper Species: The Role of MAPK and Plant Hormone Signaling Pathways. *Int. J. Mol. Sci.* **2024**, *25*, 9355. [CrossRef]
59. Kothari, K.S.; Dansana, P.K.; Giri, J.; Tyagi, A.K. Rice Stress Associated Protein 1 (OsSAP1) Interacts with Aminotransferase (OsAMTR1) and Pathogenesis-Related 1a Protein (OsSCP) and Regulates Abiotic Stress Responses. *Front. Plant Sci.* **2016**, *7*, 1057. [CrossRef]
60. Seo, P.J.; Lee, A.K.; Xiang, F.; Park, C.M. Molecular and functional profiling of Arabidopsis pathogenesis-related genes: Insights into their roles in salt response of seed germination. *Plant Cell Physiol.* **2008**, *49*, 334–344. [CrossRef]
61. Zhou, H.; Zhou, R.; Zeng, S.; Wang, B.; Zhu, P. Advance of studies on browning and antibrowning techniques in the tissue culture of horticultural plants. *Acta Hortic Sin* **2000**, *27*, 481–486.
62. Ikeuchi, M.; Sugimoto, K.; Iwase, A. Plant callus: Mechanisms of induction and repression. *Plant Cell* **2013**, *25*, 3159–3173. [CrossRef]
63. Grieneisen, V.A.; Xu, J.; Marée, A.F.; Hogeweg, P.; Scheres, B. Auxin transport is sufficient to generate a maximum and gradient guiding root growth. *Nature* **2007**, *449*, 1008–1013. [CrossRef]
64. Yang, X.; Zhang, X.; Yuan, D.; Jin, F.; Zhang, Y.; Xu, J. Transcript profiling reveals complex auxin signalling pathway and transcription regulation involved in dedifferentiation and redifferentiation during somatic embryogenesis in cotton. *BMC Plant Biol.* **2012**, *12*, 1471–2229. [CrossRef]
65. Okushima, Y.; Fukaki, H.; Onoda, M.; Theologis, A.; Tasaka, M. ARF7 and ARF19 regulate lateral root formation via direct activation of LBD/ASL genes in Arabidopsis. *Plant Cell* **2007**, *19*, 118–130. [CrossRef]
66. Lee, D.K.; Geisler, M.; Springer, P.S. LATERAL ORGAN FUSION1 and LATERAL ORGAN FUSION2 function in lateral organ separation and axillary meristem formation in Arabidopsis. *Development* **2009**, *136*, 2423–2432. [CrossRef]
67. Riou-Khamlichi, C.; Huntley, R.; Jacqmard, A.; Murray, J.A. Cytokinin activation of Arabidopsis cell division through a D-type cyclin. *Science* **1999**, *283*, 1541–1544. [CrossRef]
68. Banno, H.; Ikeda, Y.; Niu, Q.W.; Chua, N.H. Overexpression of Arabidopsis ESR1 induces initiation of shoot regeneration. *Plant Cell* **2001**, *13*, 2609–2618. [CrossRef]
69. Ikeda, Y.; Banno, H.; Niu, Q.W.; Howell, S.H.; Chua, N.H. The ENHANCER OF SHOOT REGENERATION 2 gene in Arabidopsis regulates CUP-SHAPED COTYLEDON 1 at the transcriptional level and controls cotyledon development. *Plant Cell Physiol.* **2006**, *47*, 1443–1456. [CrossRef]
70. Wu, G.; Wei, X.; Wang, X.; Wei, Y. Changes and transcriptome regulation of endogenous hormones during somatic embryogenesis in *Ormosia henryi* Prain. *Front. Plant Sci.* **2023**, *14*, 1121259. [CrossRef]
71. Zhao, H.; Nie, K.; Zhou, H.; Yan, X.; Zhan, Q.; Zheng, Y.; Song, C.P. ABI5 modulates seed germination via feedback regulation of the expression of the PYR/PYL/RCAR ABA receptor genes. *New Phytol.* **2020**, *228*, 596–608. [CrossRef] [PubMed]
72. Wang, N.N.; Yang, Y.C.; Sun, D.W.; Pu, H.; Zhu, Z. Shelf-Life Prediction of 'Gros Michel' Bananas with Different Browning Levels Using Hyperspectral Reflectance Imaging. *Food Anal. Methods* **2015**, *8*, 1173–1184. [CrossRef]
73. Dong, Y.S.; Fu, C.H.; Su, P.; Xu, X.P.; Yuan, J.; Wang, S.; Zhang, M.; Zhao, C.F.; Yu, L.J. Mechanisms and effective control of physiological browning phenomena in plant cell cultures. *Physiol. Plant* **2016**, *156*, 13–28. [CrossRef] [PubMed]

74. Yang, T.; Zhang, T.; Li, Y.; Kang, Y.; Wang, P.; Liu, W.; Wang, Y.; Tian, L.; Dai, J.; Zhou, Y. Genome-Wide Identification and Expression Analysis of the Chalcone Synthase (CHS) Gene Family in Dendrobium catenatum. *Agronomy* **2023**, *13*, 1488. [CrossRef]
75. Francisco A Tomás-Barberán, J.C.E. Phenolic compounds and related enzymes as determinants of quality in fruits and vegetables. *J. Sci. Food Agric.* **2001**, *81*, 853–876. [CrossRef]
76. Krishna, H.; Sairam, R.K.; Singh, S.K.; Patel, V.; Sharma, R.R.; Grover, M.; Nain, L.; Sachdev, A. Mango explant browning: Effect of ontogenic age, mycorrhization and pre-treatments. *Sci. Hortic.* **2008**, *118*, 132–138. [CrossRef]
77. Gao, X.; Hu, Y.; Xu, Z.; Peng, D.; Guo, Q. Expression profiling of the phenylalanine ammonia-lyase (PAL) gene family in *Ginkgo biloba* L. *Plant Signal Behav.* **2023**, *18*, 30. [CrossRef] [PubMed]
78. Gooding, P.S.; Bird, C.; Robinson, S.P. Molecular cloning and characterisation of banana fruit polyphenol oxidase. *Planta* **2001**, *213*, 748–757. [CrossRef]
79. Qin, F.; Hu, C.; Dou, T.; Sheng, O.; Yang, Q.; Deng, G.; He, W.; Gao, H.; Li, C.; Dong, T.; et al. Genome-wide analysis of the polyphenol oxidase gene family reveals that MaPPO1 and MaPPO6 are the main contributors to fruit browning in *Musa acuminate*. *Front. Plant Sci.* **2023**, *14*, 1125375. [CrossRef]
80. Dixon, R.A.; Achnine, L.; Kota, P.; Liu, C.J.; Reddy, M.S.; Wang, L. The phenylpropanoid pathway and plant defence-a genomics perspective. *Mol. Plant Pathol.* **2002**, *3*, 371–390. [CrossRef]
81. Bukh, C.; Nord-Larsen, P.H.; Rasmussen, S.K. Phylogeny and structure of the cinnamyl alcohol dehydrogenase gene family in Brachypodium distachyon. *J. Exp. Bot.* **2012**, *63*, 6223–6236. [CrossRef]
82. Ata, N.; Yusuf, N.A.; Tan, B.C.; Husaini, A.; Mohd Yusuf, Y.; Majid, N.A.; Khalid, N. Expression profiles of flavonoid-related gene, 4 coumarate: Coenzyme A ligase, and optimization of culturing conditions for the selected flavonoid production in Boesenbergia rotunda. *Plant Cell Tissue Organ Cult. (PCTOC)* **2015**, *123*, 47–55. [CrossRef]
83. Irshad, M.; Rizwan, H.M.; Debnath, B.; Anwar, M.; Li, M.; Liu, S.; He, B.; Qiu, D. Ascorbic Acid Controls Lethal Browning and Pluronic F-68 Promotes High-frequency Multiple Shoot Regeneration from Cotyldonary Node Explant of Okra (*Abelmoschus esculentus* L.). *HortScience* **2018**, *53*, 183–190. [CrossRef]
84. Sharada, M.; Ahuja, A.; Kaul, M.K. Regeneration of plantlets via callus cultures in *Celastrus paniculatus* Willd—A rare endangered, medicinal plant. *J. Plant Biochem. Biotechnol.* **2003**, *12*, 65–69. [CrossRef]
85. Kim, D.; Paggi, J.M.; Park, C.; Bennett, C.; Salzberg, S.L. Graph-based genome alignment and genotyping with HISAT2 and HISAT-genotype. *Nat. Biotechnol.* **2019**, *37*, 907–915. [CrossRef] [PubMed]
86. Liao, Y.; Smyth, G.K.; Shi, W. featureCounts: An efficient general purpose program for assigning sequence reads to genomic features. *Bioinformatics* **2014**, *30*, 923–930. [CrossRef]
87. Wu, T.D.; Watanabe, C.K. GMAP: A genomic mapping and alignment program for mRNA and EST sequences. *Bioinformatics* **2005**, *21*, 1859–1875. [CrossRef] [PubMed]
88. Chen, L.; Zhong, H.Y.; Kuang, J.F.; Li, J.G.; Lu, W.J.; Chen, J.Y. Validation of reference genes for RT-qPCR studies of gene expression in banana fruit under different experimental conditions. *Planta* **2011**, *234*, 377–390. [CrossRef]

plants

Review

Non-Invasive Micro-Test Technology in Plant Physiology Under Abiotic Stress: From Mechanism to Application

Tianpeng Zhang [1], Peipei Yin [1], Xinghong Yang [2], Yunqi Liu [3] and Ruirui Xu [1,*]

1 College of Biology and Oceanography, Weifang University, Weifang 261061, China; zhangtianpeng0718@163.com (T.Z.); yinpp1524@163.com (P.Y.)
2 State Key Laboratory of Wheat Improvement, College of Life Sciences, Shandong Agricultural University, Tai'an 271000, China; xhyang@sdau.edu.cn
3 Zhongguancun Xuyue Non-Invasive Micro-Test Technology Industrial Alliance, Beijing 100080, China; yunqi@nmtia.org.cn
* Correspondence: xuruirui2006@163.com

Abstract: Non-invasive Micro-test Technology (NMT) represents a pioneering approach in the study of physiological functions within living organisms. This technology possesses the remarkable capability to monitor the flow rates and three-dimensional movement directions of ions or molecules as they traverse the boundaries of living organisms without sample destruction. The advantages of NMT are multifaceted, encompassing real-time, non-invasive assessment, a wide array of detection indicators, and compatibility with diverse sample types. Consequently, it stands as one of the foremost tools in contemporary plant physiological research. This comprehensive review delves into the applications and research advancements of NMT within the field of plant abiotic stress physiology, including drought, salinity, extreme temperature, nutrient deficiency, ammonium toxicity, acid stress, and heavy metal toxicity. Furthermore, it offers a forward-looking perspective on the potential applications of NMT in plant physiology research, underscoring its unique capacity to monitor the flux dynamics of ions/molecules (e.g., Ca^{2+}, H^+, K^+, and IAA) in real time, reveal early stress response signatures through micrometer-scale spatial resolution measurements, and elucidate stress adaptation mechanisms by quantifying bidirectional nutrient transport across root–soil interfaces. NMT enhances our understanding of the spatiotemporal patterns governing plant–environment interactions, providing deeper insights into the molecular mechanism of abiotic stress resilience.

Keywords: Non-invasive Micro-test Technology; plant physiology; abiotic stress; ion transport; real-time assessment

1. Introduction

Plant adversity encompasses a broad array of environmental factors that are detrimental to plant growth and survival. Through extensive adaptation and evolution in nature, plants have acquired specific resilience and adaptability to harsh conditions, including extreme temperatures, drought, salinity, and heavy metal contamination [1–3]. Upon exposure to these environmental stressors, plants perceive and transmit environmental signals through multiple core physiological processes. These include regulating nutrient absorption via root membrane transporters, transducing stress signals through calcium and ROS waves, modulating stomatal aperture via guard cell ion dynamics, and coordinating developmental and stress adaptations through the auxin transport and signaling pathway [4–6]. These processes, operating at both organismal and cellular levels, elicit the coordinated transmission and regulation of endogenous and exogenous signals to maintain growth and metabolic homeostasis.

Non-invasive Micro-test Technology (NMT) is a key technology that uses selective microelectrodes to detect the ion and molecular transmembrane transport of living samples without touching or damaging them. NMT can monitor a wide range of samples in vivo, including bacteria, single cells, vacuoles, tissues, and organs. NMT has emerged as a pivotal tool in plant physiology research, particularly in elucidating ionic flux dynamics under abiotic stress conditions. Its adoption has generated paradigm-shifting discoveries in ionic dynamic analysis [7], stress signaling transduction [8], and root growth and development [9], as evidenced by publications in top-tier journals including *Science*, *Cell*, and *Nature*. NMT's unique capability for non-invasive, real-time monitoring of micron-level plant materials can accurately capture ions' spatial–temporal movement patterns in plant tissues [10]. This temporal resolution advantage, achieving detection in seconds, proves critical for capturing transient stress responses like rapid calcium signaling response and auxin transport during stress perception [11–13]. This technological advancement fosters a better understanding of nature through elucidating fundamental mechanisms underlying plant–environment interactions.

2. An Overview of NMT

NMT represents a diverse array of microelectrode technologies and methodologies. These methodologies include, but are not limited to, Scanning Ion-selective Electrode Technology (SIET), Scanning Vibrating Electrode Technique (SVET), Scanning Polarographic Electrode Technique (SPET), Self-Referencing Ion-Selective Electrode Technique (SERIS), Self-Referencing Polarographic Electrode Technique (SERP), Self-Referencing Enzyme-Assisted Electrode Technique (SERE), Scanning Reference Electrode Technique (SRET), and Microelectrode Ion Flux Estimation Technique (MIFE), alongside other pioneering technologies, as documented by Kunkel and Shabala [14,15].

The development of NMT can be traced back to the pioneering work of Lionel F. Jaffe from the Marine Biological Laboratory (MBL), who first introduced the initial concept in the 1970s. Ian A. Newman conducted the first experiment of H^+ and K^+ flux measurements in corn roots by using MIFE methodology in 1987 [16]. Subsequently, NMT was successfully employed to measure the Ca^{2+} flux rate of cells in 1990, addressing numerous scientific challenges [17,18]. In 1995, Smith, also from MBL, further elaborated on the physical principles, mathematical foundations, and application modes of NMT, significantly enhancing the theoretical framework of this technology [19]. By 2001, NMT achieved the capabilities of programmable sensor motion-tracking and three-bit automatic measurement, marking the emergence of modern NMT. The non-invasive micro-measurement system developed by Younger USA LLC (Amherst, MA, USA) and Xuyue (Beijing) Science and Technology Co., Ltd., (Beijing, China) currently represents the ninth generation of automated NMT products. This system received the "internationally leading" achievement evaluation from the Ministry of Science and Technology of China and has obtained certifications from the NMT Industry Alliance in Zhongguancun and ISO9001 quality assurance [20].

2.1. Technical Principles

NMT utilizes flow-rate microsensors to capture the signals of ions and molecules. Based on the Nernst Equation and Fick's First Law of Diffusion, it calculates the concentrations and flow rates of ions and molecules, enabling the detection of extremely subtle signals with flow rates reaching magnitudes as low as 10^{-12} mol·cm^{-2}·s^{-1}. The ion-selective microsensor of the NMT system comprises a glass microsensor, a Ag/AgCl wire, an electrolyte, and a liquid ion exchanger. It selectively recognizes specific ions via a class of organic compounds, specifically large neutral molecular carriers, within the liquid ion exchanger. Taking the measurement of the flux rate of Na^+ as an illustrative example, the Na^+-selective flux rate microsensor achieves selectivity for Na^+ through the liquid ion exchanger (LIX) filled at its tip (Figure 1) [21]. The flux rate microsensor measures two points at a predefined distance, dx, within the ion concentration gradient to be assessed, yielding voltages V_1 and V_2, respectively. The concentration difference, dc, between these two points can be calculated using V_1, V_2, and the known voltage/concentration correction curve of the microsensor, which is based on the Nernst Equation. D denotes the diffusion constant of the ion (units: $cm^{-2} \cdot s^{-1}$). Substituting these parameters into Fick's First Law of Diffusion and the Nernst Equation ($J = -D\ (dc/dx)$) yields the flux rate of the ion (units: pico $mol \cdot cm^{-2} \cdot s^{-1}$), representing the number of moles passing through the ion/molecule per square centimeter per second [22,23]. It is worth noting that NMT calculates ion/molecule flux rates with a sensitivity of 10^{-13} $mol \cdot cm^{-2} \cdot s^{-1}$ for ions and 10^{-15} $mol \cdot cm^{-2} \cdot s^{-1}$ for molecules. The measurement distance between two sensor points is in the range of 5–50 μm, and NMT offers a time resolution of 4–6 s. Calibration can be achieved by solution preparation via dilution, and fluorescent dyes/optical fibers, carbon nanowires, enzyme electrodes, metals/alloys, and other materials can be utilized to carry out the selective measurement of certain ions/molecules.

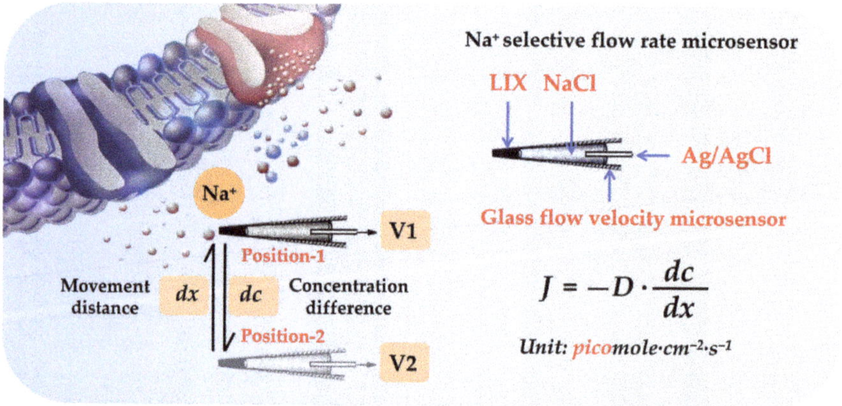

Figure 1. An illustration of the working principle of ion-selective flux rate microsensors in NMT, showing a case using Na^+ concentration gradient and a Na^+-selective flux rate microsensor. The Na^+-selective flow rate microsensor achieves ion selectivity for Na^+ by filling a Liquid Ion Exchanger (LIX) into its front-end tip. This flow rate microsensor measures the voltage (V_1 and V_2) at two points separated by a known distance (dx) within the concentration gradient of the target ion. The concentration difference (dc) between these two points can then be calculated from V_1, V_2, and the known voltage/concentration calibration curve of the microsensor (based on the Nernst Equation). D represents the ion's diffusion constant (unit: $cm^{-2} \cdot s^{-1}$). Substituting these values into Fick's First Law of Diffusion ($J = -D\ (dc/dx)$) yields the flow rate of the ion (unit: pico $mol \cdot cm^{-2} \cdot s^{-1}$). This unit represents the number of moles of the ion/molecule passing through per square centimeter per second.

2.2. Features, Advantages, and Application Fields

NMT is characterized by in vivo real-time measurements with high sensitivity (up to femtomoles $cm^{-2}\ s^{-1}$) and spatial resolution (0.5–10 μm for ions, 2–25 μm for molecules). Its non-invasive nature allows for the simultaneous quantification of over 50 parameters, including Ca^{2+}, H^+, K^+, Na^+, $NH_4{}^+$, heavy metals (e.g., Cd^{2+}), reactive oxygen species (e.g., H_2O_2), and phytohormones (e.g., IAA), without sample extraction [10,22,24]. One technical advantage is its unique three-dimensional gradient measurement capability based on Fick's diffusion laws, enabling the dynamic tracking of ion/molecule fluxes across plant membranes under stress conditions [22,25].

While NMT has broad applications spanning biomedical and pharmacological research (e.g., neural activity detection and drug efficacy evaluation), its significant contributions to plant stress physiology merit particular emphasis. In plant physiology research, NMT has revolutionized our understanding of stress adaptation mechanisms through some key applications, such as ion homeostasis regulation, heavy metal detoxification, oxidative stress response, and auxin transport and signal transduction [10]. The real-time quantification of K^+/Na^+ flux dynamics in root cells under salinity stress revealed compartmentalization strategies in halophytes [26,27]. NMT's high resolution enables the detection of H^+-ATPase-driven proton gradients crucial for maintaining membrane potential under abiotic stress [28,29]. Furthermore, NMT's multi-parameter capability (measuring two ions simultaneously) contributes to detecting the transmembrane ion flow rate and external ion concentration, and its high resolution can capture rapid signaling response events in the early stages of visible stress symptoms [10,22]. These developments facilitate the whole-plant analysis of ion redistribution patterns under abiotic stress treatments, significantly advancing our ability to cultivate stress-tolerant genotypes in crops.

3. NMT in Plant Physiology Under Abiotic Stress

The transport of mineral ions into and out of tissues and cells is central to the lives of plants [30]. Auxin plays a critical role in regulating plant adaptation and resilience to abiotic stresses and has agricultural potential for improving crop productivity [31]. NMT revolutionizes our understanding of plant stress response mechanisms by quantifying key ion and molecule fluxes. This includes quantifying ion homeostasis in real time by analyzing K^+/Na^+ transport kinetics under salinity stress [32], revealing stress signaling pathways by tracking the spatiotemporal patterns of Ca^{2+} waves under cold stress [13], and elucidating stomatal regulation by dynamically analyzing K^+/NO_3^- redistribution in guard cells and altering polar auxin transport in the root system under drought stress [12,33]. Thus, NMT offers a novel perspective by harnessing dynamic ion/molecular flux information in and out of plants, thereby unveiling more nuanced alterations in the physiological functions of diverse living plant samples. This technique provides a technological pathway for exploring plant functionalities and physiological mechanisms across multiple scales.

3.1. Salt Stress

Soil salinization poses an increasingly serious threat to agricultural and forestry productivity as well as environmental sustainability. It exerts diverse detrimental impacts on plants, encompassing ion toxicity, osmotic stress, and nutrient imbalances, ultimately hampering plant growth [34,35]. In response to a salt stress environment, plants leverage the transmembrane flux of ions across their cellular membranes. Sustaining an optimal K^+/Na^+ balance ratio is indispensable for plants to acclimate to high-salinity environments [36]. Tissue-specific K^+ retention and ROS-specific regulation confer enhanced salinity tolerance in the halophyte quinoa. Tolerant accession specifically exhibited enhanced root plasma membrane integrity and improved K^+ retention capacity in the mature root zone under saline stress [37]. NMT and molecular experiment results demonstrate that the superior leaf mesophyll K^+ retention capacity in quinoa is conferred by intrinsically lower H^+-ATPase activity, the reduced sensitivity of K^+ transporters to ROS, and the enhanced sensitivity of ROS-activated Ca^{2+}-permeable channels [38]. Similarly, NMT was employed to conduct a comparative analysis of the salt stress adaptation strategies between barley and triticale, two salt-tolerant cereal species. The results reveal that the desensitization of K^+ and Ca^{2+}-ROS-induced cation channels emerged as a key characteristic conferring salt tolerance. Compared with barley genotypes, triticale genotypes demonstrated more effective cytoplasmic K^+ retention in the root elongation zone under salt stress conditions [39]. Calmodulin (CaM), a highly conserved calcium-binding protein, plays a crucial role in plants' responses to salt stress. MIFE measurements revealed that, compared with WT plants, *OsCaM1-1* knockout mutants exhibited significantly higher Na^+ concentrations and Na^+/K^+ ratios in both shoots and roots under salt stress, along with reduced transient K^+ and Ca^{2+} fluxes in roots. These findings indicate that OsCaM1-1 positively regulates salt tolerance in rice by mediating Ca^{2+} signaling to maintain Na^+ and K^+ homeostasis [40].

The Salt Overly Sensitive (SOS) signaling module, consisting of the sodium transporter SOS1 and regulatory proteins SOS2 and SOS3, represents a well-established central mechanism for salt excretion, aiding plants in combating salt accumulation [5,41]. The findings demonstrate that the salt-induced calcium signal is decoded by 14-3-3 and SOS3/SCaBP8 proteins, which selectively activate/inactivate downstream protein kinases SOS2 and PKS5. These proteins regulate Na^+ homeostasis by co-mediating the activity of the Na^+/H^+ antiporter and H^+-ATPase in the plasma membrane. NMT was utilized to ascertain how PKS5 influences plant salt tolerance by monitoring the root Na^+ flux in *Arabidopsis* [42]. Analogously, VPS23A constitutes an essential component of the internal separation complex required for transport, playing a crucial role in the functionality of the SOS module to confer salt tolerance to plants. The determination of the Na^+ efflux rate at the *Arabidopsis* root tip via NMT unveiled that VPS23A positively impacts the secretion of Na^+ in plants under high-salinity conditions and underscores the significance of SOS2 sorting in functioning at the cell membrane [41]. In the salt-tolerant woody species *Populus euphratica*, high salinity (200 mM NaCl) increases the transcription of phospholipase Dδ (PLDδ) in roots and stems. Using NMT and biochemical experiments it was confirmed that PeGLABRA3, a basic helix–loop–helix (bHLH) transcription factor, activates the transcription of *AtPLDδ* by binding to its promoter region under salt stress. This activation confers Na^+ and ROS homeostasis through the PLDδ- and phosphatidic acid-mediated SOS1 signaling pathway, thereby enhancing salt adaptation in *Populus euphratica* [43].

In plants, glycine-rich RNA-binding proteins (GRPs) have been found to be expressed under various environmental stresses. Following long-term NaCl stress, PeGRP2 negatively affects mRNA stability. Using NMT to measure the Na^+ efflux rate in the root meristem zone of *Populus euphratica* under salt stress, it was observed that NaCl significantly increased Na^+ efflux in WT poplar. In transgenic lines, reduced Na^+ efflux capacity led to the greater accumulation of salt in the roots [44]. The HKT-type proteins play vital roles in long-distance Na^+ transport, maintenance of ion homeostasis, and improvement of salt tolerance in plants. In *Limonium bicolor*, by integrating NMT results, HKT-type proteins have been demonstrated to be involved in the salt secretion of salt glands for the first time, which provides a new perspective on the functions of HKT-type proteins under salt-stress conditions [45]. Regarding stress mitigation analysis, NMT revealed that exogenous methyl jasmonate (MeJA) promotes Na^+ efflux in *Nitraria tangutorum* roots, reduces the Na^+/K^+ ratio, and mitigates salt stress [46]. Similarly, boron (B) is an essential micronutrient for plants and plays a significant role in alleviating the inhibitory effects of soil salt stress on plant growth. NMT revealed that exogenous B induces Cl^- efflux in roots under NaCl stress, providing critical evidence for exploring the mechanism of boron-mediated alleviation of salt toxicity in sugar beet [47]. Using MIFE, researchers investigated the real-time transport rates of K^+ and H^+ in barley roots under 200 mM of NaCl stress with boron supplementation. The results reveal that under B-sufficient conditions, both root and shoot Na^+ contents were significantly reduced, while the K^+ content increased, leading to an elevated K^+/Na^+ ratio. This physiological adjustment consequently enhanced barley's tolerance to saline stress [48].

Autofluorescent inclusion (AFI), specifically accumulated in the mesophyll cells of non-secretory mangroves, is associated with salt adaptation. As a representative non-secretory mangrove species, *Kandelia obovata* exhibits exceptional salt tolerance. NMT and spatial metabolomics confirmed that AFI is structurally identified as condensed tannin (CT). CT biosynthesis showed a positive correlation with Na^+ accumulation in leaves, where chloroplast-synthesized CT is cytoplasmically transported to vacuoles, acting as molecular carriers to facilitate the compartmentalized storage of excess Na^+, thereby enhancing salt tolerance in *Kandelia obovata* [49]. Similarly, in studies on *Kandelia obovata* using NMT, it was confirmed that salicylic acid (SA)-induced salt tolerance relies on H_2O_2 generated by NADPH oxidases, which modulates Na^+/K^+ and redox homeostasis under high-salinity conditions [50]. In contrast, brassinosteroid (BR) enhances salt tolerance by reducing oxidative damage and regulating Na^+/K^+ homeostasis through the S-nitrosoglutathione reductase (GSNOR) and NO signaling pathways, thereby mediating the adaptation of *Kandelia obovata* to salt stress [51]. Integrated with NMT, elucidating the mechanisms of ion homeostasis regulation and osmoregulation in plants under salt stress will contribute to addressing the critical issues of soil salinity excess and Na^+ toxicity in agricultural production and ecosystems (Table 1).

Table 1. NMT application in plant abiotic stress.

Stress	Samples	Detection Site	Ions/Molecules	References
	Rice	Seed embryo	K^+, Na^+	[36]
	Quinoa	Root elongation and mature zone	K^+	[37]
	Quinoa	Leaf mesophyll cells	Ca^{2+}, K^+, H^+	[38]
	Barley and triticale	Root elongation and mature zone	Ca^{2+}, K^+, H^+	[39]
	Rice	Root mature zone	K^+, Ca^{2+}	[40]
	Arabidopsis	Root meristematic zone	Na^+	[41]
	Arabidopsis	Roots	Na^+, H^+	[42]
	Populus euphratica	Roots	Na^+	[43]
	Populus euphratica	Root meristematic zone	Na^+	[44]
	Limonium bicolor	Salt glands	Na^+	[45]
	Nitraria tangutorum	Root tip	Na^+, K^+	[46]
	Sugar beet	Roots	Cl^-	[47]
	Barley	Roots	K^+, H^+	[48]
	Kandelia obovata	Leaves	Na^+	[49]
	Kandelia obovata	Roots	Na^+, K^+, H^+, Ca^{2+}	[50]
	Kandelia obovata	Roots	Na^+, K^+	[51]
Alkali stress	Maize	Root meristematic zone	Na^+, H^+	[52]
	Arabidopsis	Leaf mesophyll cells	H^+	[53]
	Wheat	Roots	H^+	[54]
	Arabidopsis	Root elongation zone	H^+	[55]
	Rapeseed	Leaf mesophyll cells	Na^+, K^+, Ca^{2+}	[56]

Table 1. *Cont.*

Stress	Samples	Detection Site	Ions/Molecules	References
Low-temperature stress	Rice	Root meristematic zone	Ca^{2+}	[8]
	Watermelon	Intracellular	Ca^{2+}	[66]
	Rice	Roots	Ca^{2+}	[67]
	Rice	Roots	Ca^{2+}	[68]
	Cucumber	Roots	Ca^{2+}	[69]
	Arabidopsis	Roots	O_2	[70]
High-temperature stress	*Arabidopsis*	Leaf mesophyll cells	H^+, K^+, Ca^{2+}	[71]
	Rice	Leaf mesophyll cells	Ca^{2+}	[72]
	Lettuce	Guard cell	K^+, Ca^{2+}	[73]
	Rice	Root and aboveground parts	Ca^{2+}	[74]
	Poplar	Roots	Ca^{2+}	[75]
Nutrition stress	Apple	Stock root elongation zone	H^+	[29]
	Apple	Stock roots	H^+, NO_3^-	[76]
	Wheat	Roots	IAA, H^+	[77]
	Rice	Root meristem zone	NH_4^+	[78]
	White spruce	Roots	H^+, NH4$^+$, NO_3^-	[79]
	Lodgepole pine	Root and aboveground parts	H^+, NH4$^+$, NO_3^-	[80]
	Apple rootstock	Root mature zone	H^+	[81]
Ammonium toxicity	Wheat	Roots	K^+	[82]
	Arabidopsis	Roots	NH4$^+$, NO_3^-	[83]
	Arabidopsis	Roots	NO_3^-	[84]
Acid stress	Tea	Root mature area	H^+	[85]
	Arabidopsis	Root meristem zone, elongation zone and mature zone	H^+	[86]
	Arabidopsis	Roots	Cl^-, NO_3^-	[87]
	Malus hupehensis	Roots	H^+	[88]
	Populus	Roots	Mg^{2+}, IAA	[89]
Heavy metal toxicity	*Sedum plumbizincicola*	Roots	Cd^{2+}	[90]
	Cathay poplar	Roots	Cd^{2+}	[91]
	Populus euphratica	Root tip	Cd^{2+}	[92]
	Eichhornia crassipes	Roots, stem, leaves	H_2O_2, O_2	[93]
	Pak choi	Roots	Cd^{2+}	[94]
	Diatom	Frustule	Cd^{2+}	[95]
	Barley	Roots	Cd^{2+}, K^+, H^+, Cl^-, Ca^{2+}	[96]
	Populus euphratica	Roots	Cd^{2+}	[97]
	Apple rootstocks	Roots	Cd^{2+}	[98]

This table summarizes stress types, plant samples, detection sites, target ions/molecules, and corresponding references, highlighting NMT's utility in monitoring ion/molecule dynamics under abiotic stress conditions. Stress types: categories of abiotic stress (e.g., salt stress, alkali stress, water stress, extreme temperature stress, nutrition stress, ammonium toxicity, acid stress, heavy metal stress). Samples: plant species or tissues analyzed (e.g., *Arabidopsis*, rice, apple). Detection site: specific plant organs or subcellular regions measured (e.g., root tips, leaf mesophyll cells). Ions/molecules: ions or signaling molecules (e.g., K^+, Na^+, Ca^{2+}, NH_4^+, H^+, Cd^{2+}, NO_3^-, IAA) tracked using NMT. References: key studies demonstrating NMT's application in abiotic stress physiology.

3.2. *Alkali Stress*

Excessive salt is detrimental to plant growth and development because high soil salinity levels lead to severe degradation of plant ecosystems and negatively impact agricultural productivity. Notably, soil salinity often co-occurs with alkalinity [52]. Natural maize inbred lines exhibit variations in the Na^+ content of their aboveground parts and in their tolerance to saline and alkaline conditions ($NaHCO_3$). Specifically, a genome-wide association study revealed that the *ZmNSA1* gene influenced alterations in the Na^+ content of the aboveground parts of plants under $NaHCO_3$ treatment. Concurrently, NMT was employed to monitor changes in Na^+ and H^+ transmembrane transport, further elucidating the role of the *ZmNSA1* gene in plant saline-alkali tolerance [52]. The plasma membrane (PM) H^+-ATPase (AHA) provides energy for critical physiological and biochemical processes in plant cells and plays a crucial role in plant growth and development. Research suggests that calcineurin B-like protein 10 (CBL10) may participate in the regulation of AHA. The NMT results reveal that the proton influx rate in *cbl10* mutants was significantly lower than that in WT *Arabidopsis* plants when exposed to high external pH stress, indicating that CBL10 may act as an interconnected regulator coordinating plant responses to saline and alkaline stresses [53]. Similarly, the calcium-binding protein TaCCD1 in wheat is crucial for regulating the PM H^+-ATPase-mediated alkaline stress response. The NMT results show that *CCD1-OE* lines exhibited a significantly higher H^+ efflux rate compared to WT, while *CCD1-RNAi* lines displayed a markedly lower rate, indicating that TaCCD1 enhances proton extrusion to mediate alkaline stress adaptation [54].

Early steps in the endoplasmic reticulum (ER) lumen and cis-Golgi involve the trimming of N-glycans by Class I α-mannosidases (MNSs), which play critical roles in root growth and stress responses. Under normal pH 6.0 conditions, the *mns1 mns2 mns3 Arabidopsis* mutant exhibited proton influx in the elongation zone of root tips, whereas alkaline pH 8.2 induced a switch to proton efflux, indicating that alkaline treatment triggers a reversal in the proton flux direction from influx to efflux. In this study, NMT revealed that alkaline stress promotes H^+ efflux in mutant roots, providing indirect evidence that N-glycosylation influences IAA synthesis and transport [55]. Rapeseed (*Brassica napus* L.), a widely cultivated oilseed crop, is susceptible to ion toxicity under saline-alkali conditions. NMT analysis revealed that foliar application of 0.5% $CaCl_2$ solution under $NaHCO_3$ stress significantly alleviated membrane damage caused by ion imbalance, suppressed Na^+ influx and K^+/Ca^{2+} efflux in mesophyll cells, and maintained K^+/Na^+ homeostasis across tissues, thereby promoting seedling growth and development [56]. Along with NMT, elucidating the mechanism of pH homeostasis regulation in plants under alkaline stress will contribute to addressing the critical issues of soil alkalization and high-pH-induced nutrient deficiency in agricultural production and ecosystems (Table 1).

3.3. Water Stress

Water stress is detrimental to plant growth, and the investigation into enhancing plant adaptability to water scarcity and drought stress has garnered significant attention in recent years [12,99,100]. Maintaining root growth during the adaptation process to drought-affected soil is crucial for plants [101,102]. NMT and other methods have confirmed that, under moderate water stress, plants require auxin to stimulate proton secretion at the root tip, thereby regulating the abscisic acid-mediated adaptive processes of plants in response to drought stress [57]. Furthermore, the absorption and transport of K^+, H^+, and Ca^{2+} play a pivotal role in the drought tolerance of crops adapted to arid agricultural environments. MIFE results reveal significant differences between long-term K^+, H^+, and Ca^{2+} fluxes under drought conditions and those under short-term PEG-induced drought stress. The substantial K^+ efflux, apoplastic pH alkalinization (H^+ influx), and early Ca^{2+} efflux response in the mesophyll may serve as chemical signals and critical indicators of drought stress intensity in soybean plants [58]. By employing NMT, the flow rates of K^+, H^+, and Ca^{2+} were measured within a 24 h period across various barley lines, including XZ5, a silent line, and an overexpression line. These measurements revealed the significant roles of two key K^+ transporters, HvAKT2 and HvHAK1, in the drought tolerance of barley [59]. Drought is one of the primary limiting factors leading to reduced yield and quality in tea production. NMT and plant physiological experimental results reveal that higher ROS scavenging capacity and PM H^+-ATPase activity are the main reasons for the superior mesophyll K^+ retention capacity of drought-tolerant tea cultivars compared to drought-sensitive cultivars under drought stress [60].

Whole-genome duplication (or polyploidization) events may enhance plant adaptability to harsh environments. Using NMT to measure K^+ transport rates in the guard cells of transgenic and wild-type cotton under long-term water stress revealed that GhCIPK6D1 weakens drought resistance by positively regulating K^+ uptake, whereas GhCIPK6D3 enhances drought resistance by promoting K^+ efflux. This reveals the functional differentiation mechanism of these genes in regulating stomatal movement [61]. Moreover, soil waterlogging creates a hypoxic environment in the root zone, severely affecting plant growth and productivity. Using MIFE, researchers investigated the K^+ transport rate in wheat roots under hypoxic stress. The results reveal that hypoxia-sensing in tolerant cultivars triggers early accumulation of reactive oxygen species (ROS), which subsequently activates K^+ efflux channels. This leads to decreased cytosolic K^+ levels, ultimately inducing caspase-like protease-mediated programmed cell death (PCD). Conversely, elevated K^+ levels inhibit caspase-like protease activity, thereby enhancing plant hypoxia tolerance. This study provides novel insights for plant breeders to improve crop waterlogging resistance by modulating K^+ homeostasis mechanisms [62].

Auxin biosynthesis and transport are implicated in drought stress response [31]. NMT enables effective real-time monitoring of IAA transport in plant surface tissues while allowing simultaneous observation with existing live imaging techniques [103]. NMT was used with the newly developed IAA sensor to detect real-time auxin transport within rice roots under drought stress. The results reveal that the interaction between RoLe1 and OsAGAP disrupts OsAGAP function, altering polar auxin transport in the root system. This leads to reduced IAA influx into root tip cells, thereby modulating root system development and enhancing drought resistance in rice [12]. In *Poncirus trifoliata*, phenotypic analysis and NMT demonstrated that seedlings inoculated with the arbuscular mycorrhizal fungus (*Funneliformis mosseae*) exhibited significantly higher root hair density, length, diameter, and root IAA levels while displaying lower total root IAA efflux. These adaptations collectively enhance the drought resistance of host plants [63].

N^6-methyladenosine (m^6A) is the most prevalent internal modification in mRNAs, and drought response is highly regulated at the genomic, transcriptional, and post-transcriptional levels. In cotton, Virus-Induced Gene Silencing (VIGS) of Ca^{2+}-related genes, namely *GhECA1* and *GhCNGC4*, reduced the drought resistance of cotton, which was accompanied with a decrease in the enrichment of m^6A on the silenced genes, leading to alterations in the Ca^{2+} content [64]. Hydrogen sulfide (H_2S) is a novel gas signal molecule that can enhance plant drought resistance by inducing stomatal closure while simultaneously improving photosynthetic efficiency. Under drought stress, the fluxes of Cl^-, K^+, and H^+ in the guard cells of Chinese cabbage were detected using NMT. The results show that H_2S signaling induced the transmembrane efflux of Cl^- and H^+ while inhibiting K^+ influx, indicating that Cl^- channels serve as the primary responders for H_2S-mediated stomatal movement regulation [65]. This research underscores the critical roles of m^6A modification and gas signal molecules such as H_2S in plant drought response and provides insights into potential drought-tolerant strategies for crop improvement. Integrated with NMT, elucidating the physiological dynamics and molecular regulatory networks underlying plant water stress responses will contribute to addressing the increasing normalization of agricultural drought under global climate change (Table 1).

3.4. Low- and High-Temperature Stress

Low temperature constitutes a pivotal environmental constraint that hinders plant growth and agricultural productivity [66,102]. Rice, being a low-temperature-sensitive crop, is restricted to specific climatic zones for cultivation. Research has demonstrated the significant role of calmodulin B-like interacting protein kinases (CIPKs) under low-temperature stress conditions. Specifically, a point mutation in *OsCIPK7* in rice leads to a conformational alteration in the activation loop of its kinase domain, subsequently augmenting protein kinase activity and enhancing plant cold tolerance. Furthermore, NMT has elucidated the correlation between rice cold tolerance phenotypes and alterations in Ca^{2+} flux rates, indicating that enhanced cold tolerance in rice is associated with an augmented Ca^{2+} influx capacity [67]. Moreover, human selection for japonica rice varieties has facilitated their adaptation to low-temperature environments. The molecular basis of this adaptability is intimately linked to *COLD1*, a quantitative trait locus (QTL) associated with cold tolerance in japonica rice. The overexpression of $COLD1^{jap}$ markedly improves plant cold tolerance, whereas rice lines with $COLD1^{jap}$ deletions or downregulation exhibit increased sensitivity to low temperatures. NMT analysis of Ca^{2+} flux rates in rice roots under cold stress further validates that COLD1 interacts with G protein α, activates Ca^{2+} channels in response to low temperatures, accelerates G protein GTPase activity, and thereby enhances cold tolerance in japonica rice [8].

Additionally, OsCNGC9, a cyclic nucleotide-gated channel, augments cold tolerance in rice by modulating cold-induced Ca^{2+} influx and the subsequent elevation of cytosolic calcium levels, as detected using NMT [68]. Analogously, in cucumber, CsGPA1 interacts with CsCOR413PM2, a plasma membrane-localized protein. The inhibition of either CsGPA1 or CsCOR413PM2 results in decreased Ca^{2+} influx under low-temperature conditions, further suppressing the expression of *CsICE* and *CsCBF*. These findings provide a foundation for future investigations into the mechanism underlying cold tolerance mediated by the Gα subunit in cucumber [69]. In the Columbia ecotype of *Arabidopsis thaliana*, regardless of whether the plants are grown at 15 °C or 25 °C, the roots produce cells at the same rate and maintain consistent growth zone lengths. An analysis of whole-root oxygen consumption rates revealed that the meristematic zone exhibits steady-state Q_{10} values ranging between 0.7 and 2.1, whereas the elongation zone demonstrates higher Q_{10} values of 2.6 to 3.3, indicating that the metabolic cost of rapid cell elongation significantly exceeds that of cell division [70].

Heat stress constitutes a significant environmental challenge impacting crop growth and productivity [71,104]. The *HTS1* gene encodes a β-ketoacyl carrier protein reductase (KAR) localized in the thylakoid membrane, participating in de novo fatty acid biosynthesis. In comparison with WT plants, *hts1* mutants exhibit elevated heat-induced accumulation of H_2O_2, increased Ca^{2+} influx in mesophyll cells, and exacerbated membrane and chloroplast damage. This underscores the pivotal role of *HTS1* in maintaining membrane stability, chloroplast integrity, and stress signaling, which are fundamental for heat tolerance in rice [72]. Lettuce thrives in cooler environments, and high temperatures adversely affect its yield and quality. According to NMT results, exogenous spermidine augments the concentrations of Ca^{2+}, K^+, Fe^{3+}, Mn^{2+}, Zn^{2+}, and NO_3^- in lettuce leaves. It also facilitates K^+ efflux, enhances Ca^{2+} influx, and diminishes the relative stomatal aperture under high-temperature stress. This suggests that exogenous spermidine mitigates lettuce damage induced by high-temperature stress by modulating the ion content and altering stomatal morphology [73].

The phenomenon of global warming poses a threat to crop production. A natural quantitative trait locus, designated as TT2 (THERMOTOLERANCE 2), encodes a Gγ subunit that confers heat tolerance to rice during both vegetative and reproductive growth stages without compromising yield. The disruption of TT2 leads to a heat-triggered reduction in Ca^{2+}, which weakens the interaction between the transcription factor SCT1 (Sensing Ca^{2+} Transcription Factor 1) and calmodulin. This, in turn, influences heat tolerance and variations in the waxy content in rice [74]. Furthermore, long non-coding RNAs (lncRNAs) play a role in plant stress responses. Among the heat-responsive lncRNAs in *Populus simonii*, TCONS_00202587 binds to upstream sequences via its secondary structure, interfering with target gene transcription. Additionally, using NMT, TCONS_00260893 enhances Ca^{2+} influx following heat treatment by disrupting specific variants or isoforms of the target gene. These observations indicate that lncRNAs can regulate their target genes by functioning as potential RNA scaffolds or through RNA interference pathways [75]. Integrated with NMT, elucidating the physiological regulatory networks and molecular signaling pathways of extreme temperature stress responses in plants will contribute to addressing agricultural challenges posed by the increasing frequency of extreme high/low temperatures under global climate change (Table 1).

3.5. Nutrition Stress

Nitrogen (N) and phosphorus (P) constitute indispensable macronutrients for plant growth and development. In seedlings of the dwarfing rootstock "M9-T337", N or P deficiency, when compared to conditions of adequate nutrient supply, suppressed aboveground growth. Additionally, such deficiencies augmented the partitioning of total N and P contents towards the root system [76]. Consequently, there was an increase in the total number of root tips, root length, root volume, and root surface area, accompanied by an elevation in the root-to-shoot ratio. Furthermore, both phosphorus deficiency and nitrogen deficiency hindered the influx of NO_3^- into the root system, with H^+ pumping playing a pivotal role in the plant's response to these deficiencies [76]. In wheat, NMT results reveal that low N stress promotes lateral root development and nitrogen assimilation by regulating plant hormone signaling (such as increasing IAA), enhancing H^+-ATPase activity and H^+ efflux to expand N uptake capacity. This adaptation elevates the activity of key enzymes (nitrate reductase, glutamine synthetase, and glutamate synthase), stimulates protein synthesis, and drives root growth under N-deficient conditions [77]. Modern semi-dwarf rice varieties, which are a hallmark of the "Green Revolution", necessitate considerable quantities of N fertilizers to attain high yields. The interaction between strigolactones and gibberellins is advantageous in modulating the adaptation of rice root metabolism and development under low-N conditions, thereby ensuring the efficient absorption and translocation of available nitrogen. This synergistic effect facilitates the formulation of strategies aimed at enhancing nitrogen use efficiency in high-yielding crops [78].

Understanding the physiological processes underlying N assimilation sheds light on how boreal coniferous ecosystems develop adaptation strategies to their environmental conditions. The study revealed that white spruce roots exhibited the highest N uptake and proton efflux near the root tip, with both fluxes decreasing gradually at greater distances from the tip under treatment with 50 μM of N. In contrast, exposure to 1500 μM of N triggered significant ammonium (NH_4^+) efflux in certain root segments, highlighting differential physiological responses to varying nitrogen availability [79]. Moreover, ectomycorrhizal (EM) roots of lodgepole pine exhibited net NH_4^+ uptake, while nonmycorrhizal roots showed NH_4^+ efflux, with EM-associated seedlings displaying a higher N content in the roots and shoots, particularly when colonized by *Laccaria bicolor*. The study revealed EM fungi's role in reducing futile NH_4^+ cycling and demonstrated lodgepole pine's preference for NH_4^+ over NO_3^-, with NH_4^+ uptake rates increasing under NH_4^+-starved conditions [80].

Iron (Fe) deficiency exerts a profound impact on the growth, development, fruit productivity, and overall quality of apples. In response to Fe deficiency stress, apple roots exhibit an adaptive mechanism by augmenting the secretion of H^+, ultimately leading to soil acidification. The phosphorylation process, facilitated by the MAP kinase MxMPK6-2, exerts both direct and indirect regulatory effects on the activity of the plasma membrane H^+-ATPase MxHA2. This regulation occurs at both the post-translational and transcriptional levels, thereby synergistically intensifying H^+ secretion and enhancing root acidification in apple rootstocks subjected to Fe deficiency stress [29]. Additionally, under Fe deficiency conditions in the apple rootstock *Malus xiaojinensis*, the induced kinase MxMPK4-1 demonstrates a synergistic interaction with the IQ motif-containing protein 3 (MxIQM3). This interaction forms a functional complex that actively modulates the activity of the plasma membrane H^+-ATPase during the Fe deficiency response [81]. Aquaporins are indispensable transmembrane proteins responsible for the transport of water and several neutral solutes. In cassava, the targeted knockdown of the aquaporin gene *MePIP2;7* results in magnesium (Mg) deficiency in basal mature leaves, manifested by yellowing and necrotic spots at the leaf edges, accompanied by excessive starch accumulation. Protein interaction studies have elucidated that *MePIP2;7* is implicated in Mg^{2+} absorption and transport through its interaction with the low-affinity Mg^{2+} transporter MeMGT9 [105]. Along with NMT, deciphering the physiological and molecular response mechanisms to plant nutrient stress provides solutions for soil nutrient imbalance or deficiency in agroecosystems (Table 1).

3.6. Ammonium Toxicity and Acid Stress

N is an essential macronutrient for plant growth, with NH_4^+ and NO_3^- serving as the primary inorganic N sources utilized by plants. However, when NH_4^+ becomes the dominant nitrogen source, plants exhibit severe toxicity symptoms. For instance, the inhibition of root growth is one of the hallmark manifestations of NH_4^+ toxicity in plants [106,107]. High NH_4^+ levels impair wheat culm strength, vascular bundle integrity, nitrogen remobilization, and grain filling by competitively inhibiting K^+ uptake, but supplemental K^+ alleviates these effects by restoring transmembrane K^+ influx and tissue K^+ contents. This study links NH_4^+ toxicity in wheat to disrupted cation balance, demonstrating K^+ supplementation as a mitigation strategy for maintaining structural and physiological functions under excessive NH_4^+ [82].

The nitrate transporter NRT1.1 is involved in mediating the effects of NH_4^+ toxicity. In *Arabidopsis*, nitrate transporter NRT1.1 exacerbates NH_4^+ toxicity by enhancing NH_4^+ uptake (potentially measurable via NMT) and disrupting assimilation, leading to ethylene-driven senescence, while *NRT1.1* mutants mitigate toxicity through improved NH_4^+ metabolism and reduced accumulation [83]. Moreover, the nitrate transporter NRT1.1 interacts with the nitrate efflux channel SLAH3 to form a functional unit, which alleviates NH_4^+ toxicity by coordinating NO_3^- transport and balancing rhizosphere pH via H^+ uptake regulation, as demonstrated using NMT to monitor H^+ flux under high-NH_4^+/low-NO_3^- conditions [84]. Along with NMT, clarifying the physiological and molecular mechanisms of NH_4^+ toxicity in plants can help address the excessive accumulation of NH_4^+ in agricultural soils (Table 1).

H^+ in acidic soil can hinder plant growth [108]. However, the mechanism by which plants optimize their biological processes to reduce the adverse effects of H^+ stress still needs further exploration. Tea (*Camellia sinensis*) plants grow in acidic soil, and NMT revealed that Fe-sufficient conditions enhanced acidic stress tolerance in tea plants by promoting Fe plaque formation on the roots, which increased plasma membrane H^+-ATPase activity and H^+ efflux at pH 4.0–5.0, ultimately improving N accumulation compared to pH

6.0 conditions. The findings demonstrate that Fe plaque-mediated H^+-ATPase activation, quantified through NMT measurements of proton flux, is a key mechanism underlying tea plant adaptation to acidic environments [85]. In *Arabidopsis thaliana*, the transcription factor STOP1 enhances low pH tolerance by directly activating *NRT1.1* expression, which increases nitrate uptake to improve N use efficiency and reduces rhizospheric H^+ levels, thereby promoting root growth in acidic soils. The STOP1-NRT1.1 module optimizes plant adaptation to acidic stress by coordinating nitrate transport and H^+ homeostasis [86].

Climate change brings alternating patterns of severe drought and intense flooding events. These waterlogged conditions induce cytoplasmic acidification through oxygen deprivation in plant cells, ultimately inhibiting vital biological processes in plants. NMT and molecular biology experimental results demonstrate that *Arabidopsis* S-type anion channel AtSLAH3 directly senses cytosolic acidosis via the protonation of histidine residues, triggering structural activation and anion efflux to mediate flood stress tolerance, with wild-type plants outperforming *slah3* mutants under flooding conditions [87]. Aluminum (Al), a common element in the Earth's crust, negatively impacts vegetation in acidic soils by impairing root system expansion and hindering normal plant growth processes [109]. Soil acidification in apple orchards leads to the release of root-toxic Al^{3+} into the soil. Melatonin can alleviate Al toxicity in apple roots by activating the MdSTOP1-MdNAC2 transcriptional complex, which upregulates *MdALS3* and *MdNHX2* to enhance vacuolar H^+/Al^{3+} exchange and H^+ homeostasis, with NMT confirming the ion flux dynamics critical for Al stress mitigation [88]. Interestingly, an appropriate Mg supply can alleviate the toxic effects of high-concentration Al on poplar root growth. Al toxicity inhibits polar auxin transport and distribution in the root transition zone, but Mg supplementation partially mitigates this effect. Further analysis using NMT on the auxin transporter mutant *pin-formed2* (*pin2*) revealed that Mg alleviates Al toxicity by regulating root surface alkalinization in the transition zone through PIN2-mediated polar auxin transport [89]. Along with NMT, resolving the physiological dynamics and molecular regulatory networks of plant acid stress responses provides theoretical foundations for addressing crop growth inhibition, root developmental disorders, and nutrient imbalance caused by H^+ toxicity in acidic soils (Table 1).

3.7. Heavy Metal Toxicity

Heavy metals constitute a significant category of environmental contaminants [90]. Upon exposure to heavy metals, plants undergo alterations in morphogenesis, cell membrane permeability, photosynthesis, respiratory metabolism, enzymatic processes, and genetic impacts. When the concentration of heavy metals surpasses the threshold tolerance level of plants, it induces toxicity, disrupts metabolic processes, and inhibits plant growth [110]. Cadmium (Cd) is a particularly prominent pollutant in agricultural land, which not only severely restricts crop production but also poses a grave risk to human health through bioaccumulation in the food chain [2]. The investigation revealed that female *Populus cathayensis* exhibited heightened Cd absorption and translocation from roots to aboveground parts, whereas male *Populus cathayensis* demonstrated substantial Cd accumulation in roots, enhanced antioxidant capacity, and the effective sequestration of Cd within cells and bark. Furthermore, NMT was employed to monitor the Cd absorption rate in *Populus cathayensis* roots. The results indicate that the net Cd^{2+} influx in female *Populus cathayensis* was greater than that in male individuals, suggesting that male *Populus cathayensis* possesses higher Cd tolerance, thereby offering novel insights into the potential mechanisms underlying gender-specific responses to Cd stress [91]. NMT and biochemical experimental results confirm that in *Populus euphratica*, calcium-dependent protein kinase 21 (PeCPK21) interacts with the *Arabidopsis* nuclear transcription factor YC3

(AtNF-YC3) to reduce Cd accumulation and enhance the ROS scavenging system, thereby positively regulating the plant's adaptive capacity to Cd-contaminated environments [92].

Eichhornia crassipes is an effective ecological remediation plant that can alleviate Cd stress. Under prolonged Cd stress treatment, NMT was employed to measure the transport rates of O_2 and H_2O_2 in the roots and leaves of *Eichhornia crassipes* under stress induced by treatment with 4 mg L^{-1} $CdCl_2$. The results show that, compared to the control, the Cd treatment significantly inhibited the O_2 uptake rates in both the roots and leaves, markedly enhanced H_2O_2 efflux in the roots, and suppressed H_2O_2 efflux in the leaves. This indicates that Cd stress suppresses cellular respiration in the roots and leaves and disrupts ROS homeostasis in *Eichhornia crassipes* [93]. Additionally, NMT was utilized to measure the Cd^{2+} flux rate on the root surface of Chinese cabbage, revealing that hemin treatment reduced Cd accumulation in Chinese cabbage seedlings by decreasing plant Cd absorption rather than by influencing Cd translocation within the plant [94]. Analogously, NMT provided direct evidence that silicon (Si) treatment can enhance Cd tolerance in marine diatoms and maintain their metal homeostasis [95]. In barley, chloride (Cl^-) enhances Cd mobility and phytotoxicity by increasing Cd^{2+} uptake and disrupting ion homeostasis in roots, with Cd-sensitive genotypes exhibiting higher Cl^--mediated Cd accumulation and photosynthetic impairment. The findings underscore the critical role of soil Cl^- in Cd toxicity and propose breeding low-Cl^--uptake barley varieties to mitigate Cd contamination risks, supporting safer agricultural production for global food and beverage industries [96]. Regarding stress mitigation analysis, NMT revealed that overexpression of *PeANN1* (an annexin encoding gene facilitating Cd enrichment) enhances Cd^{2+} accumulation in transgenic *Arabidopsis*, and it can serve as a candidate gene for phytoremediation to alleviate cadmium stress [97]. In apple rootstocks, the NMT and plant physiology experiment results showed that exogenous melatonin reduces Cd accumulation in the aboveground parts of apple plants and alleviates Cd toxicity. This effect is likely attributed to melatonin-mediated compartmentalization of Cd within tissues, as well as its induction of the antioxidant defense system and upregulation of key genes involved in detoxification-related transcriptional regulation [98]. When integrated with NMT, analyzing ion transport dynamics and the detoxification of regulatory pathways under heavy metal stress addresses escalating ecological risks from soil contamination in mining/agricultural areas (Table 1).

4. Conclusions and Prospects

The application of NMT in plant stress physiology has reached substantial maturity, providing a comprehensive platform for ion/molecule flux analysis that significantly advances our understanding of plant stress resistance mechanisms. NMT should be used to study major abiotic stress types (e.g., water stress, salt stress, alkali stress, extreme temperature stress, ammonium toxicity, acid stress, heavy metal toxicity, and nutrient imbalance) to monitor real-time dynamic fluxes (influx and efflux) of different ions (e.g., Na^+, K^+, Cd^{2+}, and Ca^{2+}) within specific plant cells and tissues (e.g., guard cell, mesophyll cell, and root tip). This non-invasive approach with high sensitivity and spatial resolution enables the elucidation of plant stress defense mechanisms and provides deeper insights into plant adaptation strategies and stress signaling pathways (Figure 2).

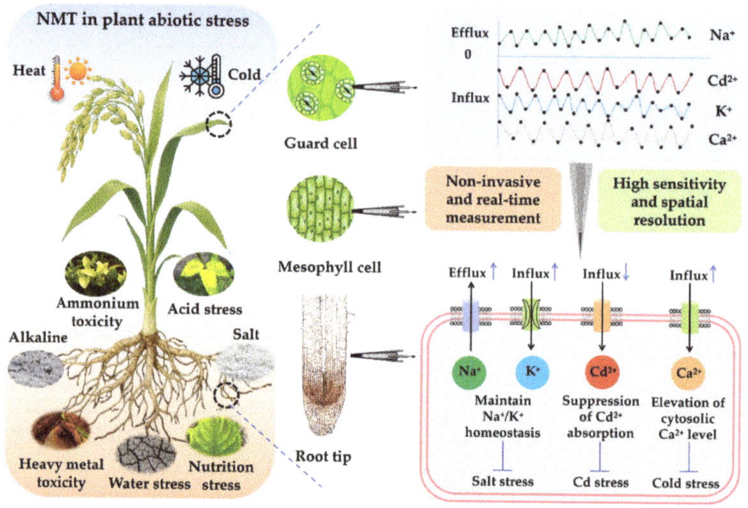

Figure 2. A comprehensive model of NMT application in elucidating plant responses to abiotic stress conditions. This schematic illustrates the integration of NMT with various major abiotic stress types, including but not limited to water stress, salt stress, alkali stress, extreme temperature stress, ammonium toxicity, acid stress, heavy metal toxicity, and nutrient imbalance. The model demonstrates how NMT can be used to monitor real-time dynamic fluxes (influx and efflux) of different ions (e.g., Na^+, K^+, Cd^{2+}, and Ca^{2+}) within different plant cells and tissues (e.g., guard cell, mesophyll cell, and root tip), thereby revealing the stress defense mechanisms of plants. By non-invasive, real-time, high sensitivity and spatial resolution measuring ion transport rate, this model provides a deeper mechanistic understanding of plant adaptation strategies and stress signaling pathways. The circles represent the measurement points for NMT. Black arrows indicate the direction of ion flux. Blue arrows represent increases or decreases in the magnitude of ion influx or efflux. Blue blunt arrowheads represent defense responses against abiotic stress.

Abiotic stress effects are dynamic and influenced by factors such as plant age, genotype, stress duration, and application methods. NMT addresses this challenge through several technical advantages. Its non-invasive nature allows for repeated measurements on the same sample over extended periods without causing tissue damage. The system incorporates environmental control chambers that maintain stable measurement conditions (humidity, temperature, and CO_2) during prolonged experiments. For genotype/age comparisons, we recommend using standardized protocols including synchronized plant growth stages, controlled stress application gradients, and multiple biological replicates. Furthermore, regarding stress mitigation analysis, NMT can quantitatively evaluate the efficacy of mitigation strategies (e.g., osmoprotectants) by comparing pre-/post-treatment ion flux patterns. NMT detects early recovery signatures (e.g., K^+ flux restoration in roots post-drought) before visible phenotypic recovery occurs. And spatial resolution enables a localized analysis of mitigation effects, such as differentiated responses in root apical and mature zones.

However, some critical challenges persist in optimizing this technology, such as hypersaline soils (inducing signal noise from competing ions like Na^+/Cl^-), severe heavy metal contamination (e.g., Cd^{2+} saturation overwhelming sensor resolution), and temperature extremes (altering membrane integrity and flux kinetics). For instance, in acidic soils, H^+ efflux measurements may conflate abiotic proton gradients with biologically regulated fluxes, while freezing conditions disrupt microelectrode stability. These limitations risk overlooking transient but biologically significant signaling events during stress adaptation.

Moreover, the reliability of NMT data hinges on rigorous sample preparation and environmental control, as subtle discrepancies in these steps can profoundly affect reproducibility. For example, sample preparation variability, including root hair density differences and microelectrode positioning accuracy, could induce flux measurement deviations. The inconsistent growth medium composition (e.g., Ca^{2+} or chelator concentrations) could unpredictably modify ion availability and membrane potential dynamics, complicating data analysis. Environmental control parameters, such as temperature fluctuations and solution convection artifacts, may cause error interference in the measurement and analysis of NMT data.

Looking forward, some strategic developments are promising in transforming NMT applications. Its integration with multi-omics platforms could enable unprecedented mechanistic insights, for instance, correlating Ca^{2+} flux dynamics with stress-responsive gene clusters (e.g., SOS pathway genes, auxin transport, and response factors) identified through single-cell RNA sequencing or validating ABA-mediated stomatal regulation through parallel proteomic profiling of guard cell H^+-ATPases. Artificial intelligence implementation may show promise in two operational domains: machine vision algorithms could automate microelectrode navigation with submicron precision, and deep learning models trained on historical flux datasets may predict stress response thresholds, effectively reducing human interpretation bias.

In conclusion, NMT possesses considerable potential to occupy a central position in agricultural breeding endeavors and ecological environment preservation. It can provide scientific underpinnings for the cultivation of stress-tolerant crops, the enhancement of ecological conditions, and the attainment of sustainable development objectives.

Author Contributions: Conceptualization, T.Z. and R.X.; writing—original draft preparation, T.Z. and P.Y.; writing—review and editing, X.Y. and Y.L.; funding acquisition, R.X., T.Z. and P.Y. All authors have read and agreed to the published version of the manuscript.

Funding: This research was financially supported by the National Natural Science Foundation of China (Grant No. 32372643), the General Program of Shandong Natural Science Foundation (Grant No. ZR2022MC064), and the Doctoral Startup Foundation of Weifang University (Grant No. 2023BS15 and 2024BS21) in China.

Conflicts of Interest: The authors declare no conflicts of interest.

References

1. Zhu, J.K. Abiotic stress signaling and responses in plants. *Cell* **2016**, *167*, 313–324. [CrossRef] [PubMed]
2. Gong, Z.; Xiong, L.; Shi, H.; Yang, S.; Herrera-Estrella, L.R.; Xu, G.; Chao, D.Y.; Li, J.; Wang, P.Y.; Qin, F.; et al. Plant abiotic stress response and nutrient use efficiency. *Sci. China Life Sci.* **2020**, *63*, 635–674. [CrossRef] [PubMed]
3. Moustakas, M. Molecular mechanisms of plant abiotic stress tolerance. *Int. J. Mol. Sci.* **2025**, *26*, 2731. [CrossRef] [PubMed]
4. Waadt, R.; Seller, C.A.; Hsu, P.K.; Takahashi, Y.; Munemasa, S.; Schroeder, J.I. Plant hormone regulation of abiotic stress responses. *Nat. Rev. Mol. Cell Biol.* **2022**, *23*, 680–694. [CrossRef]
5. Zhang, H.; Zhu, J.; Gong, Z.; Zhu, J.K. Abiotic stress responses in plants. *Nat. Rev. Genet.* **2022**, *23*, 104–119. [CrossRef]
6. Salehin, M. Emerging roles of auxin in plant abiotic stress tolerance. *Physiol. Plant.* **2024**, *176*, e14601. [CrossRef]
7. Michard, E.; Lima, P.T.; Borges, F.; Silva, A.C.; Portes, M.T.; Carvalho, J.E.; Gilliham, M.; Liu, L.H.; Obermeyer, G.; Feijó, J.A. Glutamate receptor-like genes form Ca^{2+} channels in pollen tubes and are regulated by pistil D-serine. *Science* **2011**, *332*, 434–437. [CrossRef]
8. Ma, Y.; Dai, X.; Xu, Y.; Luo, W.; Zheng, X.; Zeng, D.; Pan, Y.; Lin, X.; Liu, H.; Zhang, D.; et al. *COLD1* confers chilling tolerance in rice. *Cell* **2015**, *160*, 1209–1221. [CrossRef]
9. Li, L.; Verstraeten, I.; Roosjen, M.; Takahashi, K.; Rodriguez, L.; Merrin, J.; Chen, J.; Shabala, L.; Smet, W.; Ren, H.; et al. Cell surface and intracellular auxin signalling for H^+ fluxes in root growth. *Nature* **2021**, *599*, 273–277. [CrossRef]
10. Liu, B.; Zhang, J.; Ye, N. Noninvasive micro-test technology: Monitoring ion and molecular flow in plants. *Trends Plant Sci.* **2023**, *28*, 123–124. [CrossRef]

11. Thor, K.; Jiang, S.; Michard, E.; George, J.; Scherzer, S.; Huang, S.; Dindas, J.; Derbyshire, P.; Leitão, N.; DeFalco, T.A.; et al. The calcium-permeable channel OSCA1.3 regulates plant stomatal immunity. *Nature* **2020**, *585*, 569–573. [CrossRef] [PubMed]
12. Han, S.; Wang, Y.; Li, Y.; Zhu, R.; Gu, Y.; Li, J.; Guo, H.; Ye, W.; Nabi, H.G.; Yang, T.; et al. The OsNAC41-RoLe1-OsAGAP module promotes root development and drought resistance in upland rice. *Mol. Plant* **2024**, *17*, 1573–1593. [CrossRef] [PubMed]
13. Luo, W.; Xu, Y.; Cao, J.; Guo, X.; Han, J.; Zhang, Y.; Niu, Y.; Zhang, M.; Wang, Y.; Liang, G.; et al. COLD6-OSM1 module senses chilling for cold tolerance via 2′,3′-cAMP signaling in rice. *Mol. Cell* **2024**, *84*, 4224–4238.e9. [CrossRef]
14. Kunkel, J.G.; Cordeiro, S.; Xu, Y.; Shipley, A.M.; Feijo, J.A. The use of non-invasive ion-selective microelectrode techniques for the study of plant development. In *Plant Electrophysiology-Theory and Methods*; Volkov, A.G., Ed.; Springer: Berlin/Heidelberg, Germany, 2005; Chapter V, pp. 109–137. [CrossRef]
15. Shabala, S. Non-invasive microelectrode ion flux measurements in plant stress physiology. In *Plant Electrophysiology*; Volkov, A.G., Ed.; Springer: Berlin/Heidelberg, Germany, 2006; Chapter III, pp. 35–71. [CrossRef]
16. Newman, I.A.; Kochian, L.V.; Grusak, M.A.; Lucas, W.J. Fluxes of H^+ and K^+ in corn roots: Characterization and stoichiometries using ion-selective microelectrodes. *Plant Physiol.* **1987**, *84*, 1177–1184. [CrossRef]
17. Kühtreiber, W.M.; Jaffe, L.F. Detection of extracellular calcium gradients with a calcium-specific vibrating electrode. *J. Cell Biol.* **1990**, *110*, 1565–1573. [CrossRef]
18. Jaffe, L.F. Fast calcium waves. *Cell Calcium* **2010**, *48*, 102–113. [CrossRef]
19. Smith, P.J. Non-invasive ion probes-tools for measuring transmembrane ion flux. *Nature* **1995**, *378*, 645–646. [CrossRef]
20. *ISO 9001:2015*; Quality Management Systems–Requirements. International Organization for Standardization: Geneva, Switzerland, 2015.
21. Sun, K.; Liu, Y.; Pan, Y.; Di, D.; Li, J.; Xu, F.; Li, L.; Mimata, Y.; Chen, Y.; Xie, L.; et al. Non-invasive micro-test technology and applications. *Biophys. Rep.* **2025**, *11*, 96–111. [CrossRef]
22. Han, M.; Yang, H.; Yu, G.; Jiang, P.; You, S.; Zhang, L.; Lin, H.; Liu, J.; Shu, Y. Application of Non-invasive Micro-test Technology (NMT) in environmental fields: A comprehensive review. *Ecotoxicol. Environ. Saf.* **2022**, *240*, 113706. [CrossRef]
23. Jiang, W.; You, S.; Shi, Y.; Jiang, P.; Chen, M.; Yang, X. Application progress of Non-invasive Micro-test Technology in environmental bioremediation and stress physiology research. *J. Plant Interact.* **2023**, *18*, 2268131. [CrossRef]
24. Li, L.Z.; Yu, S.Y.; Peijnenburg, W.J.G.M.; Luo, Y.M. Determining the fluxes of ions (Pb^{2+}, Cu^{2+} and Cd^{2+}) at the root surface of wetland plants using the scanning ion-selective electrode technique. *Plant Soil* **2017**, *414*, 1–12. [CrossRef]
25. Xu, Y.; Sun, T.; Yin, L.P. Application of non-invasive microsensing system to simultaneously measure both H^+ and O_2 fluxes around the pollen tube. *J. Integr. Plant Biol.* **2006**, *48*, 823–831. [CrossRef]
26. Tang, X.; Zhang, H.; Shabala, S.; Li, H.; Yang, X.; Zhang, H. Tissue tolerance mechanisms conferring salinity tolerance in a halophytic perennial species *Nitraria sibirica* Pall. *Tree Physiol.* **2021**, *41*, 1264–1277. [CrossRef] [PubMed]
27. Huang, Z.; Yao, L.; Li, B.; Ma, X.; Si, E.; Yang, K.; Zhang, H.; Meng, Y.; Wang, J.; Wang, H. *HgS2*, a novel salt-responsive gene from the Halophyte *Halogeton glomeratus*, confers salt tolerance in transgenic *Arabidopsis*. *Physiol. Plant.* **2024**, *176*, e14356. [CrossRef]
28. Stéger, A.; Hayashi, M.; Lauritzen, E.W.; Herburger, K.; Shabala, L.; Wang, C.; Bendtsen, A.K.; Nørrevang, A.F.; Madriz-Ordeñana, K.; Ren, S.; et al. The evolution of plant proton pump regulation via the R domain may have facilitated plant terrestrialization. *Commun. Biol.* **2022**, *5*, 1312. [CrossRef]
29. Sun, Q.; Zhao, D.; Gao, M.; Wu, Y.; Zhai, L.; Sun, S.; Wu, T.; Zhang, X.; Xu, X.; Han, Z.; et al. MxMPK6-2-mediated phosphorylation enhances the response of apple rootstocks to Fe deficiency by activating PM H^+-ATPase MxHA2. *Plant J.* **2023**, *116*, 69–86. [CrossRef]
30. Newman, I.A. Ion transport in roots: Measurement of fluxes using ion-selective microelectrodes to characterize transporter function. *Plant Cell Environ.* **2001**, *24*, 1–14. [CrossRef]
31. Jing, H.; Wilkinson, E.G.; Sageman-Furnas, K.; Strader, L.C. Auxin and abiotic stress responses. *J. Exp. Bot.* **2023**, *74*, 7000–7014. [CrossRef]
32. Xu, H.; Chen, H.; Halford, N.G.; Xu, R.; He, T.; Yang, B.; Zhou, L.; Guo, H.; Liu, C. Ion homeostasis and coordinated salt tolerance mechanisms in a barley (*Hordeum vulgare* L.) doubled haploid line. *BMC Plant Biol.* **2025**, *25*, 52. [CrossRef]
33. Kou, Y.; Su, B.; Yang, S.; Gong, W.; Zhang, X.; Shan, X. Phosphorylation of *Arabidopsis* NRT1.1 regulates plant stomatal aperture and drought resistance in low nitrate condition. *BMC Plant Biol.* **2025**, *25*, 95. [CrossRef]
34. Yu, Z.; Duan, X.; Luo, L.; Dai, S.; Ding, Z.; Xia, G. How plant hormones mediate salt stress responses. *Trends Plant Sci.* **2020**, *25*, 1117–1130. [CrossRef] [PubMed]
35. Zhao, S.; Zhang, Q.; Liu, M.; Zhou, H.; Ma, C.; Wang, P. Regulation of plant responses to salt stress. *Int. J. Mol. Sci.* **2021**, *22*, 4609. [CrossRef] [PubMed]
36. Zeng, P.; Xie, T.; Shen, J.; Liang, T.; Yin, L.; Liu, K.; He, Y.; Chen, M.; Tang, H.; Chen, S.; et al. Potassium transporter OsHAK9 regulates seed germination under salt stress by preventing gibberellin degradation through mediating OsGA2ox7 in rice. *J. Integr. Plant Biol.* **2024**, *66*, 731–748. [CrossRef]

37. Bazihizina, N.; Vita, F.; Balestrini, R.; Kiferle, C.; Caparrotta, S.; Ghignone, S.; Atzori, G.; Mancuso, S.; Shabala, S. Early signalling processes in roots play a crucial role in the differential salt tolerance in contrasting *Chenopodium quinoa* accessions. *J. Exp. Bot.* **2022**, *73*, 292–306. [CrossRef]
38. Tanveer, M.; Wang, L.; Huang, L.; Zhou, M.; Chen, Z.H.; Shabala, S. Understanding mechanisms for differential salinity tissue tolerance between quinoa and spinach: Zooming on ros-inducible ion channels. *Crop J.* **2024**, *12*, 1357–1368. [CrossRef]
39. Rasouli, F.; Yun, P.; Kiani-Pouya, A.; Movahedi, A.; Rasouli, M.; Salehi, M.; Shabala, S. One size does not fit all: Different strategies employed by triticale and barley plants to deal with soil salinity. *Environ. Exp. Bot.* **2024**, *218*, 105585. [CrossRef]
40. Wei, S.; Chen, M.; Wang, F.; Tu, Y.; Xu, Y.; Fu, L.; Zeng, F.; Zhang, G.; Wu, D.; Shen, Q. OsCaM1-1 is responsible for salt tolerance by regulating Na^+/K^+ homoeostasis in rice. *Plant Cell Environ.* **2025**, *48*, 1393–1408. [CrossRef]
41. Lou, L.; Yu, F.; Tian, M.; Liu, G.; Wu, Y.; Wu, Y.; Xia, R.; Pardo, J.M.; Guo, Y.; Xie, Q. ESCRT-I component VPS23A sustains salt tolerance by strengthening the SOS module in *Arabidopsis*. *Mol. Plant* **2020**, *13*, 1134–1148. [CrossRef]
42. Yang, Z.; Wang, C.; Xue, Y.; Liu, X.; Chen, S.; Song, C.; Yang, Y.; Guo, Y. Calcium-activated 14-3-3 proteins as a molecular switch in salt stress tolerance. *Nat. Commun.* **2019**, *10*, 1199. [CrossRef]
43. Zhang, Y.; Yin, K.; Yao, J.; Zhao, Z.; Liu, Z.; Yan, C.; Zhang, Y.; Liu, J.; Li, J.; Zhao, N.; et al. *Populus euphratica* GLABRA3 binds PLDδ promoters to enhance salt tolerance. *Int. J. Mol. Sci.* **2023**, *24*, 8208. [CrossRef]
44. Li, J.; Zhao, R.; Liu, J.; Yao, J.; Ma, S.; Yin, K.; Zhang, Y.; Liu, Z.; Yan, C.; Zhao, N.; et al. *Populus euphratica* GRP2 interacts with target mRNAs to negatively regulate salt tolerance by interfering with photosynthesis, Na^+, and ROS homeostasis. *Int. J. Mol. Sci.* **2024**, *25*, 2046. [CrossRef] [PubMed]
45. Zhu, Z.; Liu, X.; Meng, F.; Jiang, A.; Zhou, Y.; Yuan, F.; Chen, M. LbHKT1;1 negatively regulates salt tolerance of *Limonium bicolor* by decreasing salt secretion rate of salt glands. *Plant Cell Environ.* **2025**, *48*, 3544–3558. [CrossRef] [PubMed]
46. Gao, Z.; Gao, S.; Li, P.; Zhang, Y.; Ma, B.; Wang, Y. Exogenous methyl jasmonate promotes salt stress-induced growth inhibition and prioritizes defense response of *Nitraria tangutorum* Bobr. *Physiol. Plant.* **2021**, *172*, 162–175. [CrossRef]
47. Dong, X.; Sun, L.; Guo, J.; Liu, L.; Wang, B. Exogenous boron alleviates growth inhibition by NaCl stress by reducing Cl uptake in sugar beet (*Beta vulgaris*). *Plant Soil* **2021**, *464*, 423–439. [CrossRef]
48. Qu, M.; Havshøi, N.W.; Huang, X.; Shabala, L.; Yu, M.; Fuglsang, A.T.; Shabala, S. Understanding the mechanistic basis of ameliorative effects of boron on salinity in barley (*Hordeum vulgare*). *Environ. Exp. Bot.* **2024**, *220*, 105690. [CrossRef]
49. Huang, H.; Zhuang, L.; Tang, H.; Guo, Z.; Li, Q.; Lin, Z.; Dai, M.; Wang, X.; Wang, Y.; Zheng, H.; et al. Biosynthesis-based spatial metabolome of condensed tannin reveals its role in salt tolerance of non-salt-secretor mangrove *Kandelia obovata*. *Plant Cell Environ.* **2025**, *48*, 1874–1889. [CrossRef]
50. Wu, X.; Li, J.; Song, L.Y.; Zeng, L.L.; Guo, Z., J.; Ma, D.N.; Wei, M.Y.; Zhang, L.D.; Wang, X.X.; Zheng, H.L. NADPH oxidase-dependent H_2O_2 production mediates salicylic acid-induced salt tolerance in mangrove plant *Kandelia obovata* by regulating Na^+/K^+ and redox homeostasis. *Plant J.* **2024**, *118*, 1119–1135. [CrossRef]
51. Zeng, L.L.; Song, L.Y.; Wu, X.; Ma, D.N.; Song, S.W.; Wang, X.X.; Zheng, H.L. Brassinosteroid enhances salt tolerance via S-nitrosoglutathione reductase and nitric oxide signaling pathway in mangrove *Kandelia obovata*. *Plant Cell Environ.* **2024**, *47*, 511–526. [CrossRef]
52. Cao, Y.; Zhang, M.; Liang, X.; Li, F.; Shi, Y.; Yang, X.; Jiang, C. Natural variation of an EF-hand Ca^{2+}-binding-protein coding gene confers saline-alkaline tolerance in maize. *Nat. Commun.* **2020**, *11*, 186. [CrossRef]
53. Xie, Q.; Yang, Y.; Wang, Y.; Pan, C.; Hong, S.; Wu, Z.; Song, J.; Zhou, Y.; Jiang, X. The calcium sensor CBL10 negatively regulates plasma membrane H^+-ATPase activity and alkaline stress response in *Arabidopsis*. *Environ. Exp. Bot.* **2022**, *194*, 104752. [CrossRef]
54. Cui, M.; Li, Y.; Li, J.; Yin, F.; Chen, X.; Qin, L.; Wei, L.; Xia, G.; Liu, S. Ca^{2+}-dependent TaCCD1 cooperates with TaSAUR215 to enhance plasma membrane H^+-ATPase activity and alkali stress tolerance by inhibiting PP2C-mediated dephosphorylation of TaHA2 in wheat. *Mol. Plant* **2023**, *16*, 571–587. [CrossRef] [PubMed]
55. Xia, T.; Zhan, Y.; Mu, Y.; Zhang, J.; Xu, W. MNSs-mediated N-glycan processing is essential for auxin homeostasis in *Arabidopsis* roots during alkaline response. *iScience* **2022**, *25*, 104298. [CrossRef] [PubMed]
56. Cao, X.; Sun, L.; Wang, W.; Zhang, F. Exogenous calcium application mediates K^+ and Na^+ homeostasis of different salt-tolerant rapeseed varieties under $NaHCO_3$ stress. *Plant Growth Regul.* **2024**, *102*, 367–378. [CrossRef]
57. Xu, W.; Jia, L.; Shi, W.; Liang, J.; Zhou, F.; Li, Q.; Zhang, J. Abscisic acid accumulation modulates auxin transport in the root tip to enhance proton secretion for maintaining root growth under moderate water stress. *New Phytol.* **2013**, *197*, 139–150. [CrossRef]
58. Mak, M.; Babla, M.; Xu, S.C.; O'Carrigan, A.; Liu, X.H.; Gong, Y.M.; Holford, P.; Chen, Z.H. Leaf mesophyll K^+, H^+ and Ca^{2+} fluxes are involved in drought-induced decrease in photosynthesis and stomatal closure in soybean. *Environ. Exp. Bot.* **2014**, *98*, 1–12. [CrossRef]
59. Feng, X.; Liu, W.; Qiu, C.W.; Zeng, F.; Wang, Y.; Zhang, G.; Chen, Z.H.; Wu, F. HvAKT2 and HvHAK1 confer drought tolerance in barley through enhanced leaf mesophyll H^+ homoeostasis. *Plant Biotechnol. J.* **2020**, *18*, 1683–1696. [CrossRef]

60. Zhang, X.; Wu, H.; Chen, J.; Chen, L.; Chang, N.; Ge, G.; Wang, X. Higher ROS scavenging ability and plasma membrane H^+-ATPase activity are associated with potassium retention in drought tolerant tea plants. *J. Plant Nutr. Soil Sci.* **2020**, *183*, 406–415. [CrossRef]
61. Sun, W.; Xia, L.; Deng, J.; Sun, S.; Yue, D.; You, J.; Wang, M.; Jin, S.; Zhu, L.; Lindsey, K.; et al. Evolution and subfunctionalization of CIPK6 homologous genes in regulating cotton drought resistance. *Nat. Commun.* **2024**, *15*, 5733. [CrossRef]
62. Huang, X.; Shabala, L.; Zhang, X.; Zhou, M.; Voesenek, L.A.C.J.; Hartman, S.; Yu, M.; Shabala, S. Cation transporters in cell fate determination and plant adaptive responses to a low-oxygen environment. *J. Exp. Bot.* **2022**, *73*, 636–645. [CrossRef]
63. Liu, C.Y.; Zhang, F.; Zhang, D.J.; Srivastava, A.K.; Wu, Q.S.; Zou, Y.N. Mycorrhiza stimulates root-hair growth and IAA synthesis and transport in trifoliate orange under drought stress. *Sci. Rep.* **2018**, *8*, 1978. [CrossRef]
64. Li, B.; Zhang, M.; Sun, W.; Yue, D.; Ma, Y.; Zhang, B.; Duan, L.; Wang, M.; Lindsey, K.; Nie, X.; et al. N^6-methyladenosine RNA modification regulates cotton drought response in a Ca^{2+} and ABA-dependent manner. *Plant Biotechnol. J.* **2023**, *21*, 1270–1285. [CrossRef] [PubMed]
65. Zhang, W.; Wang, L.; Zhang, L.; Kong, X.; Zhang, J.; Wang, X.; Pei, Y.; Jin, Z. H_2S-mediated balance regulation of stomatal and non-stomatal factors responding to drought stress in Chinese cabbage. *Hortic. Res.* **2022**, *10*, uhac284. [CrossRef] [PubMed]
66. Chang, J.; Guo, Y.; Li, J.; Liu, L.; Liu, J.; Yuan, L.; Wei, C.; Ma, J.; Zhang, Y.; Ahammed, G.J.; et al. Cyclic nucleotide-gated ion channel 20 regulates melatonin-induced calcium signaling and cold tolerance in watermelon. *Plant Physiol.* **2024**, *26*, kiae630. [CrossRef]
67. Zhang, D.; Guo, X.; Xu, Y.; Li, H.; Ma, L.; Yao, X.; Weng, Y.; Guo, Y.; Liu, C.M.; Chong, K. OsCIPK7 point-mutation leads to conformation and kinase-activity change for sensing cold response. *J. Integr. Plant Biol.* **2019**, *61*, 1194–1200. [CrossRef]
68. Wang, J.; Ren, Y.; Liu, X.; Luo, S.; Zhang, X.; Liu, X.; Lin, Q.; Zhu, S.; Wan, H.; Yang, Y.; et al. Transcriptional activation and phosphorylation of OsCNGC9 confer enhanced chilling tolerance in rice. *Mol. Plant* **2021**, *14*, 315–329. [CrossRef]
69. Yan, Y.; Sun, M.; Ma, S.; Feng, Q.; Wang, Y.; Di, Q.; Zhou, M.; He, C.; Li, Y.; Gao, L.; et al. Mechanism of CsGPA1 in regulating cold tolerance of cucumber. *Hortic. Res.* **2022**, *9*, uhac109. [CrossRef]
70. Zimmermann, M.J.; Bose, J.; Kramer, E.M.; Atkin, O.K.; Tyerman, S.D.; Baskin, T.I. Oxygen uptake rates have contrasting responses to temperature in the root meristem and elongation zone. *Physiol. Plant.* **2022**, *174*, e13682. [CrossRef]
71. Guo, Z.; Zuo, Y.; Wang, S.; Zhang, X.; Wang, Z.; Liu, Y.; Shen, Y. Early signaling enhance heat tolerance in *Arabidopsis* through modulating jasmonic acid synthesis mediated by HSFA2. *Int. J. Biol. Macromol.* **2024**, *267*, 131256. [CrossRef]
72. Chen, F.; Dong, G.; Wang, F.; Shi, Y.; Zhu, J.; Zhang, Y.; Ruan, B.; Wu, Y.; Feng, X.; Zhao, C.; et al. A β-ketoacyl carrier protein reductase confers heat tolerance via the regulation of fatty acid biosynthesis and stress signaling in rice. *New Phytol.* **2021**, *232*, 655–672. [CrossRef]
73. Huang, H.; Han, Y.; Hao, C.F.S. Exogenous spermidine modulates osmoregulatory substances and leaf stomata to alleviate the damage to lettuce seedlings caused by high temperature stress. *J. Plant Growth Regul.* **2023**, *42*, 1236–1255. [CrossRef]
74. Kan, Y.; Mu, X.R.; Zhang, H.; Gao, J.; Shan, J.X.; Ye, W.W.; Lin, H.X. TT2 controls rice thermotolerance through SCT1-dependent alteration of wax biosynthesis. *Nat. Plants* **2022**, *8*, 53–67. [CrossRef] [PubMed]
75. Song, Y.; Chen, P.; Liu, P.; Bu, C.; Zhang, D. High-temperature-responsive poplar lncRNAs modulate target gene expression via RNA interference and act as RNA scaffolds to enhance heat tolerance. *Int. J. Mol. Sci.* **2020**, *21*, 6808. [CrossRef] [PubMed]
76. Xie, B.; Chen, Y.; Zhang, Y.; An, X.; Li, X.; Yang, A.; Kang, G.; Zhou, J.; Cheng, C. Comparative physiological, metabolomic, and transcriptomic analyses reveal mechanisms of apple dwarfing rootstock root morphogenesis under nitrogen and/or phosphorus deficient conditions. *Front. Plant Sci.* **2023**, *14*, 1120777. [CrossRef] [PubMed]
77. Lv, X.; Zhang, Y.; Hu, L.; Zhang, Y.; Kong, L. Low-nitrogen stress stimulates lateral root initiation and nitrogen assimilation in wheat: Roles of phytohormone signaling. *J. Plant Growth Regul.* **2021**, *40*, 436–450. [CrossRef]
78. Sun, H.; Guo, X.; Zhu, X.; Gu, P.; Zhang, W.; Tao, W.; Wang, D.; Wu, Y.; Zhao, Q.; Xu, G.; et al. Strigolactone and gibberellin signaling coordinately regulate metabolic adaptations to changes in nitrogen availability in rice. *Mol. Plant* **2023**, *16*, 588–598. [CrossRef]
79. Alber, A.; Ehlting, B.; Ehlting, J.; Hawkins, B.J.; Rennenberg, H. Net NH_4^+ and NO_3^- flux, and expression of NH_4^+ and NO_3^- transporters in roots of *Picea glauca*. *Trees* **2012**, *26*, 1403–1411. [CrossRef]
80. Hawkins, B.J.; Robbins, S. Comparison of ammonium, nitrate, and proton fluxes in mycorrhizal and nonmycorrhizal roots of lodgepole pine in contrasting nitrogen treatments. *Can. J. For. Res.* **2022**, *52*, 1245–1253. [CrossRef]
81. Sun, Q.; Zhai, L.; Zhao, D.; Gao, M.; Wu, Y.; Wu, T.; Zhang, X.; Xu, X.; Han, Z.; Wang, Y. Kinase MxMPK4-1 and calmodulin-binding protein MxIQM3 enhance apple root acidification during Fe deficiency. *Plant Physiol.* **2023**, *191*, 1968–1984. [CrossRef]
82. Kong, L.; Sun, M.; Wang, F.; Liu, J.; Feng, B.; Si, J.; Zhang, B.; Li, S.; Li, H. Effects of high NH_4^+ on K^+ uptake, culm mechanical strength and grain filling in wheat. *Front. Plant Sci.* **2014**, *5*, 703. [CrossRef]
83. Jian, S.; Liao, Q.; Song, H.; Liu, Q.; Lepo, J.E.; Guan, C.; Zhang, J.; Ismail, A.M.; Zhang, Z. NRT1.1-related NH_4^+ toxicity is associated with a disturbed balance between NH_4^+ uptake and assimilation. *Plant Physiol.* **2018**, *178*, 1473–1488. [CrossRef]

84. Xiao, C.; Sun, D.; Liu, B.; Fang, X.; Li, P.; Jiang, Y.; He, M.; Li, J.; Luan, S.; He, K. Nitrate transporter NRT1.1 and anion channel SLAH3 form a functional unit to regulate nitrate-dependent alleviation of ammonium toxicity. *J. Integr. Plant Biol.* **2022**, *64*, 942–957. [CrossRef] [PubMed]
85. Zhang, X.; Wu, H.; Chen, L.; Li, Y.; Wan, X. Efficient iron plaque formation on tea (*Camellia sinensis*) roots contributes to acidic stress tolerance. *J. Integr. Plant Biol.* **2019**, *61*, 155–167. [CrossRef] [PubMed]
86. Ye, J.Y.; Tian, W.H.; Zhou, M.; Zhu, Q.Y.; Du, W.X.; Zhu, Y.X.; Liu, X.X.; Lin, X.Y.; Zheng, S.J.; Jin, C.W. STOP1 activates NRT1.1-mediated nitrate uptake to create a favorable rhizospheric pH for plant adaptation to acidity. *Plant Cell* **2021**, *33*, 3658–3674. [CrossRef]
87. Lehmann, J.; Jørgensen, M.E.; Fratz, S.; Müller, H.M.; Kusch, J.; Scherzer, S.; Navarro-Retamal, C.; Mayer, D.; Böhm, J.; Konrad, K.R.; et al. Acidosis-induced activation of anion channel SLAH3 in the flooding-related stress response of *Arabidopsis*. *Curr. Biol.* **2021**, *31*, 3575–3585.e9. [CrossRef]
88. Wang, C.; Bian, C.; Li, J.; Han, L.; Guo, D.; Wang, T.; Sun, Z.; Ma, C.; Liu, X.; Tian, Y.; et al. Melatonin promotes Al^{3+} compartmentalization via H^+ transport and ion gradients in *Malus hupehensis*. *Plant Physiol.* **2023**, *193*, 821–839. [CrossRef]
89. Zhang, Z.; Liu, D.; Meng, H.; Li, S.; Li, N. Magnesium alleviates aluminum toxicity by promoting polar auxin transport and distribution and root alkalization in the root apex in populus. *Plant Soil* **2020**, *448*, 565–585. [CrossRef]
90. Lu, Z.; Yu, M.; Han, X.; Qiao, G.; Xu, J.; Wu, L.; Qiu, W.; Zhuo, R. SpbZIP60 confers cadmium tolerance by strengthening the root cell wall compartmentalization in *Sedum plumbizincicola*. *J. Hazard. Mater.* **2024**, *480*, 135936. [CrossRef]
91. Liu, M.; Liu, X.; Kang, J.; Korpelainen, H.; Li, C. Are males and females of *Populus cathayana* differentially sensitive to Cd stress? *J. Hazard. Mater.* **2020**, *393*, 122411. [CrossRef]
92. Yin, K.; Liu, Y.; Liu, Z.; Zhao, R.; Zhang, Y.; Yan, C.; Zhao, Z.; Feng, B.; Zhang, X.; An, K.; et al. *Populus euphratica* CPK21 interacts with NF-YC3 to enhance cadmium tolerance in *Arabidopsis*. *Int. J. Mol. Sci.* **2024**, *25*, 7214. [CrossRef]
93. Liu, L.; Gui, H.; Zou, D.; Jiao, W.; Wang, S.; Wan, X. Long-term adaptation of water hyacinth to low cadmium involves antioxidant enzyme and metallothionein transcriptional regulation. *Chemosphere* **2024**, *365*, 143346. [CrossRef]
94. Su, N.; Niu, M.; Liu, Z.; Wang, L.; Zhu, Z.; Zou, J.; Chen, Y.; Cui, J. Hemin-decreased cadmium uptake in pak choi (*Brassica chinensis* L.) seedlings is heme oxygenase-1 dependent and relies on its by-products ferrous iron and carbon monoxide. *Environ. Pollut.* **2021**, *274*, 115882. [CrossRef] [PubMed]
95. Ma, J.; Zhou, B.; Tan, Q.; Zhang, L.; Pan, K. The roles of silicon in combating cadmium challenge in the Marine diatom *Phaeodactylum tricornutum*. *J. Hazard. Mater.* **2020**, *389*, 121903. [CrossRef] [PubMed]
96. Mak, M.; Zhang, M.; Randall, D.; Holford, P.; Chen, Z.H. Chloride transport at plant-soil interface modulates barley cd tolerance. *Plant Soil* **2019**, *441*, 409. [CrossRef]
97. Zhang, Y.; Sa, G.; Zhang, Y.; Hou, S.; Wu, X.; Zhao, N.; Zhang, Y.; Deng, S.; Deng, C.; Deng, J.; et al. *Populus euphratica* annexin1 facilitates cadmium enrichment in transgenic *Arabidopsis*. *J. Hazard. Mater.* **2021**, *405*, 124063. [CrossRef]
98. He, J.; Zhuang, X.; Zhou, J.; Sun, L.; Wan, H.; Li, H.; Lyu, D. Exogenous melatonin alleviates cadmium uptake and toxicity in apple rootstocks. *Tree Physiol.* **2020**, *40*, 746–761. [CrossRef]
99. Kuromori, T.; Fujita, M.; Takahashi, F.; Yamaguchi-Shinozaki, K.; Shinozaki, K. Inter-tissue and inter-organ signaling in drought stress response and phenotyping of drought tolerance. *Plant J.* **2022**, *109*, 342–358. [CrossRef]
100. Liu, H.; Able, A.J.; Able, J.A. Priming crops for the future: Rewiring stress memory. *Trends Plant Sci.* **2022**, *2*, 699–716. [CrossRef]
101. Gupta, A.; Rico-Medina, A.; Caño-Delgado, A.I. The physiology of plant responses to drought. *Science* **2020**, *368*, 266–269. [CrossRef]
102. Kim, J.S.; Kidokoro, S.; Yamaguchi-Shinozaki, K.; Shinozaki, K. Regulatory networks in plant responses to drought and cold stress. *Plant Physiol.* **2024**, *195*, 170–189. [CrossRef]
103. McLamore, E.S.; Diggs, A.; Calvo Marzal, P.; Shi, J.; Blakeslee, J.J.; Peer, W.A.; Murphy, A.S.; Porterfield, D.M. Non-invasive quantification of endogenous root auxin transport using an integrated flux microsensor technique. *Plant J.* **2010**, *63*, 1004–1016. [CrossRef]
104. Zhang, X.; Zhang, X.; Fu, Y.; Cui, Y.; Wu, N.; Li, Y.; Yang, Z.; Zhang, C.; Song, H.; He, G.; et al. HTT1, a Stearoyl-Acyl Carrier Protein Desaturase involved unsaturated fatty acid biosynthesis, affects rice heat tolerance. *Plant Cell Environ.* **2025**, *48*, 3391–3405. [CrossRef] [PubMed]
105. Ma, Q.; Feng, Y.; Luo, S.; Cheng, L.; Tong, W.; Lu, X.; Li, Y.; Zhang, P. The aquaporin MePIP2;7 improves MeMGT9-mediated Mg^{2+} acquisition in cassava. *J. Integr. Plant Biol.* **2023**, *65*, 2349–2367. [CrossRef] [PubMed]
106. Liu, X.X.; Zhang, H.H.; Zhu, Q.Y.; Ye, J.Y.; Zhu, Y.X.; Jing, X.T.; Du, W.X.; Zhou, M.; Lin, X.Y.; Zheng, S.J.; et al. Phloem iron remodels root development in response to ammonium as the major nitrogen source. *Nat. Commun.* **2022**, *13*, 561. [CrossRef] [PubMed]
107. Pandey, A.; Devi, L.L.; Gupta, S.; Prasad, P.; Agrwal, K.; Asif, M.H.; Pandey, A.K.; Bandyopadhyay, K.; Singh, A.P. Jasmonate signaling modulates root growth by suppressing iron accumulation during ammonium stress. *Plant Physiol.* **2024**, *196*, 2213–2231. [CrossRef]

108. Bolan, N.; Sarmah, A.K.; Bordoloi, S.; Bolan, S.; Padhye, L.P.; Van Zwieten, L.; Sooriyakumar, P.; Khan, B.A.; Ahmad, M.; Solaiman, Z.M.; et al. Soil acidification and the liming potential of biochar. *Environ. Pollut.* **2023**, *317*, 120632. [CrossRef]
109. Dhandapani, S.; Sng, Y.H.; Agisha, V.N.; Suraby, E.J.; Park, B.S. Mitigating aluminum toxicity and promoting plant resilience in acidic soil with *Penicillium olsonii* TLL1. *Front. Plant Sci.* **2024**, *15*, 1423617. [CrossRef]
110. Feng, Z.; Ji, S.; Ping, J.; Cui, D. Recent advances in metabolomics for studying heavy metal stress in plants. *TrAC-Trend Anal. Chem.* **2021**, *143*, 116402. [CrossRef]

plants

MDPI

Review

The IAOx-Dependent IAA Biosynthesis Pathway: Acquired Insights, Paradigm Shifts, and Unresolved Questions

Ming-Kun Ma [1], Verena Kriechbaumer [2,*] and Dong-Wei Di [3,4,*]

1 Institute of Agricultural Resources and Environment, Sichuan Academy of Agricultural Sciences, Chengdu 610066, China; mamingkun@scsaas.cn
2 School of Biological and Medical Sciences, Faculty of Health, Science and Technology, Oxford Brookes University, Oxford OX3 0BP, UK
3 State Key Laboratory of Soil and Sustainable Agriculture, Institute of Soil Science, Chinese Academy of Sciences, Nanjing 210008, China
4 University of Chinese Academy of Sciences, Beijing 100049, China
* Correspondence: vkriechbaumer@brookes.ac.uk (V.K.); dwdi@issas.ac.cn (D.-W.D.)

Abstract

The auxin indole-3-acetic acid (IAA) is essential for plant growth and stress adaptation. Its biosynthesis via the indole-3-acetaldoxime (IAOx) pathway has recently undergone a paradigm shift. Recent genetic and metabolomic studies have fundamentally revised the indole-3-acetaldoxime (IAOx) pathway from a linear route (IAOx→IAN→IAM→IAA) to a dynamic network. This review synthesizes this paradigm shift by integrating evidence from key Arabidopsis studies. Crucially, mutants disrupting multiple downstream enzyme families fail to block IAA overproduction in the IAOx-accumulating *superroot 2* (*sur2*) background. Functioning as a central branching point between auxin and defense metabolism, the tryptophan-derived metabolite IAOx, along with indole-3-acetonitrile (IAN) and indole-3-acetamide (IAM), elicits auxin responses via independent, tissue-specific pathways, with no metabolic requirement for IAM as a universal intermediate. Furthermore, IAN and IAM levels do not increase with massive IAOx accumulation, indicating a bypass route from IAOx to IAA. We conclude that IAOx acts as a central metabolic hub, partitioning flux competitively between growth and defense. Resolving the unknown IAOx-converting enzyme, the signaling roles of IAN/IAM, and the logic of metabolic channeling is vital to understanding how plants integrate hormonal and stress responses.

Keywords: auxin biosynthesis; indole-3-acetaldoxime; indole-3-acetamide; indole-3-acetonitrile; growth-defense trade-off

1. Introduction

Auxin, a core class of plant hormones, plays a pivotal role in regulating plant development and stress responses [1,2]. It governs diverse physiological processes, including cell division and elongation, tropic movements, apical dominance, flowering, and senescence, while also coordinating adaptations to biotic and abiotic stresses [1,3]. Among the three naturally occurring active auxins, indole-3-acetic acid (IAA), phenylacetic acid (PAA), and 4-chloroindole-3-acetic acid (4-Cl-IAA), IAA stands out as the most ubiquitous and physiologically crucial [4,5]. To ensure normal growth and adaptation in complex environments, plants meticulously regulate IAA levels through a sophisticated homeostatic system. This system relies on a dynamic equilibrium among IAA biosynthesis, polar transport, signal transduction, and the interconversion between its active and inactive forms [6–9].

https://doi.org/10.3390/plants15030436

IAA biosynthesis serves as the foundation of this regulatory network, originating from chorismate in the chloroplast, which is the terminal product of the shikimate pathway [10]. Biosynthetic routes are primarily classified as tryptophan (Trp)-dependent (TD) or Trp-independent (TI), with both pathways converging at the key metabolic intermediate indole-3-glycerol phosphate (IGP) [10,11]. During Trp synthesis in *Arabidopsis thaliana*, chorismate is converted to anthranilate by the anthranilate synthase (AS). This rate-limiting step is mediated by the *WEI2/ASA1* and *WEI7/ASB1* genes, encoding the α and β subunits of AS enzyme, respectively [12–14]. Anthranilate is then sequentially transformed into 1-(O-carboxylphenylamino)-1-deoxyribulose-5-phosphate (CdRP) through the coordinated actions of phosphoribosyltransferase (PAT1) and phosphoribosyl anthranilate isomerase (PAI), ultimately yielding IGP via IGP synthase catalysis [15]. As a critical metabolic branch point, IGP is predominantly channeled to Trp through the tryptophan synthase complex (TSA/TSB) for entry into TD pathways, while also serving as the substrate for indole synthesis via indole synthase (INS) to initiate the TI route [16,17].

Among TD pathways, the indole-3-pyruvic acid (IPyA) pathway represents the predominant and best-characterized route for plant IAA production [2,18]. This pathway involves two core enzymatic reactions: the transamination of L-Trp to IPyA by the TAA1/TAR family of aminotransferases [19–22], followed by oxidative conversion of IPyA to IAA catalyzed by YUCCA (YUC) family flavin monooxygenases, a reaction recognized as the rate-limiting step [23–28]. The identification of the YUC gene family [27], coupled with evidence of its members' functional redundancy, distinct spatiotemporal expression patterns [18,24,26], and functional coordination with the upstream TAA1/TAR enzymes [23,28], collectively underscores the centrality of the IPyA pathway in plant auxin biosynthesis. The Arabidopsis *YUC* gene family comprises 11 members exhibiting both functional redundancy and spatial specificity, thereby facilitating precise spatiotemporal control of IAA synthesis [18]. Furthermore, this pathway is further modulated by feedback regulation through the aminotransferase REVERSAL OF SAV3 PHENOTYPE 1 (VAS1), which catalyzes IPyA reconversion to Trp to maintain metabolic homeostasis [29].

Beyond the IPyA pathway, multiple additional Trp-derived routes contribute to IAA biosynthesis. The indole-3-acetamide (IAM) pathway involves IAM hydrolysis to IAA by amidase 1 (AMI1), its homologs (such as TOC64s and FAAHs), or the recently identified IAM hydrolases (IAMH1/2), though the endogenous origins of IAM remain partially elusive, potentially deriving from IAOx conversion or unidentified sources [30,31]. Although the tryptamine (TAM) pathway has been proposed, its physiological significance remains controversial, with potential functionality restricted to specific tissues or particular conditions [32]. TAM exhibits auxin-like activity, but several lines of evidence challenge its role as a major IAA precursor: it is often found at high, non-specific levels across species, originates from distinct Trp pools, and its proposed conversion to IAA lacks robust enzymatic support [33–35]. While tissue-specific conversion (e.g., in pea roots) suggests limited, context-dependent functionality, TAM is not considered a conserved, primary route for IAA biosynthesis [2]. In contrast, the indole-3-acetaldoxime (IAOx) pathway, which is especially prominent in Brassicaceae species, constitutes a major TD branch that is intricately linked to specialized metabolism and stress responses, and will be the primary focus of this review [26,27].

The TI pathway, which bypasses Trp, is postulated to initiate from IGP through direct conversion to indole via INS, followed by poorly defined metabolic steps ultimately yielding IAA [16,17,35]. These studies, using genetic and isotopic tracing evidence, suggested a potential role for this pathway in embryogenesis [17] and maintaining basal IAA levels [16], although its precise enzymatic mechanisms remain elusive to date [35]. This pathway is hypothesized to function during embryogenesis and in maintaining basal IAA levels,

https://doi.org/10.3390/plants15030436

though its enzymatic mechanisms and regulatory networks remain substantial challenges for future investigation [16,17]. Crucially, the entire IAA biosynthetic network operates under multi-layered regulation encompassing epigenetic modifications, transcriptional control, post-transcriptional processing, post-translational modifications, and metabolic feedback, all functioning coordinately to maintain IAA dynamic balance and support plant growth and environmental adaptation [6,18,36]. This involves integrated control at multiple levels, from epigenetic and transcriptional regulation [6,36] to the precise spatiotemporal control of key biosynthetic genes like *YUC* [18].

The field of auxin biology has undergone a profound paradigm shift, decisively moving beyond the confines of the canonical nuclear Skp1-Cullin1-F-box (SCF) TIR1/AFBs-Auxin/Indole-3-Acetic Acid (Aux/IAA)- Auxin Response Factor (ARF) signaling pathway to embrace a more complex, decentralized, and multi-layered signaling network. This transformation is underscored by the establishment of a sophisticated dual-stream mechanism: one consisting of a rapid, cell-surface signaling cascade mediated by the Auxin-Binding Protein (ABP1)/ABP1-Like protein (ABL)-Transmembrane Kinase (TMK) module, which directly phosphorylates key targets like PIN-FORMED (PIN) efflux carriers and plasma membrane H^+-ATPases to coordinately regulate auxin transport and the acid growth process within minutes [37–43]; and a refined view of nuclear signaling, in which the TIR1/AFB receptors depend on their intrinsic adenylate cyclase activity to generate local cAMP pools essential for full transcriptional reprogramming of auxin-responsive genes [8,44–46]. This bifurcation elegantly resolves long-standing questions about the temporal disconnect between rapid physiological responses and slower transcriptional changes [8,37,38].

Building on this conceptual framework, recent advances have further elucidated the self-organizing properties of the auxin system. Studies on self-organizing transport modules, together with high-resolution structural insights into AUX1/LAX influx and PIN efflux carriers, have revealed fundamental biophysical principles governing the spatiotemporal control of auxin distribution [38,47–50]. These findings bridge signaling and transport, illustrating how auxin can rapidly modulate its own directional movement through non-transcriptional pathways while also generating complex distribution patterns that feed back into both signaling branches. Through intricate feedback loops and crosstalk with other hormonal and environmental signals, this integrated network provides a robust molecular basis for developmental plasticity, from root gravitropism to phyllotaxis.

Despite these advances, significant knowledge gaps persist between the well-characterized IPyA pathway and less understood routes such as TAM, IAM, and particularly IAOx [2,51]. The IAOx pathway has recently undergone significant conceptual revisions, with transformative insights from seminal work by Fenech et al. (2025) compelling a comprehensive re-evaluation of its metabolic fluxes and physiological roles [52]. Against this backdrop of renewed interest and evolving paradigms, this review systematically traces the research trajectory of the IAOx pathway, providing an overview of established conceptual frameworks, a critical analysis of emerging evidence, and an identification of key scientific questions for future investigation. It should be noted that the mechanistic insights and paradigm shifts discussed herein derive primarily from studies in Brassicaceae species, notably Arabidopsis.

2. The Established Paradigm: IAOx as the Central Precursor and Linear Pathway

The IAOx pathway represents a classic model for the biosynthesis of the plant auxin IAA, primarily characterized in Brassicaceae species. This model, systematically articulated by Sugawara et al. [34], posits a linear metabolic cascade: IAOx is converted to IAN by

https://doi.org/10.3390/plants15030436

cytochrome P450 enzymes of the CYP71A family (e.g., CYP71A12/A13) [53–55]. IAN is subsequently transformed into IAA and/or IAM by nitrilases (NITs) [51,56–60]. Finally, the conversion of IAM to IAA is mediated by a range of enzymes, encompassing amidase AMI1, its homologs (e.g., TOC64s and FAAHs), as well as the recently characterized IAM hydrolases IAMH1 and IAMH2 [30,61–65] (Figure 1A). Strong genetic support for this model comes from studies of the *sur2* mutant. By blocking the diversion of IAOx into indole glucosinolate (IG) biosynthesis, *sur2* leads to substantial IAOx accumulation, and this accumulated IAOx is then channeled into IAA production, resulting in characteristic high-auxin phenotypes such as shortened roots, disintegrated hypocotyls, and epinastic cotyledons [66,67].

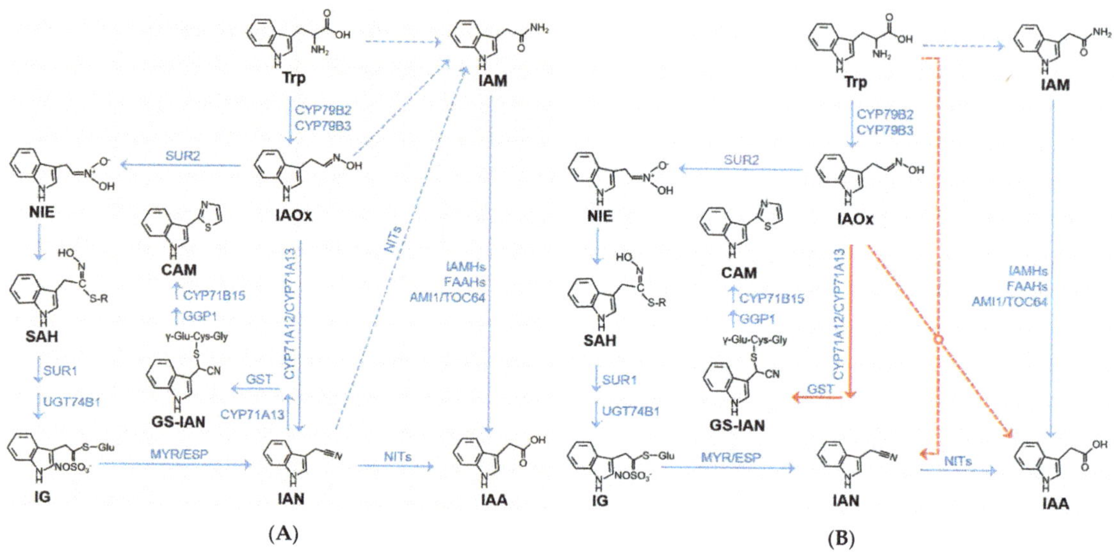

Figure 1. Paradigm shift in the IAOx-dependent auxin biosynthesis pathway: An evolving view. (**A**) The established paradigm: A linear IAOx-dependent auxin biosynthesis pathway. (**B**) An emerging view: IAOx-dependent auxin biosynthesis as a complex metabolic network. Abbreviations: CYP79B2/B3, Cytochrome P450 Family 79 Subfamily B Polypeptide 2/3; SUR1, SUPERROOT 1; SUR2, SUPERROOT 2 (CYP83B1); UGT74B1, UDP-Glycosyltransferase 74B1; MYR, Myrosinase; ESP, Epithiospecifier Protein; CYP71B15, Cytochrome P450 Family 71 Subfamily B Polypeptide 15 (PAD3); GGP1, General Regulator of Peptidase 1; GST: Glutathione S-Transferase; CYP71A12/13, Cytochrome P450 Family 71 Subfamily A Polypeptide 12/13; NIT, Nitrilase; IAMH, Indole-3-Acetamide Hydrolase; FAAH, Fatty Acid Amide Hydrolase; AMI1, Amidase 1; TOC64: Translocon of the Outer Membrane of the Chloroplasts 64; Trp, Tryptophan; IAOx, Indole-3-Acetaldoxime; NIE, 1-aci-Nitro-2-Indolyl-Ethane; SAH, S-alkyl-thiohydroximate; IAN, Indole-3-Acetonitrile; GS-IAN, Glutathione Conjugate of Indole-3-Acetonitrile; IAM, Indole-3-Acetamide; IG, Indole Glucosinolate; CAM, Camalexin; IAA, Indole-3-Acetic Acid. All depicted biochemical pathways and molecular structures were drawn based on established knowledge from the cited literature.

Early investigations detected IAOx unequivocally in a few Brassicaceae species, such as Arabidopsis, while it was generally absent in non-Brassicaceae species such as tomato, rice, maize, pea, and tobacco, suggesting potential species specificity [15,34,68]. However, the synthesis and metabolic flux of IAOx are significantly enhanced under various stress conditions, including pathogen infection (e.g., *Piriformospora indica*) [69], high temperature [70], or salt stress [71], indicating its crucial physiological role in plant stress responses. Building on this view, subsequent studies have reported significant IAOx accumulation in

https://doi.org/10.3390/plants15030436

specific contexts beyond Brassicaceae, such as in maize following herbivory [72,73] and in *Medicago truncatula* under iron deficiency [74]. These observations demonstrate that IAOx metabolism can be activated by specific environmental cues beyond the Brassicaceae family. A key question emerging from this finding is whether the associated auxin phenotypes in non-Brassicaceae species rely on CYP79 homologs or operate through entirely distinct mechanisms. Furthermore, exogenous IAOx application in *Medicago truncatula* induces typical auxin responses reminiscent of the Arabidopsis glucosinolate-deficient mutants *sur1* and *sur2*, including increased lateral root formation and suppressed primary root elongation [75,76]. These observations demonstrate that IAOx metabolism or perception can be activated in non-Brassicaceae species, yet the enzymatic basis for IAOx conversion to IAA in these species remains unclear and may not involve orthologs of the Arabidopsis CYP79B or the unknown enzyme implied by the *sur2* studies.

In Arabidopsis, the biosynthesis of IAOx, catalyzed by CYP79B2/B3, constitutes the rate-limiting step for its downstream metabolism [70,76,77] (Figure 1). Overexpression of CYP79B2 leads to IAOx accumulation and consequent IAA overproduction phenotypes, such as elongated hypocotyls, leaf epinasty, and elevated IAA levels. Conversely, the *cyp79b2/b3* double mutant lacks IAOx and exhibits reduced IAA levels and growth defects under high-temperature stress [34,70]. Beyond being a mere precursor of IAA, IAOx acts as a critical metabolic hub [78]. Its flux is precisely directed toward IG biosynthesis by CYP83B1/SUR2 [66,67,79], or, under inducing conditions, channeled into IAN by CYP71A12/CYP71A13 [53,55] (Figure 1). It is noteworthy that IG can be converted to IAN through catalysis by MYR and ESP [31,80,81]. IAN is subsequently conjugated to form GS-IAN, which is then utilized by GGP1 and CYP71B15 for the synthesis of the phytoalexin camalexin (CAM) [52,78].

The traditional model is underpinned by multiple lines of evidence consolidating IAM and IAN as key intermediates [82]. Genetically, the *cyp79b2/b3* double mutant shows significantly reduced levels of IAM and IAN, whereas glucosinolate-deficient mutants *sur1/sur2*, which accumulate IAOx, exhibit elevated IAM and/or IAN levels, driving IAA overproduction [34,76,83]. Biochemically, isotope tracing experiments demonstrated the conversion of exogenous IAOx to IAM and IAN [34], and CYP71A12/CYP71A13 have been confirmed to convert IAOx to IAN *in vitro* and *in vivo* [55]. The nitrilases NIT1-3 and the amidase hydrolases AMI1/IAMH have been identified as key enzymes catalyzing the hydrolysis of IAN and IAM to IAA, respectively [30,51,56–58]. Furthermore, cross-species studies reinforce this connection. In IAOx-treated Medicago plants, significant IAN accumulation was detected in shoots and in roots of iron-deficient plants, indicating that IAOx can be converted to IAN in roots or after translocation to shoots, a process occurring under both iron-deficient and sufficient conditions [74].

Nevertheless, as of 2025, several fundamental questions remain unresolved. First, the complete linear conversion route from IAOx to IAA is not fully established. While CYP71A13 catalyzes IAOx→IAN, the enzyme(s) responsible for the direct conversion of IAOx to IAM or other potential intermediates remain unknown [83]. Second, the necessity of IAN and IAM under basal growth conditions is questionable, as higher-order *nit* and *ami1 iamh* mutants do not exhibit significant IAA-deficient phenotypes under normal growth. This suggests functional redundancy or a primary role for these pathways outside of maintaining basal growth [30,51,56,83]. Furthermore, a conceptual gap persists between the apparent restriction of the IAOx pathway to Brassicaceae and its broader importance in stress responses [34,68]. Finally, the regulatory mechanisms governing the 'channeling' of metabolites between different pathways, such as the potential physical or metabolic separation of the IAN pool dedicated to CAM biosynthesis from the pool potentially

https://doi.org/10.3390/plants15030436

involved in IAA production, remain unclear [54,84]. These outstanding issues collectively define the critical research frontiers in the field beyond 2025.

3. Contemporary Revisions: Redefining the Pathway Through Network Regulation

Prior to the study by Fenech et al. (2025) [52], research on IAOx-mediated IAA biosynthesis was governed by a classic theoretical framework depicting a linear metabolic pathway (Figure 1A). However, Fenech et al. [52] fundamentally challenged this model through systematic genetic and metabolomic analyses (Figure 1B). Through a comprehensive genetic approach involving the generation and analysis of higher-order mutants targeting entire gene families (*cyp71a12 a13 a18*, *nit1 nit2 nit3 nit4*, *ami1 toc64-III toc64-V faah1 faah2*), combined with detailed metabolomic profiling, Fenech et al. (2025) provided compelling evidence that challenges the established linear model [52]. Notably, none of these mutants exhibited developmental abnormalities under standard conditions, in contrast to the weak auxin-deficient mutant *wei8/sav3/ckrc1* [19–21]. This absence of phenotypic defects clearly indicated that these gene families are dispensable for basal auxin biosynthesis.

A pivotal breakthrough came from systematically introducing these higher-order mutations into the *sur2* background, where the IAOx pathway is hyperactive. The linear model predicted that disrupting any downstream step in this background should suppress the high-auxin phenotypes by blocking IAOx conversion to IAA (Figure 1A). Contrary to this prediction, and inconsistent with the linear model, all mutant combinations fully retained the characteristic *sur2* phenotypes at both seedling and adult stages [52]. In stark contrast, the *cyp79b2 cyp79b3 sur2* triple mutant, which blocks IAOx production at the source, completely suppressed these phenotypes. This collective genetic evidence strongly demonstrates that the *CYP71As*, *NITs*, and *AMI/TOC64/FAAHs* gene families are not required for the excessive IAA production in *sur2* [52].

Further pharmacological and reporter gene analyses demonstrated that IAOx, IAN, and IAM engage in intricate tissue-specific metabolic interactions. Although all three compounds activated *p*DR5:GFP expression and induced auxin-like physiological responses, their effects were organ-dependent. IAOx exerted its primary influence in root tissues, IAM specifically enhanced hypocotyl elongation, and IAN was active in both roots and hypocotyls [52]. To decipher whether these distinct phenotypes reflected interconnected or parallel pathways, the researchers leveraged their mutant collection. Crucially, the *nit1 nit2 nit3 nit4* mutant was root-insensitive to IAN but responded normally to IAOx. Similarly, the *iamh1 iamh2* mutant was hypocotyl-insensitive to IAM but responded normally to IAOx and IAN. These results definitively show that the conversion of IAOx and IAN to IAA does not require IAM as an intermediate (Figure 1). Furthermore, the normal IAM response in higher-order AMI family mutants suggests that IAM-to-IAA conversion is primarily mediated by IAMHs, not the AMI family [30].

Metabolomics data provided direct quantitative evidence contradicting the linear model. In *sur2* and its higher-order mutants, despite a dramatic 60- to 120-fold accumulation of IAOx, the putative downstream intermediates IAN and IAM failed to accumulate and were even reduced. A decisive finding came from the *ami1-2 toc64-III toc64-V faah1 faah2 sur2* and *iamh1 iamh2 sur2* mutants, where IAM accumulation was anticipated if it were a central intermediate. The absence of significant IAM accumulation in these lines is entirely inconsistent with the linear model and strongly supports the existence of an unknown route from IAOx to IAA that bypasses both IAN and IAM.

In summary, the work by Fenech et al. presents a compelling rebuttal of the linear IAOx pathway model and establishes a robust, evidence-based framework for under-

 https://doi.org/10.3390/plants15030436

standing IAOx metabolism in Arabidopsis [52]. This framework is founded on three key pillars supported by integrated genetic, pharmacological and metabolomic data: (1) the existence of an efficient alternative pathway from IAOx to IAA that is hyperactivated in the *sur2* mutant; (2) the operational independence of this pathway from the *CYP71A*, *NIT*, *AMI/TOC64/FAAH* and *IAMH* gene families; and (3) the parallel and metabolically decoupled roles of IAOx, IAN and IAM as precursors, each engaging tissue-specific routes to produce IAA via distinct primary mechanisms, an unknown enzyme for IAOx, NITs for IAN and IAMHs for IAM [30,51,55,56,85]. This evidence-based model conclusively defines what is known, the existence and independence of this pathway, while starkly revealing what remains unknown: the biochemical identity of the conversion mechanism itself.

4. Future Research Perspectives: Unresolved Mysteries and Future Frontiers

The study by Fenech et al. has overturned the traditional linear model of IAOx-mediated IAA biosynthesis, representing a paradigm shift in the field (Figure 1B). Rather than concluding the research journey, this work reveals a more complex and highly networked regulatory landscape, and, most importantly, defines the precise boundaries of our current ignorance. The following directions constitute the most critical research frontiers within this revised framework. Crucially, the discussions below on mechanisms and origins are informed speculations derived from established evidence, not extensions of the evidence itself.

(i) The central enigma: The unidentified enzyme(s) catalyzing IAOx-to-IAA conversion and its regulation

The most pressing unresolved question, revealed by the work of Fenech et al. (2025) [52], is the molecular identity of the enzyme(s) responsible for the efficient conversion of accumulated IAOx to IAA in the *sur2* mutant. While its existence is robustly inferred from the inability of mutation in *CYP71A*, *NIT*, and *AMI* families to suppress *sur2* phenotypes, its molecular identity, biochemical properties, and regulatory mechanisms remain entirely unknown. While the genetic and metabolomic evidence unequivocally demands the existence of an efficient IAOx-to-IAA conversion route, its biochemical mechanism remains unresolved. Several plausible hypotheses merit consideration: a single unidentified enzyme (e.g., a novel oxidoreductase, lyase or dioxygenase) could catalyze direct conversion; a multi-enzyme metabolon, akin to the camalexin biosynthetic complex, might channel IAOx through transient association of catalytic and auxiliary subunits; strict subcellular compartmentalization (e.g., in peroxisomes or specialized ER domains) could provide both the requisite catalytic milieu and regulatory isolation; or an enzyme-triggered chemical step might generate a reactive intermediate that rapidly rearranges to IAA in the cellular context. These speculative models are grounded in the logical imperative of the genetic data and in known biochemical principles, and their experimental dissection represents the immediate biochemical frontier in the field.

Future research must focus on identifying its encoding gene(s), employing strategies such as genetic suppressor screens, genome-wide association studies, or interactome proteomics. Subsequently, profound regulatory questions emerge: Is the activity of this enzyme finely regulated by upstream signaling pathways? Does a competitive or mutually exclusive regulatory relationship exist between this enzyme and defense-related enzymes (e.g., CYP71A12/13) for the common substrate IAOx? Particularly under stress conditions such as pathogen infection, is this enzyme directly suppressed by core defense transcription factors like WRKY33 [69], thereby channeling metabolic flux preferentially toward defense pathways? Elucidating these mechanisms will not only identify a key missing

 https://doi.org/10.3390/plants15030436

component in auxin metabolism but also reveal a critical control point in the plant's stress response network.

(ii) Re-evaluating the physiological origins and functions of IAN and IAM

With IAN and IAM being excluded from the primary route of IAOx-to-IAA conversion, their physiological roles demand a critical re-examination that goes beyond their former status as pathway intermediates. Emerging evidence suggests that IAN and IAM not only contribute to systemic IAA biosynthesis but also exhibit distinct tissue-specific signaling functions in planta. IAM accumulation and AMI1 expression are particularly pronounced in reproductive and juvenile tissues such as flowers, young leaves, and developing seeds, indicating a potential role in localized auxin production during reproductive development and embryogenesis. In contrast, IAN is enriched in root tips, lateral root primordia, and leaf margins, suggesting a spatially restricted role in organogenesis and stress-responsive growth regulation. These observations highlight the nuanced, tissue-partitioned contributions of IAN and IAM to auxin-mediated signaling, beyond their canonical roles as mere biosynthetic intermediates.

The immediate task is to elucidate their endogenous biosynthetic pathways in wild-type plants. For instance, does IAM originate from a yet-unidentified, independent pathway involving the direct amidation of tryptophan? More significantly, IAN and IAM may function not merely as auxin precursors but as output signals of distinct metabolic decisions. IAN, derived from glucosinolate hydrolysis, for example, could act as a 'danger signal' that triggers specific defense responses locally or systemically. This possibility raises several pivotal questions: Does the plant cell possess sophisticated sensing mechanisms, such as NLR-like immune receptors or specific kinases, capable of discriminating IAN from different origins? Furthermore, how does the cell prevent IAN generated from defense pathways from inappropriately activating auxin signaling cascades? However, several key experimental gaps persist in fully elucidating these signaling roles. First, tissue and cell-type-specific functions remain unclear due to the lack of high-resolution localization studies and tissue-specific genetic manipulations. Second, the mechanisms by which IAN/IAM signaling integrates with canonical auxin perception pathways (e.g., TIR1/AFB-Aux/IAA-ARF) require systematic genetic and biochemical validation. Third, the environmental modulation of these pathways, particularly under drought, pathogen challenge, or nutrient stress, remains poorly mapped at the molecular level. Finally, the evolutionary conservation and functional divergence of IAN/IAM signaling across plant species warrant comprehensive comparative studies.

These questions position IAN and IAM as central players in an 'IAOx metabolic decision-making system'. Delving into these mechanisms is, therefore, essential to understand how this system governs the core life strategy of plant growth-defense trade-offs, a theme that is intrinsically linked to metabolic channeling and environmental signal integration. To address these gaps, future research should prioritize: (1) deploying CRISPR/Cas9-based tissue-specific editing and single-cell metabolomic imaging to resolve spatial dynamics of IAN/IAM accumulation and conversion; (2) conducting interactome screenings and genetic epistasis analyses to identify signaling components that interface with IAN/IAM metabolic nodes; (3) establishing multifactorial stress regimes to dissect environment-dependent regulation of IAN/IAM pathways; and (4) expanding phylogenetic-functional analyses across monocot and eudicot lineages to trace the evolution of IAN/IAM-mediated signaling modules.

(iii) The regulatory logic of metabolic channeling and the determinative role of cell identity

Substantial evidence indicates that IAOx metabolism exhibits remarkable 'channeling' or 'compartmentalization', whereby the identical precursor is precisely directed into

 https://doi.org/10.3390/plants15030436

distinct downstream pathways. For instance, pathogen infection can almost completely shut off the flux of IAOx into IG while simultaneously enhancing its conversion to the phytoalexin CAM. The molecular mechanisms underlying this precise metabolic reprogramming, particularly its upstream regulatory logic, remain elusive. A central hypothesis posits that this regulation may rely on cell type-specific 'preset metabolic programs'. In other words, particular cell types, due to their inherent transcriptomic signatures, their 'cell identity', may be intrinsically biased toward channeling IAOx into a specific metabolic route. Critically, metabolic channeling is not a static configuration but is dynamically reshaped by developmental and environmental cues. For example, under pathogen attack, transcriptional reprogramming mediated by WRKY33 may suppress the unknown IAOx-to-IAA converting enzyme while upregulating *CYP71A12/13*, thereby redirecting IAOx flux toward camalexin synthesis. Such dynamic repartitioning of metabolic flux likely involves rapid post-translational modifications, allosteric regulation, or the reversible assembly of enzyme complexes. Furthermore, this channeling is likely underpinned by precise subcellular organization and protein–protein interactions. The biosynthesis of camalexin, for instance, involves a metabolon comprising CYP71A12, CYP71A13, GGP1, and GSTU4, which spatially coordinates intermediate transfer and prevents diffusion of reactive metabolites. By analogy, the unknown route from IAOx to IAA may operate within dedicated subcellular niches, such as specialized endoplasmic reticulum domains or peroxisomes, or through transient enzyme complexes that physically segregate auxin production from defense-related pathways. Future research must leverage cutting-edge technologies such as single-cell metabolomics and spatial transcriptomics to construct a high-resolution spatiotemporal atlas of the IAOx metabolic network in planta. Integrating these approaches will be essential to map context-dependent channeling events and elucidate how plants toggle between metabolic programs in real time. This approach will allow us to move beyond tissue-level averaging and ultimately uncover the cellular basis and fundamental principles driving metabolic path decisions. Understanding this regulatory layer may also provide insights into the evolution of species-specific metabolic strategies.

(iv) Evolutionary trajectory and the mechanisms underlying species specificity

The IAOx pathway, as currently understood from genetic dissection in Arabidopsis, exhibits remarkable complexity within Brassicaceae, yet the evolutionary drivers and molecular basis for this remain incompletely understood. Does this specificity primarily stem from innovations in Brassicaceae-specific cis-regulatory elements, or is it driven by the adaptive evolution of key enzymes such as CYP79B? Conversely, in non-Brassicaceae species like maize and *Medicago truncatula*, are the enzymatic bases for the auxin responses triggered by IAOx or its analogs homologous to the newly discovered unknown IAA biosynthesis pathway in Arabidopsis? A key goal is to determine whether the unknown enzyme in Arabidopsis has functional analogs in other species, and how cell-specific regulatory logics have evolved to shape pathway diversity.

(v) Network integration and signal crosstalk under environmental stress

Within the new metabolic network framework, a central question emerges: how do diverse environmental stresses, such as high temperature, salinity, and pathogen attack, orchestrate resource allocation within the IAOx pathway? Specifically, how are stress signals perceived and transduced to prioritize the flux of IAOx toward defense metabolites like CAM, while simultaneously fine-tuning local IAA levels to coordinate adaptive growth responses? Deciphering the mechanisms governing resource partitioning and signal crosstalk between auxin biosynthesis and defense metabolism under stress conditions holds significant theoretical importance. In this context, the allocation of IAOx flux provides a mechanistic basis for classical growth–defense trade-offs. Rather than viewing

https://doi.org/10.3390/plants15030436

IAA and defense metabolites as endpoints of a single linear pathway, the revised model posits IAOx as a central resource node whose flux is partitioned through competing, and often mutually exclusive, enzymatic commitments (Figure 1B). Key determinants of this partitioning include (1) the relative activity and substrate affinity of the unknown IAOx-to-IAA enzyme versus CYP71A12/13 (for CAM) and SUR2 (for IG); (2) their spatiotemporal expression and subcellular localization; and (3) post-translational regulation that rapidly toggles these activities in response to signals. For example, pathogen-triggered suppression of the IAOx-converting enzyme (potentially via WRKY33) coupled with induction of *CYP71A12/13* would establish a hardwired metabolic switch that prioritizes defense (CAM) at the expense of local auxin biosynthesis. Conversely, developmental cues favoring growth may enhance the expression or activity of the unknown enzyme, thereby channeling IAOx toward IAA production.

This network-based view fundamentally revises traditional interpretations of hormone–defense crosstalk. The former linear model implied a direct metabolic coupling in which defense intermediates (e.g., IAN derived from IG hydrolysis or CYP71A13 activity) served as obligatory precursors for auxin, creating a paradox whereby defense activation could inadvertently stimulate growth. The new model decouples these pathways: IAN and IAM are not required intermediates for IAOx-derived IAA. Instead, they arise from parallel, independently regulated branches (e.g., IG hydrolysis for IAN, or yet-unidentified routes for IAM) and can themselves function as tissue-specific auxin precursors or signaling molecules. Thus, crosstalk is not an obligatory metabolic linkage but a competitive, regulatory relationship centered on the IAOx branch point. Hormone and defense outputs can be coordinated or antagonized through transcriptional and allosteric control of the enzymes that compete for IAOx, enabling a more flexible and context-dependent balance between growth and defense. Furthermore, it unveils potential targets for future crop improvement strategies aimed at precisely manipulating the growth-defense balance.

5. Conclusions

The IAOx-dependent IAA biosynthesis pathway, once viewed as a linear metabolic route, has now emerged as a dynamic, channeled network that bridges auxin production with defense metabolism. The paradigm shift initiated by recent genetic and metabolomic studies not only redefines the role of IAOx, IAN, and IAM but also reveals a sophisticated regulatory logic governing resource allocation between growth and defense programs.

Looking forward, several frontier challenges remain. Future research must prioritize the identification of the unknown enzyme(s) catalyzing IAOx-to-IAA conversion, elucidate the tissue-specific signaling functions of IAN and IAM, and unravel the mechanisms of metabolic channeling under environmental stress. Moreover, integrating the IAOx pathway into the broader auxin signaling network, encompassing both nuclear and cell-surface signaling modules, will be essential to understand how plants dynamically balance growth and defense in fluctuating environments. Advancing these questions will require interdisciplinary approaches, including single-cell omics, spatial metabolomics, and systems biology modeling, to move from a descriptive framework toward a predictive understanding of auxin-driven plant adaptation.

Author Contributions: D.-W.D.: conceptualization; M.-K.M. and D.-W.D.: Figure preparation; M.-K.M., V.K. and D.-W.D.: draft manuscript preparation; M.-K.M., D.-W.D. and V.K.: manuscript revision. D.-W.D.: funding acquisition. All authors have read and agreed to the published version of the manuscript.

Funding: This work was supported by grants from the Natural Science Foundation of Jiangsu Province (No. BK20240214) and the Field Frontier Program of the Institute of Soil Science (ISSAS2412).

 https://doi.org/10.3390/plants15030436

Data Availability Statement: Data are contained within the article.

Acknowledgments: We are deeply grateful to Weiming Shi for his invaluable insights and constructive suggestions on this manuscript.

Conflicts of Interest: The authors declare no conflicts of interest.

References

1. Vanneste, S.; Pei, Y.; Friml, J. Mechanisms of auxin action in plant growth and development. *Nat. Rev. Mol. Cell Biol.* **2025**, *26*, 648–666. [CrossRef] [PubMed]
2. Di, D.-W.; Zhang, C.; Luo, P.; An, C.-W.; Guo, G.-Q. The biosynthesis of auxin: How many paths truly lead to IAA? *Plant Growth Regul.* **2016**, *78*, 275–285. [CrossRef]
3. Di, D.-W.; Wu, J.; Ma, M.; Li, G.; Wang, M.; Kronzucker, H.J.; Shi, W. PIN5 is involved in regulating NH_4^+ efflux and primary root growth under high-ammonium stress via mediating intracellular auxin transport. *Plant Soil* **2024**, *505*, 25–40. [CrossRef]
4. Enders, T.A.; Strader, L.C. Auxin activity: Past, present, and future. *Am. J. Bot.* **2015**, *102*, 180–196. [CrossRef]
5. Simon, S.; Petrášek, J. Why plants need more than one type of auxin. *Plant Sci.* **2011**, *180*, 454–460. [CrossRef]
6. Wang, J.-L.; Di, D.-W.; Luo, P.; Zhang, L.; Li, X.-F.; Guo, G.-Q.; Wu, L. The roles of epigenetic modifications in the regulation of auxin biosynthesis. *Front. Plant Sci.* **2022**, *13*, 959053. [CrossRef] [PubMed]
7. Luo, P.; Li, T.-T.; Shi, W.-M.; Ma, Q.; Di, D.-W. The roles of GRETCHEN HAGEN3 (GH3)-dependent auxin conjugation in the regulation of plant development and stress adaptation. *Plants* **2023**, *12*, 4111. [CrossRef]
8. Wu, L.; Kriechbaumer, V.; Di, D.-W. Emerging roles of cAMP: A transcriptional master regulator in the canonical TIR1/AFB-mediated auxin signaling. *Plant Hormones* **2025**, *1*, 0018. [CrossRef]
9. Blakeslee, J.J.; Spatola Rossi, T.; Kriechbaumer, V. Auxin biosynthesis: Spatial regulation and adaptation to stress. *J. Exp. Bot.* **2019**, *70*, 5041–5049. [CrossRef]
10. Radwanski, E.R.; Last, R.L. Tryptophan biosynthesis and metabolism: Biochemical and molecular genetics. *Plant Cell* **1995**, *7*, 921–934. [CrossRef]
11. Chandler, J.W. Auxin as compère in plant hormone crosstalk. *Planta* **2009**, *231*, 1–12. [CrossRef]
12. Niyogi, K.K.; Last, R.L.; Fink, G.R.; Keith, B. Suppressors of Trp1 fluorescence identify a new Arabidopsis gene, *TRP4*, encoding the anthranilate synthase beta subunit. *Plant Cell* **1993**, *5*, 1011–1027. [CrossRef]
13. Stepanova, A.N.; Hoyt, J.M.; Hamilton, A.A.; Alonso, J.M. A link between ethylene and auxin uncovered by the characterization of two root-specific ethylene-insensitive mutants in Arabidopsis. *Plant Cell* **2005**, *17*, 2230–2242. [CrossRef]
14. Di, D.-W.; Wu, L.; Luo, P.; Zhang, L.; Zhang, T.-Z.; Sun, X.; Wei, S.-D.; An, C.-W.; Guo, G.-Q. Analysis the role of Arabidopsis *CKRC6/ASA1* in auxin and cytokinin biosynthesis. *J. Plant Biol.* **2016**, *59*, 162–171. [CrossRef]
15. Mano, Y.; Nemoto, K. The pathway of auxin biosynthesis in plants. *J. Exp. Bot.* **2012**, *63*, 2853–2872. [CrossRef]
16. Ouyang, J.; Shao, X.; Li, J. Indole-3-glycerol phosphate, a branchpoint of indole-3-acetic acid biosynthesis from the tryptophan biosynthetic pathway in *Arabidopsis thaliana*. *Plant J.* **2000**, *24*, 327–334. [CrossRef]
17. Wang, B.; Chu, J.; Yu, T.; Xu, Q.; Sun, X.; Yuan, J.; Xiong, G.; Wang, G.; Wang, Y.; Li, J. Tryptophan-independent auxin biosynthesis contributes to early embryogenesis in *Arabidopsis*. *Proc. Natl. Acad. Sci. USA* **2015**, *112*, 4821–4826. [CrossRef]
18. Luo, P.; Di, D.-W. Precise regulation of the TAA1/TAR-YUCCA auxin biosynthesis pathway in plants. *Int. J. Mol. Sci.* **2023**, *24*, 8514. [CrossRef] [PubMed]
19. Zhou, Z.; Zhang, C.; Wu, L.; Zhang, C.; Chai, J.; Wang, M.; Jha, A.; Jia, P.; Cui, S.; Yang, M.; et al. Functional characterization of the *CKRC1/TAA1* gene and dissection of hormonal actions in the Arabidopsis root. *Plant J.* **2011**, *66*, 516–527. [CrossRef] [PubMed]
20. Tao, Y.; Ferrer, J.-L.; Ljung, K.; Pojer, F.; Hong, F.; Long, J.A.; Li, L.; Moreno, J.E.; Bowman, M.E.; Ivans, L.J.; et al. Rapid synthesis of auxin via a new tryptophan-dependent pathway is required for shade avoidance in plants. *Cell* **2008**, *133*, 164–176. [CrossRef] [PubMed]
21. Stepanova, A.N.; Robertson-Hoyt, J.; Yun, J.; Benavente, L.M.; Xie, D.-Y.; Doležal, K.; Schlereth, A.; Jürgens, G.; Alonso, J.M. TAA1-mediated auxin biosynthesis is essential for hormone crosstalk and plant development. *Cell* **2008**, *133*, 177–191. [CrossRef] [PubMed]
22. Yamada, M.; Greenham, K.; Prigge, M.J.; Jensen, P.J.; Estelle, M. The *TRANSPORT INHIBITOR RESPONSE2* gene is required for auxin synthesis and diverse aspects of plant development. *Plant Physiol.* **2009**, *151*, 168–179. [CrossRef] [PubMed]
23. Di, D.-W.; Wu, L.; Zhang, L.; An, C.-W.; Zhang, T.-Z.; Luo, P.; Gao, H.-H.; Kriechbaumer, V.; Guo, G.-Q. Functional roles of Arabidopsis *CKRC2/YUCCA8* gene and the involvement of PIF4 in the regulation of auxin biosynthesis by cytokinin. *Sci. Rep.* **2016**, *6*, 36866. [CrossRef]
24. Kriechbaumer, V.; Wang, P.; Hawes, C.; Abell, B.M. Alternative splicing of the auxin biosynthesis gene *YUCCA4* determines its subcellular compartmentation. *Plant J.* **2012**, *70*, 292–302. [CrossRef]

https://doi.org/10.3390/plants15030436

25. Poulet, A.; Kriechbaumer, V. Bioinformatics analysis of phylogeny and transcription of *TAA/YUC* auxin biosynthetic genes. *Int. J. Mol. Sci.* **2017**, *18*, 1791. [CrossRef]
26. Kriechbaumer, V.; Botchway, S.W.; Hawes, C. Localization and interactions between Arabidopsis auxin biosynthetic enzymes in the TAA/YUC-dependent pathway. *J. Exp. Bot.* **2016**, *67*, 4195–4207. [CrossRef]
27. Zhao, Y.; Christensen, S.K.; Fankhauser, C.; Cashman, J.R.; Cohen, J.D.; Weigel, D.; Chory, J. A role for flavin monooxygenase-like enzymes in auxin biosynthesis. *Science* **2001**, *291*, 306–309. [CrossRef]
28. Won, C.; Shen, X.; Mashiguchi, K.; Zheng, Z.; Dai, X.; Cheng, Y.; Kasahara, H.; Kamiya, Y.; Chory, J.; Zhao, Y. Conversion of tryptophan to indole-3-acetic acid by TRYPTOPHAN AMINOTRANSFERASES OF *ARABIDOPSIS* and YUCCAs in *Arabidopsis*. *Proc. Natl. Acad. Sci. USA* **2011**, *108*, 18518–18523. [CrossRef]
29. Zheng, Z.; Guo, Y.; Novák, O.; Dai, X.; Zhao, Y.; Ljung, K.; Noel, J.P.; Chory, J. Coordination of auxin and ethylene biosynthesis by the aminotransferase VAS1. *Nat. Chem. Biol.* **2013**, *9*, 244–246. [CrossRef] [PubMed]
30. Gao, Y.; Dai, X.; Aoi, Y.; Takebayashi, Y.; Yang, L.; Guo, X.; Zeng, Q.; Yu, H.; Kasahara, H.; Zhao, Y. Two homologous *INDOLE-3-ACETAMIDE (IAM) HYDROLASE* genes are required for the auxin effects of IAM in Arabidopsis. *J. Genet. Genom.* **2020**, *47*, 157–165. [CrossRef]
31. Fu, L.; Wang, M.; Han, B.; Tan, D.; Sun, X.; Zhang, J. Arabidopsis myrosinase genes *AtTGG4* and *AtTGG5* are root-tip specific and contribute to auxin biosynthesis and root-growth regulation. *Int. J. Mol. Sci.* **2016**, *17*, 892. [CrossRef]
32. Tivendale, N.D.; Ross, J.J.; Cohen, J.D. The shifting paradigms of auxin biosynthesis. *Trends Plant Sci.* **2014**, *19*, 44–51. [CrossRef]
33. Cooney, T.P.; Nonhebel, H.M. Biosynthesis of indole-3-acetic acid in tomato shoots: Measurement, mass-spectral identification and incorporation of ^2H from ^2H_2O into indole-3-acetic acid, d- and l-tryptophan, indole-3-pyruvate and tryptamine. *Planta* **1991**, *184*, 368–376. [CrossRef]
34. Sugawara, S.; Hishiyama, S.; Jikumaru, Y.; Hanada, A.; Nishimura, T.; Koshiba, T.; Zhao, Y.; Kamiya, Y.; Kasahara, H. Biochemical analyses of indole-3-acetaldoxime-dependent auxin biosynthesis in *Arabidopsis*. *Proc. Natl. Acad. Sci. USA* **2009**, *106*, 5430–5435. [CrossRef]
35. Brumos, J.; Alonso, J.M.; Stepanova, A.N. Genetic aspects of auxin biosynthesis and its regulation. *Physiol. Plant* **2014**, *151*, 3–12. [CrossRef]
36. Di, D.; Sun, L.; Wang, M.; Wu, J.; Kronzucker, H.J.; Fang, S.; Chu, J.; Shi, W.; Li, G. WRKY46 promotes ammonium tolerance in Arabidopsis by repressing *NUDX9* and indole-3-acetic acid-conjugating genes and by inhibiting ammonium efflux in the root elongation zone. *New Phytol.* **2021**, *232*, 190–207. [CrossRef]
37. Friml, J.; Gallei, M.; Gelová, Z.; Johnson, A.; Mazur, E.; Monzer, A.; Rodriguez, L.; Roosjen, M.; Verstraeten, I.; Živanović, B.D.; et al. ABP1-TMK auxin perception for global phosphorylation and auxin canalization. *Nature* **2022**, *609*, 575–581. [CrossRef] [PubMed]
38. Rodriguez, L.; Fiedler, L.; Zou, M.; Giannini, C.; Monzer, A.; Vladimirtsev, D.; Randuch, M.; Yu, Y.; Gelová, Z.; Verstraeten, I.; et al. ABP1/ABL3-TMK1 cell-surface auxin signaling targets PIN2-mediated auxin fluxes for root gravitropism. *Cell* **2025**, *188*, 6138–6150.e17. [CrossRef] [PubMed]
39. Li, L.; Verstraeten, I.; Roosjen, M.; Takahashi, K.; Rodriguez, L.; Merrin, J.; Chen, J.; Shabala, L.; Smet, W.; Ren, H.; et al. Cell surface and intracellular auxin signalling for H^+ fluxes in root growth. *Nature* **2021**, *599*, 273–277. [CrossRef]
40. Lin, W.; Zhou, X.; Tang, W.; Takahashi, K.; Pan, X.; Dai, J.; Ren, H.; Zhu, X.; Pan, S.; Zheng, H.; et al. TMK-based cell-surface auxin signalling activates cell-wall acidification. *Nature* **2021**, *599*, 278–282. [CrossRef] [PubMed]
41. Wang, J.-L.; Wang, M.; Zhang, L.; Li, Y.-X.; Li, J.-J.; Li, Y.-Y.; Pu, Z.-X.; Li, D.-Y.; Liu, X.-N.; Guo, W.; et al. WAV E3 ubiquitin ligases mediate degradation of IAA32/34 in the TMK1-mediated auxin signaling pathway during apical hook development. *Proc. Natl. Acad. Sci. USA* **2024**, *121*, e2314353121. [CrossRef] [PubMed]
42. Yu, Y.; Tang, W.; Lin, W.; Li, W.; Zhou, X.; Li, Y.; Chen, R.; Zheng, R.; Qin, G.; Cao, W.; et al. ABLs and TMKs are co-receptors for extracellular auxin. *Cell* **2023**, *186*, 5457–5471.e17. [CrossRef]
43. Wang, J.; Jin, D.; Deng, Z.; Zheng, L.; Guo, P.; Ji, Y.; Song, Z.; Zeng, H.Y.; Kinoshita, T.; Liao, Z.; et al. The apoplastic pH is a key determinant in the hypocotyl growth response to auxin dosage and light. *Nat. Plants* **2025**, *11*, 279–294. [CrossRef] [PubMed]
44. Chen, H.; Qi, L.; Zou, M.; Lu, M.; Kwiatkowski, M.; Pei, Y.; Jaworski, K.; Friml, J. TIR1-produced cAMP as a second messenger in transcriptional auxin signalling. *Nature* **2025**, *640*, 1011–1016. [CrossRef]
45. Qi, L.; Kwiatkowski, M.; Chen, H.; Hoermayer, L.; Sinclair, S.; Zou, M.; Del Genio, C.I.; Kubeš, M.F.; Napier, R.; Jaworski, K.; et al. Adenylate cyclase activity of TIR1/AFB auxin receptors in plants. *Nature* **2022**, *611*, 133–138. [CrossRef]
46. Luo, P.; Luan, C.-S.; Li, T.-T.; Kriechbaumer, V.; Di, D.-W. 3′,5′-cAMP in plants: An integrated view of homeostasis, effectors, and physiological functions. *J. Exp. Bot.* **2025**, eraf517. [CrossRef]
47. Yang, Z.; Wei, H.; Gan, Y.; Liu, H.; Cao, Y.; An, H.; Que, X.; Gao, Y.; Zhu, L.; Tan, S.; et al. Structural insights into auxin influx mediated by the Arabidopsis AUX1. *Cell* **2025**, *188*, 3960–3973.e15. [CrossRef] [PubMed]
48. Ung, K.L.; Winkler, M.; Schulz, L.; Kolb, M.; Janacek, D.P.; Dedic, E.; Stokes, D.L.; Hammes, U.Z.; Pedersen, B.P. Structures and mechanism of the plant PIN-FORMED auxin transporter. *Nature* **2022**, *609*, 605–610. [CrossRef]

https://doi.org/10.3390/plants15030436

49. Pérez-Henríquez, P.; Nagawa, S.; Liu, Z.; Pan, X.; Michniewicz, M.; Tang, W.; Rasmussen, C.; Cui, X.; Van Norman, J.; Strader, L.; et al. PIN2-mediated self-organizing transient auxin flow contributes to auxin maxima at the tip of Arabidopsis cotyledons. *Nat. Commun.* **2025**, *16*, 1380. [CrossRef]
50. Huang, R.; Wang, J.; Chang, M.; Tang, W.; Yu, Y.; Zhang, Y.; Peng, Y.; Wang, Y.; Guo, Y.; Lu, T.; et al. TMK-PIN1 drives a short self-organizing circuit for auxin export and signaling in Arabidopsis. *Dev. Cell* **2025**, *61*, 73–84.e6. [CrossRef]
51. Vorwerk, S.; Biernacki, S.; Hillebrand, H.; Janzik, I.; Müller, A.; Weiler, E.W.; Piotrowski, M. Enzymatic characterization of the recombinant *Arabidopsis thaliana* nitrilase subfamily encoded by the *NIT 2*/*NIT 1*/*NIT 3*-gene cluster. *Planta* **2001**, *212*, 508–516. [CrossRef]
52. Fenech, M.; Brumos, J.; Pěnčík, A.; Edwards, B.; Belcapo, S.; DeLacey, J.; Patel, A.; Kater, M.; Li, X.; Ljung, K.; et al. The *CYP71A*, *NIT*, *AMI*, and *IAMH* gene families are dispensable for indole-3-acetaldoxime-mediated auxin biosynthesis in *Arabidopsis*. *Plant Cell* **2025**, *37*, koaf242. [CrossRef]
53. Pastorczyk, M.; Kosaka, A.; Piślewska-Bednarek, M.; López, G.; Frerigmann, H.; Kułak, K.; Glawischnig, E.; Molina, A.; Takano, Y.; Bednarek, P. The role of CYP 71A12 monooxygenase in pathogen-triggered tryptophan metabolism and Arabidopsis immunity. *New Phytol.* **2020**, *225*, 400–412. [CrossRef]
54. Müller, T.M.; Böttcher, C.; Morbitzer, R.; Götz, C.C.; Lehmann, J.; Lahaye, T.; Glawischnig, E. TRANSCRIPTION ACTIVATOR-LIKE EFFECTOR NUCLEASE-mediated generation and metabolic analysis of camalexin-deficient *cyp71a12 cyp71a13* double knockout lines. *Plant Physiol.* **2015**, *168*, 849–858. [CrossRef] [PubMed]
55. Nafisi, M.; Goregaoker, S.; Botanga, C.J.; Glawischnig, E.; Olsen, C.E.; Halkier, B.A.; Glazebrook, J. *Arabidopsis* Cytochrome P450 monooxygenase 71A13 catalyzes the conversion of indole-3-acetaldoxime in camalexin synthesis. *Plant Cell* **2007**, *19*, 2039–2052. [CrossRef]
56. Song, M.; Fan, X.; Chen, J.; Qu, H.; Luo, L.; Xu, G. OsNAR2.1 interaction with OsNIT1 and OsNIT2 functions in root-growth responses to nitrate and ammonium. *Plant Physiol.* **2020**, *183*, 289–303. [CrossRef]
57. Normanly, J.; Grisafi, P.; Fink, G.R.; Bartel, B. Arabidopsis mutants resistant to the auxin effects of indole-3-acetonitrile are defective in the nitrilase encoded by the *NIT1* gene. *Plant Cell* **1997**, *9*, 1781–1790. [CrossRef]
58. Park, W.J.; Kriechbaumer, V.; Müller, A.; Piotrowski, M.; Meeley, R.B.; Gierl, A.; Glawischnig, E. The nitrilase *ZmNIT2* converts indole-3-acetonitrile to indole-3-acetic acid. *Plant Physiol.* **2003**, *133*, 794–802. [CrossRef] [PubMed]
59. Lehmann, T.; Janowitz, T.; Sánchez-Parra, B.; Alonso, M.-M.P.; Trompetter, I.; Piotrowski, M.; Pollmann, S. Arabidopsis NITRILASE 1 contributes to the regulation of root growth and development through modulation of auxin biosynthesis in seedlings. *Front. Plant Sci.* **2017**, *8*, 36. [CrossRef] [PubMed]
60. Wójcikowska, B. Involvement of nitrilases in auxin biosynthesis and their role in plant development and stress response. *Plant Cell Physiol.* **2025**, pcaf152. [CrossRef]
61. Pollmann, S.; Neu, D.; Weiler, E.W. Molecular cloning and characterization of an amidase from *Arabidopsis thaliana* capable of converting indole-3-acetamide into the plant growth hormone, indole-3-acetic acid. *Phytochemistry* **2003**, *62*, 293–300. [CrossRef] [PubMed]
62. Pollmann, S.; Müller, A.; Piotrowski, M.; Weiler, E. Occurrence and formation of indole-3-acetamide in *Arabidopsis thaliana*. *Planta* **2002**, *216*, 155–161. [CrossRef]
63. Pollmann, S.; Neu, D.; Lehmann, T.; Berkowitz, O.; Schäfer, T.; Weiler, E.W. Subcellular localization and tissue specific expression of *Amidase 1* from *Arabidopsis thaliana*. *Planta* **2006**, *224*, 1241–1253. [CrossRef]
64. Aronsson, H.; Boij, P.; Patel, R.; Wardle, A.; Töpel, M.; Jarvis, P. Toc64/OEP64 is not essential for the efficient import of proteins into chloroplasts in *Arabidopsis thaliana*. *Plant J.* **2007**, *52*, 53–68. [CrossRef] [PubMed]
65. Keereetaweep, J.; Blancaflor, E.B.; Hornung, E.; Feussner, I.; Chapman, K.D. Ethanolamide oxylipins of linolenic acid can negatively regulate *Arabidopsis* seedling development. *Plant Cell* **2013**, *25*, 3824–3840. [CrossRef]
66. Bak, S.; Tax, F.E.; Feldmann, K.A.; Galbraith, D.W.; Feyereisen, R. CYP83B1, a cytochrome P450 at the metabolic branch point in auxin and indole glucosinolate biosynthesis in Arabidopsis. *Plant Cell* **2001**, *13*, 101–111. [CrossRef]
67. Delarue, M.; Prinsen, E.; Va, H.; Onckelen; Caboche, M.; Bellini, C. *Sur2* mutations of *Arabidopsis thaliana* define a new locus involved in the control of auxin homeostasis. *Plant J.* **1998**, *14*, 603–611. [CrossRef]
68. Quittenden, L.J.; Davies, N.W.; Smith, J.A.; Molesworth, P.P.; Tivendale, N.D.; Ross, J.J. Auxin biosynthesis in pea: Characterization of the tryptamine pathway. *Plant Physiol.* **2009**, *151*, 1130–1138. [CrossRef]
69. Nongbri, P.L.; Johnson, J.M.; Sherameti, I.; Glawischnig, E.; Halkier, B.A.; Oelmüller, R. Indole-3-acetaldoxime-derived compounds restrict root colonization in the beneficial interaction between *Arabidopsis* roots and the endophyte *Piriformospora indica*. *Mol. Plant-Microbe Interact* **2012**, *25*, 1186–1197. [CrossRef]
70. Zhao, Y.; Hull, A.K.; Gupta, N.R.; Goss, K.A.; Alonso, J.; Ecker, J.R.; Normanly, J.; Chory, J.; Celenza, J.L. Trp-dependent auxin biosynthesis in *Arabidopsis*: Involvement of cytochrome P450s CYP79B2 and CYP79B3. *Genes Dev.* **2002**, *16*, 3100–3112. [CrossRef] [PubMed]

https://doi.org/10.3390/plants15030436

71. Van Zelm, E.; Koevoets, I.T.; Meyer, A.J.; Van Der Velde, K.; De Zeeuw, T.A.J.; Verstappen, F.; Holmer, R.; Kohlen, W.; Willemsen, V.; Gommers, C.M.M.; et al. CYP79B2 and CYP79B3 contribute to root branching through production of the auxin precursor indole-3-acetonitrile. *bioRxiv* **2023**, 559630. [CrossRef]
72. Irmisch, S.; Zeltner, P.; Handrick, V.; Gershenzon, J.; Köllner, T.G. The maize cytochrome P450 CYP79A61 produces phenylacetaldoxime and indole-3-acetaldoxime in heterologous systems and might contribute to plant defense and auxin formation. *BMC Plant Biol.* **2015**, *15*, 128. [CrossRef]
73. Perez, V.C.; Dai, R.; Bai, B.; Tomiczek, B.; Askey, B.C.; Zhang, Y.; Rubin, G.M.; Ding, Y.; Grenning, A.; Block, A.K.; et al. Aldoximes are precursors of auxins in Arabidopsis and maize. *New Phytol.* **2021**, *231*, 1449–1461. [CrossRef]
74. Roman, A.; Montenegro, J.; Fraile, L.; Urra, M.; Buezo, J.; Cornejo, A.; Moran, J.F.; Gogorcena, Y. Indole-3-acetaldoxime delays root iron-deficiency responses and modify auxin homeostasis in *Medicago truncatula*. *Plant Sci.* **2023**, *332*, 111718. [CrossRef]
75. Buezo, J.; Esteban, R.; Cornejo, A.; López-Gómez, P.; Marino, D.; Chamizo-Ampudia, A.; Gil, M.J.; Martínez-Merino, V.; Moran, J.F. IAOx induces the SUR phenotype and differential signalling from IAA under different types of nitrogen nutrition in *Medicago truncatula* roots. *Plant Sci.* **2019**, *287*, 110176. [CrossRef]
76. Mikkelsen, M.D.; Naur, P.; Halkier, B.A. *Arabidopsis* mutants in the *C–S* lyase of glucosinolate biosynthesis establish a critical role for indole-3-acetaldoxime in auxin homeostasis. *Plant J.* **2004**, *37*, 770–777. [CrossRef]
77. Hull, A.K.; Vij, R.; Celenza, J.L. *Arabidopsis* cytochrome P450s that catalyze the first step of tryptophan-dependent indole-3-acetic acid biosynthesis. *Proc. Natl. Acad. Sci. USA* **2000**, *97*, 2379–2384. [CrossRef]
78. Glawischnig, E.; Hansen, B.G.; Olsen, C.E.; Halkier, B.A. Camalexin is synthesized from indole-3-acetaldoxime, a key branching point between primary and secondary metabolism in *Arabidopsis*. *Proc. Natl. Acad. Sci. USA* **2004**, *101*, 8245–8250. [CrossRef] [PubMed]
79. Douglas Grubb, C.; Zipp, B.J.; Ludwig-Müller, J.; Masuno, M.N.; Molinski, T.F.; Abel, S. Arabidopsis glucosyltransferase UGT74B1 functions in glucosinolate biosynthesis and auxin homeostasis. *Plant J.* **2004**, *40*, 893–908. [CrossRef] [PubMed]
80. Lambrix, V.; Reichelt, M.; Mitchell-Olds, T.; Kliebenstein, D.J.; Gershenzon, J. The Arabidopsis epithiospecifier protein promotes the hydrolysis of glucosinolates to nitriles and influences *Trichoplusia Ni Herbivory*. *Plant Cell* **2001**, *13*, 2793. [CrossRef] [PubMed]
81. Searle, L.M.; Chamberlain, K.; Rausch, T.; Butcher, D.N. The conversion of 3-indolylmethylglucosinolate to 3-indolylacetonitrile by myrosinase, and its relevance to the clubroot disease of the Cruciferae. *J. Exp. Bot.* **1982**, *33*, 935–942. [CrossRef]
82. Lehmann, T.; Hoffmann, M.; Hentrich, M.; Pollmann, S. Indole-3-acetamide-dependent auxin biosynthesis: A widely distributed way of indole-3-acetic acid production? *Eur. J. Cell Biol.* **2010**, *89*, 895–905. [CrossRef]
83. Kasahara, H. Current aspects of auxin biosynthesis in plants. *Biosci. Biotechnol. Biochem.* **2016**, *80*, 34–42. [CrossRef]
84. Mucha, S.; Heinzlmeir, S.; Kriechbaumer, V.; Strickland, B.; Kirchhelle, C.; Choudhary, M.; Kowalski, N.; Eichmann, R.; Hueckelhoven, R.; Grill, E.; et al. The formation of a camalexin-biosynthetic metabolon. *Plant Cell* **2019**, *31*, 2697–2710. [CrossRef] [PubMed]
85. Sui, J.; Tian, H.; Ding, Z.; Kong, X. Crop designs: The ideal root architecture for future crop breeding. *New Crop.* **2024**, *1*, 100030. [CrossRef]

 https://doi.org/10.3390/plants15030436

MDPI AG
Grosspeteranlage 5
4052 Basel
Switzerland
Tel.: +41 61 683 77 34

Plants Editorial Office
E-mail: plants@mdpi.com
www.mdpi.com/journal/plants

www.ingramcontent.com/pod-product-compliance
Lightning Source LLC
Chambersburg PA
CBHW061628210926
51850CB00009B/2475
* 9 7 8 3 7 2 5 8 7 7 4 1 6 *